D0944669

SALT DREAMS

William deBuys

Photographs by Joan Myers

University of New Mexico Press Albuquerque

Salt Dreams

LAND & WATER IN LOW-DOWN CALIFORNIA

FIRST EDITION

Earlier versions of the following material from *Salt Dreams* have been previously published:
Portions of chapter 1 as "Aerial Reconnaissance," *Northern Lights,* 9: 3 (Fall 1993), 27–30, and as "Aerial Reconnaissance," in *Northern Lights: A Selection of New Writing from the American West,* Deborah Clow and Donald Snow, editors (New York: Vintage, 1994), 207–218.
Portions of chapter 1 as "You Can Be in My Dream if I Can Be in Yours," *Head/Waters,* Linny Stovall, editor (Hillsboro, Ore.: Blue Heron Press, 1994), 12–20.
Chapter 3 as "Dreams of Earth," in *Reopening the American West,* Hal Rothman, editor (Tucson: University of Arizona, 1998), 3–23.
Chapter 13 as "The Theory and Practice of Borders," *DoubleTake* 2: 2 (Spring 1996), 106–118.
Side Trip: Calipatria as "Here: Flagpole—Calipat, California," *DoubleTake* 2: 1 (Winter 1996), 10–11.

Library of Congress Cataloging-in-Publication Data

DeBuys, William Eno.
Salt dreams : land and water in low-down California / William deBuys ; photographs by Joan Myers. — 1st ed.
p. cm.
Includes bibliographical references (p.) and index.
ISBN 0-8263-2126-7 (alk. paper)
1. Salton Sea (Calif.) — Environmental conditions. 2. Water resources development—Colorado Desert (Calif. and Mexico)—History—20th century. 3. Water salinization—California—Salton Sea—History—20th century. I. Myers, Joan, 1944– . II. Title.
GE155.S35 D4 1999
333.91′64—dc21 99-6422
CIP

For Anne—WILLIAM DEBUYS

For my mother, Florence Smith—JOAN MYERS

Contents

PART III: CONSEQUENCES

Photographer's Preface

In 1986, the Salton Sea was to me no more than a blue splotch on a map of southern California. But it made me curious: why was so much water in the middle of the desert? I frequently drove between Los Angeles and Phoenix, and on one trip I decided to turn south at Indio and drive along the west side of the sea and take a look. I headed past the strip malls in Indio, through Coachella and the orchards and vineyards beyond. Small roadside markets and fruit stands with signs in both English and Spanish displayed dates, oranges, nuts, and avocados. I stopped for a date shake at Valerie Jean's Dates in Oasis and, looking around at the hand-lettered signs on the produce, was surprised by how quickly and completely I had left behind the sprawling outskirts of Los Angeles. Farther south, I got my first glimpse of the Salton Sea, a flat but lustrous far-off sheen of water.

Curious to see more, I drove to Salton City. The road from the main highway descended toward the sea past the town's commercial center, little more than a dingy restaurant advertising burgers, a small market with gas pump, and a Chamber of Commerce building. It curved past a couple of small homes and acres of empty lots. As it approached the shore of the sea, the road became a broad, divided avenue lined with palms, an impressive boulevard now in disrepair. Palm fronds littered the edges, and weeds grew in the cracks of the pavement. The road ended abruptly in the parking lot of a large round building, its white paint peeling and windows gaping. A sign placed grandly over its entrance door proclaimed it to be the Salton Bay Yacht Club.

As I got out of my car, a disagreeable smell wafted toward me from the water. Inside the building, electric wires dangled from ripped conduit. Bird droppings spattered the scraps of the carpet on the cement slab. The glass windows of the curved dining room facing the sea lay shattered on the floor. A pelican watched me through the floor-to-ceiling window frames from a broken chunk of concrete a few yards offshore. The sea paled in the morning heat. The horizon between sea and sky trembled in the haze. Off to one side, a man stood fishing, oblivious to all else.

At the beginning, it was the eerie beauty of the sea and the strangeness of the failed development around it that captured me. I began photographing the yacht club with its swimming pool, exhausted gardens, and empty motel. I wandered through nearby trailer camps and along the canals where egrets and herons lived. Gradually I widened my exploration to the surrounding area—the emigrant route through Anza Borrego, the Mexican border, the sand dunes west of Yuma, and the Colorado River and its delta where the river empties into the Sea of Cortez. I began to see the region as all of a piece.

Soon, I met some of the people who made the sea their home: snowbirds at the Slabs, wildlife workers, residents of the trailer camps. Everywhere, I listened to stories. When I set up my view camera to photograph the yacht club, couples would drive into the cracked parking lot, lean out of their cars, and tell me about the time back in the sixties when they drove down from Indio for a special dinner, about the white tablecloths, attentive waiters, and the view of yachts moored on moonlit water, about their vision of a pleasure city on the edge of a languid sea.

Almost involuntarily, I found myself returning year after year to photograph. A vacation spot it was not. My car bottomed out on desert roads, skidded into slick caliche puddles when it rained. My tripod came back from each trip covered with a salty muck. Although I chose the coolest months of the year for my visits, the temperature often approached 100 degrees. In this land of vast agricultural fields, fine dining was something to dream of, not to experience. Salads, it seemed, were pathetic heaps of iceberg lettuce, and El Centro was the nearest place offering a choice of decent food. Until a motel was built on the east side of the sea to service the new prison, passable lodging was also nearly an hour's drive away.

In 1991, historian and author Bill deBuys joined me for a short visit to the sea to consider collaborating on a book. We visited Salton City, where we found the yacht club locked behind barbed wire and heard rumors that a new owner was about to restore it to its former glory. (As the years have passed, the talk continues but the building becomes ever more covered with graffiti.) We circled the sea, scouting the enfolding desert and the canals and drains of the great fields of onions and alfalfa. On the east side, we stopped at Bombay Beach to look at buildings, fences, and vehicles that lay partially submerged after a recent rise in the sea's level. Strewn along the beach were wood remnants from a disintegrated pier, wire and concrete slabs, and desiccated, salt-encrusted fish. We sensed a story larger than what we saw. Years passed as we individually responded to the necessities of gainful employment and family responsibilities while nudging the project forward. Our association was periodic but enduring. No one could ask for a collaborator with a better sense of humor or greater powers of observation, synthesis, and storytelling.

When we started we hardly recognized the extent of our undertaking. So many themes of the American West flowed into the Salton basin—water issues, the border, agriculture, flimflam development, environmental degradation, and Native rights. As time passed and more of such issues surfaced, the scope of the book broadened and the text grew. Concurrently, my interests radiated outward. I began to photograph Imperial Valley agriculture with its impressive plumbing and broad flat fields, the lively border towns of Calexico and Mexicali, and the social life of small communities around the sea like Salton City, Bombay Beach, and the Slabs. I also photographed early historical sites, including Port Isabel, which required an eight-mile hike through water and quicksand in the eerie mirage-land of the lowermost Colorado River delta.

The more I saw, the stranger the place seemed. The contrast grew between the seductive beauty of the sea and the reality beneath its surface; between the majesty of a flock of pelicans in flight and the pollution of the New River draining into the sea; between the sacredness of earth-art sites like that of the horse intaglio and the profane devastation of the Salton Sea Test Base. I was fascinated by the mud volcanoes and the nearby arching pipes of the geothermal plants sucking heat and energy from deep below the earth's surface. I was lured by the forbidden bombing ranges in the Chocolate Mountains and by the awesome bleakness of the Yuha desert. One year, I photographed the sea basin from a small helicopter.

For years I searched for the soul of the region, trying to find the sacred amid desolation and to capture it in my photographic frames. Perversely, the more intensely I photographed, the less able I was to distill the essence of the sea and its surroundings. Repeatedly, I found my frames too small to contain the breadth of what I perceived, the range between poetry and obscenity all around me. Finally, on my last trip, I gave up the effort. I shot panoramas and concentrated on the emptiness of desert and sea, allowing the space itself to fill the frame. In the end, these photographs brought me an acceptance of what I had first seen twelve years earlier—a sea and a region of harsh contrasts and grave uncertainties, struggling for survival.

This project complete, I will miss the Salton Sea. Each year I have migrated there, like the wild birds, and photographed for a few days. I have come to love its unfriendly exterior and to marvel at the unremitting and largely unsuccessful human attempts to tame it. Human failure is cruelly evident in this place of naked clarity. At the beginning of a new millennium, it spreads lessons before us. We ignore them at our risk.

ON A PROJECT of more than ten years' duration, many people provided me with information and assistance. I would like to thank Norm Niver; Homer Townsend, head ranger at Anza Borrego State Park; Ron Hensen and Barbara Moore of Earthrise Company (a spirulina algae farm near the sea); Johnny Hernandez, who showed me a personal side of Imperial Valley agriculture and of the border; Bureau of Land Management archaeologist Boma Johnson; E. O. (Mac) McDaniel, now, sadly, deceased; and Tad Nichols, for his advice and notes on the location of Port Isabel. Leonard Knight not only showed me the details of his painted mountain and posed for me on several occasions but also showed me, by example, a truly happy man.

I am especially grateful to those who accompanied and assisted me on one or another trip: Jim Londo took the door off his small helicopter so that I could take photographs; William (Rudy) Arruda from the California Department of Fish and Game drove me into the gunnery range in the Chocolate Mountains (where bombs fall at any hour, day or night) to photograph historic sites and an Indian waterhole; Mick St. John and his son Gabe accompanied me on an exploration by land and sea of the Colorado delta, the strangest landscape I have ever seen. *Mi amor* Bernie Lopez shared more bad meals and lugged more camera gear than I like to remember; without his help I'm not sure that I would have survived one especially nasty trip in the wilds of Anza Borrego where a wrong turn on a rough road led to a slashed tire, and where a hike up a small hill caused a knee injury.

Kristina Kachele, the book's designer, was a pleasure to work with. She deserves considerable credit for integrating a lengthy text and numerous photographs into a handsome and readable book. My studio assistant, Mike Webb, deserves special thanks for printing the silver prints for book reproduction. Finally, I would like to thank Bill deBuys for his unwavering enthusiasm and encouragement throughout our lengthy collaboration and Beth Hadas of the University of New Mexico Press for her patience, support, and belief that someday a book would come.

Author's Acknowledgments

A saying from the rural South comes to mind: "If you see a turtle on a fencepost, you know he didn't get there by himself." Like such a turtle, *Salt Dreams* had a lot of help getting where it is.

And it needed every bit. The book turned out to be far more ambitious than Joan Myers or I imagined at the outset. For want of funds, its completion was delayed for years. At last help appeared, and my thanks to Jessica Hobby Catto cannot be more emphatic or sincere. Without her kindness and generosity, this book might never have been finished. And it might never have been so handsomely produced without the generous assistance of Eugene V. Thaw, Susan Herter, and the Thaw Charitable Trust.

Many others helped, too, including scores of people from around the Salton Sea and throughout the Imperial, Coachella, and Mexicali valleys who shared their knowledge and insight. From among the water agencies and boards, I would particularly like to thank Rich Thiery, Randall Stocker, Don Cox, Bill Condit, Paul Cunningham, and Tom Kirk. Larry Vavra, Danny and Reba Miller, Juan Ulloa, Olivia Doughty, Manuel Abundis, and especially Johnny and Maria Hernandez could not have been more hospitable or helpful in introducing me to their communities. Mary Belardo, Tom Luebben, and Dick Young kindly shared both knowledge and documents concerning the Torres-Martinez land claim. Where the environmental troubles of the sea are concerned, Norm Niver is in a class by himself, and I am proud to know him. Another one-of-a-kind, in a completely different way, is Leonard Knight; my thanks go also to him. Around Salton City, Lester Murrah, Walt Lindsay, the late Helen Burns, and many others gave generously of their time. So did Kim Nichols and Maurice Cardenas of the California Department of Fish and Game. Over a period of years, the staff at the Salton Sea National Wildlife Refuge were unfailingly helpful, and I owe special thanks to Clark Bloom, Ken Sturm, Bill Radke, and Ken Voget. In the same way, I am indebted to Patricia Rice and Alec Rosenberg, who for long periods covered the environmental beat for the Imperial Valley Press, and to Boma Johnson in the Yuma office of the Bureau of Land Management.

My thanks go to Don Snow and Deb Clow at *Northern Lights* and Dennis and Linny Stovall at *Blue Heron Press*, who published early versions of two chapters, and especially to Robert Coles and Alex Harris at *DoubleTake,* who commissioned the research and writing that became chapter 12.

Closer to home, Daniel Morper and Rob Leutheuser provided valuable contacts and introductions; Robin Abell and John Herron helped with research; Juliana Henderson and Darla Sather transcribed many hours of tapes while maintaining their sense of humor and mine. Ellen Bradbury and Recursos de Santa Fe helped with the administration of precious funding. Dan Voll was a wonderful traveling companion and a provocative co-interviewer.

Dale Pontius read various chapters in early stages and later the whole manuscript; he also steered me to sources, meetings, and conversations that were vitally important. Another good friend, Hal Rothman, likewise plowed through inches of manuscript, gave needed criticism and encouragement, and generously included an early version of chapter 2 in his collection *Reopening the American West.* Lois Rudnick was the ideal "naive" reader; she gave the manuscript a close and thoughtful going over and broke the news about needed revisions with clarity and gentleness. Jane Kepp likewise lent a deft hand to the improvement of the text; her edit of the final copy mended innumerable errors.

The manuscript for *Salt Dreams* was delivered to UNM Press four and a half years after the deadline set in its contract. I do not know whether I am more grateful to press director Beth Hadas for her patience with the years of delay or for her expert (and badly needed) editing of the manuscript. All writers should be lucky enough to work with such an editor and to have such a friend.

In the friendship department, Roger and Frances Kennedy have few peers. They cheered this project on when it most needed cheering, and Frances, especially, helped it in more ways, large and small, than can be counted. No amount of thanks is enough. The same goes for Joan Myers, whose brilliant photography speaks for itself, and whose patience, research instincts, book collection, and good cheer made the endeavor possible. Finally, thanks to all those who put up with me during the long process of producing this work, especially my children, Kate and David, and my wife Anne, to whom I dedicate this work.

INVOCATIONS

The white pelican and the crow, whose feathers were used on the stabbing pike (akwil) were both tipai (persons).

Other birds which were tipai were the roadrunner and the mockingbird. The latter was called tuwilaú. Birds which were tipai were not eaten. Mammals which were tipai were the coyote and the wild cat (nyimet). The latter, however, was eaten at times.—E. W. GIFFORD, 1931

The last three and the northernmost of the Jesuit missions in California were made possible by a Borgian heiress. The tale is told that when she made the gift she was asked in what country she wished the missions established. "In the most outlandish place in the world," she replied. The Jesuits consulted their atlases and returned the answer: "The most outlandish place in all the world is California." So in California the three missions were founded. Of course this is just a story.
—HERBERT EUGENE BOLTON, 1936

In the evening Glanton and the judge and a detail of five men rode downriver into the Yuma encampment. . . . The leader was a man named Caballo en Pelo and this old mogul wore a belted wool overcoat that would have served a far colder climate and beneath it a woman's blouse of embroidered silk and a pair of pantaloons of gray cassinette. He was small and wiry and he had lost an eye to the Maricopas and he presented the Americans with a strange priapic leer that may have at one time been a smile. At his right rode a lesser chieftain named Pascual in a frogged coat out at the elbows and who wore in his nose a bone hung with small pendants. The third man was Pablo and he was clad in a scarlet coat with tarnished braiding and tarnished epaulettes of silver wire. He was barefooted and bare of leg and he wore on his face a pair of round green goggles. In this attire they arranged themselves before the Americans and nodded austerely.—CORMAC MCCARTHY, 1985

It is easy for any one, who knows what miracles have been wrought in our far Western deserts during the last few years, to harbor the suspicion that the great, brown waste which lies on the borders of two republics, as voiceless now as the Mississippi Basin at the close of the Revolution, will some time be as densely populous as the lands of the Nile, as rich in industry as the Kingdom of Holland.
—WILLIAM E. SMYTHE, 1900

As our little sloop floated down the Red River on our way to the Vermillion Sea, on a February afternoon in 1904, the lazy current of low water carried us past a wing dam of stakes and brush thrown out from the western shore of the Colorado, a few miles below Yuma, and this slender barrage was coaxing a reluctant and shallow stream toward a low opening in the bank where it entered a channel of

ditches and sloughs leading to distant dry plains which it was converting to fertile farm lands. Locks, control or headworks there were none, and our party, not inexperienced in engineering emergencies nor unlearned in the ways of the river, spent a lengthened evening in camp nearby, in discussion and conjecture as to what pressure of necessity or overbold haste could lead to such unguarded opening of the cage of a sleeping tiger.
—D. T. MACDOUGAL, 1908

I stopped Kaweah and glanced back at the Salton Sea, which I was now leaving for a time. It is at best a rather cheerless object, beautiful in a pale, placid way, but the beauty is like that of the mirage, the placidity that of stagnation and death. Charm of color it has, but none of sentiment; mystery, but not romance. Loneliness has its own attraction and it is a deep one; but this is not so much loneliness as abandonment, not a solitude sacred but a solitude shunned. Even the gulls that drift and flicker over it seem to have a spectral air, like bird-ghosts banished from the wholesome ocean.
—J. SMEATON CHASE, 1919

After more than two months of observation and investigation in Imperial Valley, it is my conviction that a group of growers have exploited a "communist" hysteria for the advancement of their own interests; that they have welcomed labor agitation, which they could brand as "Red," as a means of sustaining supremacy by mob rule, thereby preserving what is so essential to their profits—Cheap labor; that they have succeeded in drawing into their conspiracy certain county officials who have become the principal tools of their machine.
—BRIG. GENERAL PELHAM D. GLASSFORD, 1934

The border is the juncture, not the edge.—GUILLERMO GÓMEZ-PEÑA, 1993

Americans love ballyhoo. We always have and we always will. It takes different forms at different periods, but it's one of our national predilections. We love to be swindled, even when it's done clumsily and without finesse. We choose to believe in the dream even though we know, somehow, it's a lie. We cherish and reward our flimflam artists, those dream peddlers who can convince us of almost anything.
—RICHARD SHELTON, 1980

The Salton Sea looks good from afar, it really does. Water, it's pretty. But you go down there, forget it.
—MARY BELARDO, 1994

1 | HEAD WATERS

We Americans may be the only people on earth who speak of a national dream. There is no French Rêve Nationale nor a Sueño Mexicano, so far as I know, nor a Senegalese or Iranian or Laotian Dream. And there may never be. It took the extraordinary conjunction of a perception of new lands, free for the taking, with crescent economic and political individualism to launch the idea of an American Dream. World events have not seen the like again. One wonders whether the planet could bear it if they did.

From the day Europeans first landed on Atlantic shores, down through the centuries to the golf resorts and gated communities of the present, the theme of American experience north of Mexico has never been to lower one's conceit of attainable felicity, as Herman Melville once advised, but continually to raise and revise it, year by year and generation by generation. In the United States, the hope has been not just to do better but to *be* better—to be happier, richer, wittier, bustier, more powerful, less balding. Americans expect life's options to be profuse, if not prodigal, and in each generation we reinvent ourselves in terms of the desiderata, economic, personal, and spiritual, of our era.

Perhaps no force has shaped our society more than this collective yearning. It fueled the nation's territorial expansion and shaped its economic culture; it has inspired natives and immigrants alike; it has infused our arts and letters with a brashness and sometimes a hopefulness that is distinctly American. Its history is our autobiography as a people.

But its history is local as well as national. The details of its influence on places and communities are always site-specific. On occasion, Americans have adapted their dreams to the realities of the places where they dwelled—or what they perceived those realities to be. More often, they altered their places to conform to their dreams. In either case, the process of alteration and accommodation never ended, for both dreams and places are restless things and change continually. This interplay of dreams with land, of reciprocal change and adaptation, becomes a kind of biography of the continent.

This book is about that interplay. It traces the interaction through time of American dreams with an extraordinary American place—one of the most naturally austere and barren deserts in the world. In the nineteenth century, westering North Americans called it the Colorado Desert, a term they applied loosely to lands along the lower Colorado River, especially on the California side. This book deals not with the totality of the Colorado Desert but with its busiest and most history-afflicted portion: the long, low strip of territory stretching southward from Palm Springs, California, through the desert basin of the Salton Sea and the Imperial Valley, then through the border city of Mexicali and the farmlands of the Mexicali Valley, and down to saltwater at the head of the Gulf of California. The rugged jumble of California's Coastal Range walls off the desert on the west, and more mountains, dunes, and the Colorado River limit the area on the east.

These lands acquired their unity through the agency of two powerful forces. One was geological: the famed San Andreas fault combined with other faults to shape a long topographical depression, much of it lying below sea level. We call this the Salton Trough. In essence, it is a northward extension of the same geologic trench that shapes the Gulf of California. Nearly all of our story is contained within the trough, with the exception of a few excursions to the transportation and trading center of Yuma, from which so much activity affecting the region was launched. The second great force was the Colorado River, which, for as long as it has existed, has poured into the trough both its water and the sediments it collected from the nearly quarter million square miles of its watershed. The subset of the Colorado Desert on which we focus was—and is—the land where the Colorado River comes to an end.

The unity of the region is no longer obvious. An international border stretches across its belly, and the waters of the river have been made to transform this hottest and driest of deserts into one of the great agricultural regions of the continent. The purpose of this book is to examine such transformations, their place within the national dramas of the United States and, to a lesser degree, Mexico, and the considerable challenges they bequeath to the present.

The book takes a narrative and thematic approach that might be called an "archaeology of place." The idea is to seek in each successive stratum of the region's occupation a narrative that conveys the story of that time. A number of recurrent themes emerge. One concerns the way in which the vast and empty desert served as a geographic tabula rasa, an empty stage on which successive actors strove to impose their dreams and desires. A second theme holds that in low places consequences collect—that the hydraulic and geophysical realities of the region produce effects that flow inexorably downward to the trough, where they intrude upon the imaginings of its dreamers.

By intention, this book fails to observe the dictum of one of southern California's most memorable fictional citizens. Sergeant Joe Friday of the television series *Dragnet* demanded, "Just the facts, ma'am." But the facts, for our purposes, are not enough. We have tried to capture the flavor as well as the facts of events and places, and so the reader may find that the following pages depart in style and content from conventional history writing. Moreover, if anything offered here can capture the dreamlike quality of the places this book explores, it is the photographs. They are illuminations, not illustrations. One might expect a work like this to include photographs drawn from history, but we have instead sought to illuminate past events with contemporary images, not to erase the gulf of time but to feel its distance and depth while glimpsing its far shore.

A WORD OF ADVICE. If you have occasion to travel by airplane above the deserts of the Southwest, do not fail to note the color of the region's greatest river. Looking down from thirty thousand feet, you will see that the formerly great Colorado is a *blue* ribbon, the same baby blue beloved by cartographers for rivers and creeks of all kinds. That cheery color is no less than an epitaph for the natural West. The river you gaze upon is no longer brown, as would befit a stream formerly as silt laden as any on earth, nor still less red, which is what Juan de Oñate and his lieutenants had in mind in 1604 when they called a side stem *colorado*.

Today, assuredly, the river is blue, and so are the tepid lakes behind the colossal dams that block its canyons. Hoover and Glen Canyon dams, which retain lakes Mead and Powell, respectively, are among the largest man-made structures on earth. Beneath the houseboaters who putt-putt up the side canyons and the jet-skiers who roar across the lakes' domesticated surfaces, the earthen harvest of the immense, eroding intermountain West settles invisibly. The red of the river falls out, whole deltas sifting down to the lake bottoms, forming a series of geologic Ellis Islands where immigrant grains of soil arrive and arrive and arrive, never departing.

If you fly above the Colorado where it exits its final canyon, you can look down and see a pool of blue water backed up behind the concrete geometry of Imperial Dam. Beside it, the "resort" of Imperial Oasis glares upward, hurling daggers of reflection from acre upon acre of trailer and RV sheet metal. Here at Imperial Dam is the end of a river and the beginning of a story.

WHEN I FIRST visited Imperial Dam, I drove up from the south across the Gila River and Yuma Proving Grounds. It did not surprise me, crossing a low bridge, that the Gila, a lesser river, had no water. In the Southwest a river, in order to be a river, need not carry water but only provide it for irrigation, which the Gila does to the ultimate drop. What surprised me was that the dry bed of the Gila had been plowed, a phenomenon that approached Homeric strangeness—like the great sailor Odysseus carrying an oar into the deserts of Africa or Arabia until he should come to a place where no one knew the oar's purpose. Here, the plowed river, as puzzling to me as an oar to an inland Bedouin, may have had more to do with floodway maintenance than with placating gods, but the sight of it still did not prepare me for misplacing the Colorado.

I knew I was close to the big river, after crossing the bestirred desert of the proving grounds, when I came to a series of green-water ditches in a marsh where blackbirds trilled. I drove across a causeway and up a low ridge of sand, expecting any second to see the great river of Wyoming, Colorado, and Utah, of the Grand Canyon itself, diminished but vibrant, threading through its plain. But at the top of the sand ridge the road turned south, and I marveled that even here, downstream of so much monumental plumbing, the Colorado still forced highways and human plans to bend. On I drove several miles, ever southward, and saw no river nor any chance to turn west, where the river ought to be. Only slowly did I begin to suspect that the river had not forced the road from its logical path so much as it had somehow evaded me. I was traveling alone, a condition in which one entertains thoughts that do not occur in company. Had I crossed the river and missed it? Had I blacked out? Was I even now in a twilight of consciousness? Worse had happened in other times and places, and I had been on the road without rest since Show Low, nearly four hundred miles back.

Anxious to get my bearings, I pulled over where a dirt road met the highway, and there encountered a barrier and a sign proclaiming, "All-American Canal, Property of Imperial Irrigation District." I got out of the car and heard the whispering suck of great waters moving fast. A hundred steps forward and I stood at the canal's concrete bank. At my feet ran the brawny, unimpeded flush of the mighty, canyon-carving River of the West.

Now I realized that the stagnant marsh where blackbirds trembled with desire was the old main channel of the Colorado—and a mapmaker's lie. Atlases innumerable notwithstanding, the blue line of the Colorado reaches no saltwater outlet in the Gulf of California. Except when El Niño is up to mischief and runoff is greater than the reservoirs can handle, the Colorado fails even to visit its former delta, which once was a wildlife area deserving comparison to the undisturbed Everglades. Today the delta is for the most part a desiccated mudflat, a desert of salt cedar, iodine bush, and pickleweed.

The tamed Colorado flows by way of one aqueduct to Los Angeles and San Diego, by way of another to Phoenix and Tucson. It flows to the farms of greater Yuma by various siphons and canals. A modest portion detours through hydroelectric generators before returning to the natural riverbed in time to cross the international border and fetch up against a final dam that diverts it to the fields and cities of northern Baja. But miles upstream, the last strong pulse of the Colorado had already departed the old path of the river. This last pulse, a stream of water consisting of more than a fifth of the river's native flow, pours westward across the driest, hottest desert in the United States to the Imperial Valley and Salton Sink of southeastern California. It accomplishes this unlikely journey by way of what is today the lower river's de facto main channel, the All-American Canal, the name of which provides full answer to anyone south of the border who wonders where the river went.

FROM THE AIR there is no mystery. The green-water ditch, crowded on either side by a gauze of tamarisk, trickles down toward Yuma and the Mexican line. Terraces of cotton fields and mesquite woodlands separate it from the dun vastness of the desert. What draws the eye is the perfect and unnatural geometry of the blue All-American Canal branching gracefully from its lesser parent and arcing sinuously through gravel hills toward unseen destinations.

If you are airborne, flying westward to San Diego, your jet will follow. The canal snakes into a wilderness of sand—the Algodones dunes, once a menace to travelers but today a noisy and ravaged playground for the off-road-vehicle and dirtbike tribe. The canal contours around the shifting slopes, twisting in long parabolic curves, blue on buff. Amid the dunes the canal divides. The smaller portion, the Coachella Canal, angles northward toward purple mountains. The larger part, still the All-American, veers to the border and then runs laser-straight along it as far as the eye can see.

Your jet drones westward, the canal and border under its wing. Now a haze lies on the land, a thickening murk of moisture, smoke, and dust, and through it emerges an apparition of monumental cultivation. You see checkerboard lines and quilted greens on a scale to match the cotton fields of Texas, or Iowa buried in corn. What lies below is an agricultural sea: field after field, square and rectangle, fallow and full, Nile green and bile green, emerald and jade. The twill of crop rows runs here with the sun, there athwart, everywhere at different angles, and each presents a new weave of shadow, dirt, and leaf.

This is the Imperial Valley, where the last waters of the Colorado River feed nearly half a million acres of cropland and, by extension, the people of the United States. In its fields grow dozens of varieties of head and leaf lettuce, carrots and

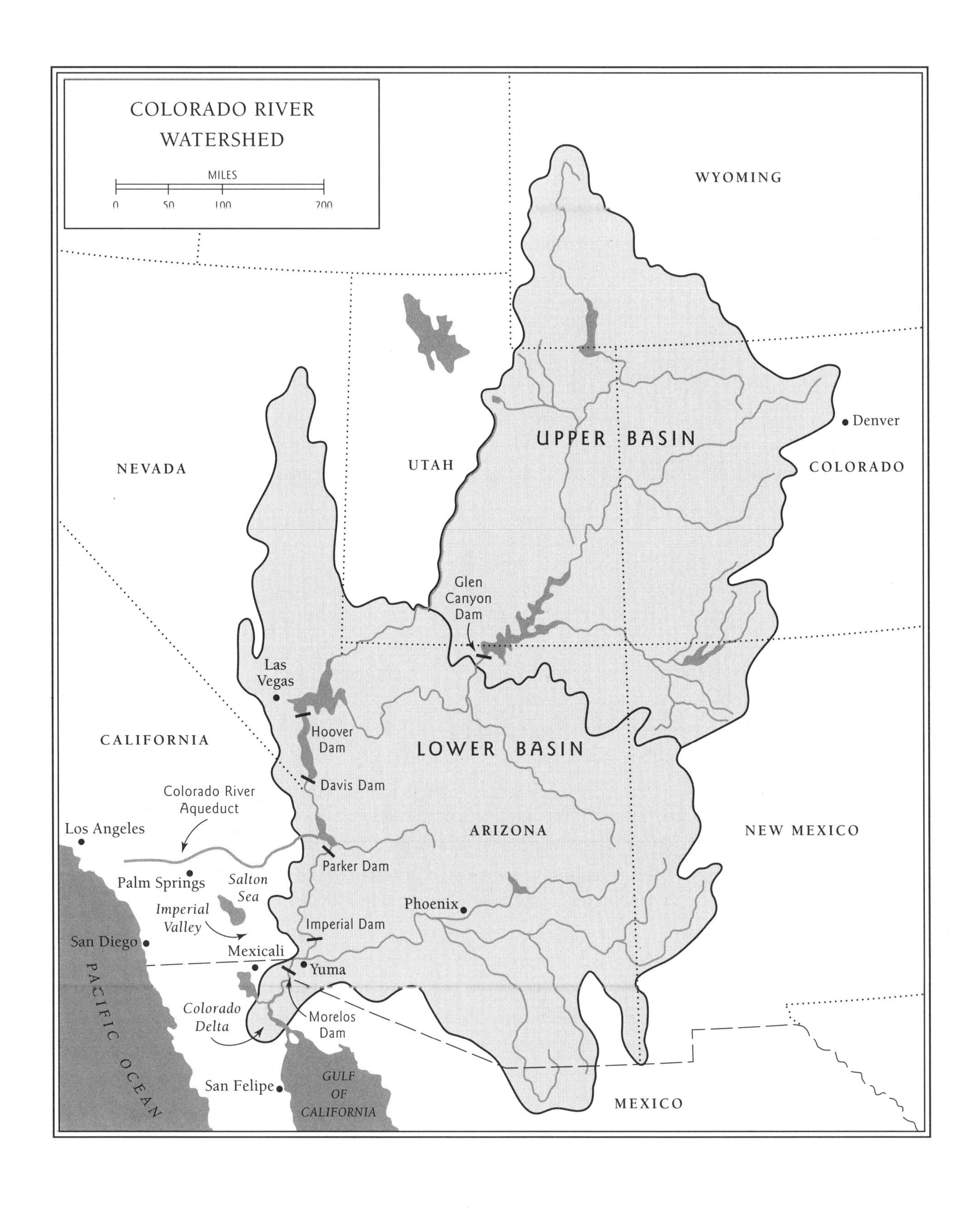

COLORADO RIVER
WATERSHED
MILES
0
50
100
200
WYOMING
NEVADA
UTAH
UPPER BASIN
Denver
COLORADO
Glen
Canyon
Dam
Las
Vegas
Hoover
Dam
CALIFORNIA
LOWER BASIN
Davis Dam
Colorado River
Aqueduct
Los Angeles
ARIZONA
NEW MEXICO
Parker Dam
Palm Springs
Salton
Sea
Imperial
Valley
Phoenix
Imperial Dam
San Diego
Mexicali
Yuma
PACIFIC OCEAN
Colorado
Delta
Morelos
Dam
San Felipe
GULF
OF
CALIFORNIA
MEXICO

artichokes, asparagus, beans, beets, and broccoli. There are bok choy and celery, cilantro and cucumbers, eggplant, peppers, and okra. There are cabbages and kale, collards and cauliflower. But the list is just beginning. One must not omit the onions, garlic, parsnips, and squashes, or the potatoes and tomatoes, the watermelons, muskmelons, honeydews, cantaloupes, and casabas. Nor should one overlook the wheat, barley, and sugar beets, sorghum and oil seeds, sweet corn and feed corn, the square miles of alfalfa. There are also pistachios, cashews, and nuts you never heard of, and fruits including dates, lemons, oranges, grapefruit, tangerines, nectarines, and hybrid-*ines* of varying description. The valley grows fifty thousand acres of grasses for pasture and seed and plants another eighty thousand in Sudan grass, much of which goes to Japan to fatten Kobe beef. It grows cotton and other fibers, waterlilies and turnips, fennel and kohlrabi—all told, nearly the entire complex of cultigens supporting North American civilization. But plants provide only part of the feast that graces our national table, as the people of the valley well know. Animals grace that table too, and they are here represented by roughly a million sheep and feedlot cattle, plus dairy cows, swine, farmed catfish, and enough commercially tended bees to keep the organs of the plants and the air humming.[1]

Here, beyond the reach of frost and chill, the growing season attains a state of nearly perpetual motion: discing, planting, irrigating, harvesting, discing, planting, and on again, restlessly and efficiently, thanks to armies of work-hungry, brown-skinned pickers and packers, thanks to boxcars, tankercars, and truck caravans of fertilizer, herbicide, and pesticide, the pesticide alone totaling eight million pounds a year.[2] And water, *gracias á dios,* without which nothing can live, a great continental river delivering the equivalent of an inland sea, all of it originating in distant lands with different climates, all making possible in this place the environmental semblance—from a seedling's point of view—of forty inches of annual rainfall, where less than five actually come from the sky. (And the farmer hardly welcomes those few natural inches, for they make the fields hard to work and mar the perfection of absolute control.) The result is the apotheosis of industrial agriculture: here food is not grown so much as manufactured. From a pragmatic perspective, one can argue that it must be so: if we are to have cities like New York and Los Angeles, if our markets are to offer year-round selection and unending abundance, we must have farms like these.

Farther to the north, beyond the limit of the fields, shines a mirror to the sky. It is thirty-six miles long and over fifteen wide, a mirror large enough to reflect the vanity of a powerful and prosperous nation. This watery mirror is the Salton Sea, California's largest lake, which receives the leachate and dross of the Imperial Valley, just as the valley receives in the Colorado's water the leachate of countless fields upstream in the river's watershed. Selenates washed from Wyoming rangelands end their travels here, as do the progressively sharper-tasting effluents of fields and towns in Colorado, Utah, New Mexico, Arizona, and Nevada, to say nothing of the raw sewage that flows north across the international boundary from Mexicali along the New River, which for decades has borne the unhappy distinction of being the nation's filthiest waterway.

This desert basin, half food factory, half sump, is what the intermountain West boils down to—or leaches or evaporates or otherwise reduces to. The verb is variable, but the process is as immutable as the laws of gravity and evaporation, which are perhaps the only laws that the hydraulic West has not amended to its purposes.

Gravity decrees that in low places consequences collect, and here is the lowest of the low: Salton Sea, growing saltier by the day and stewing with the waste of the upstream world. The sea's fluctuating surface lies roughly 227 feet below sea level, and the deepest hollow of its unseen, nacreous bottom lies still 51 feet deeper than that. In the days before floods formed the sea, when the Salton Sink was a dry bowl, geographers and explorers speculated whether it or Death Valley was the lowest location on the continent. Probably the sink held that distinction, by a foot or two, but before the question might be settled, floods came to the desert and rendered it moot.[3]

The floods were the result of the most spectacularly bungled development scheme of the century, perhaps of all time. The developers of the Imperial Valley brought water to the valley by means of a jury-rigged irrigation heading on the main stem of the Colorado, a dozen or so miles below Yuma. The Colorado flooded, blew out the heading, and poured its waters into the developers' canal system, which effectively became the main channel of the river for the next two years. The river water collected in the sink—and collected and collected, filling and rising, until the lake that had formed was too big to be called a lake. It was the Salton Sea.

MY INTRODUCTION to the sea—and to the surrounding lands that are its hydraulic co-dependents—came by way of a photograph. The image, black and white, shows an empty swimming pool yawning like a gutted melon beneath a useless diving board. A solitary, drought-murdered palm stands guard in the minimal distance, and the entire tableau is reflected in the liquid filth at the bottom of the pool. This ruin, I learned, was part of the once-vaunted Salton Bay Yacht Club, chief jewel of a giant real-estate promotion staged in vacant desert beside the Salton Sea. Brochures, film clips, and other come-ons, replete with leggy, sun-bronzed models lounging beside this very pool, promised a Palm Springs life style for people of ordinary means. It was to be the land of summum bonum—life without work, and golf forever.

The chant of hyperbole was not hard to imagine: Year-round sunshine! Fishing, sailing, no urban congestion! The lowest greens fees you ever heard of, plus bingo, drinks, and dancing after dark.

Problems? Forget 'em. Imagine yourself in a chair by the pool, sipping a frozen margarita under the shade of the palm. Never mind the goosefart stench at the water's edge or the river of Mexican sewage flowing up from the south. Never mind that house lots were sold from airplanes and tents in a rush to grab the dollars of the guileless. Never mind that the developers would one day turn off the irrigation of their make-believe and pull out, leaving yacht clubs, golf courses, and palm trees to wither in the solar wind.

The photograph completes the tale. In the putrescence at the bottom of the pool you can see what happened when the ad-man's appeals to greed and indolence turned belly up. Suffice it to say that the image of the pool—the sump within the sump at the end of the West—suggested that more rivers than one ended in the Salton Sink. Clearly, the Colorado subsided into nothingness there, but possibly another river did as well, a river of spirit and dreams. This other river, I thought, might flow from historical instead of geographic headwaters; it might rise from notions, born centuries ago, of free land and westward migration. It might, flowing through time, change character as long rivers do, reflecting not so much the country from which it departed as the country in which it endlessly arrives.

Perhaps the river of dreams might metamorphose into something unforeseen at its headwaters. Perhaps it might become an honest-to-god New River, transformed by gold rush, Hollywood, and postwar defense bonanzas, flowing past the frozen, face-lift smiles of Palm Springs, where kisses taste like piña coladas and golf clubs rattle timelessly, flowing downward past the shacks of migrant crop pickers and unemployed Indians in the Coachella Valley, down to the loneliest of deserts, flowing ever toward the mirage, barely out of reach, of the pool in the sun and the girl by the pool. Her beckoning smile is as bright as the white linen suit of the master of ceremonies, who repeats, over and over, that this deal, played right, is the only jackpot you'll ever need: you can buy one lot to build on and another for investment; the second will pay for the first. His chant sings you onward, palm trees swaying, toward the lounge at the country club, where faces turn in welcome, toward the quiet house on the cul de sac with its wet bar, climate control, and carpets soft as beds, toward the promise—*deal again! fifty on black!*—of getting something for nothing, and then doubling that.

All this, the photograph seemed to say, lay in the sump within a sump at the end of the West.

THE DREAM OF real-estate bonanzas in the Colorado Desert is perhaps as strong today as it has ever been. The Salton Sea has become a long-running ecological disaster featuring repeated wildlife epidemics and promising still more. In Congress and elsewhere, advocates for remedying the environmental ills of the Salton Sea justify their call for expensive cures by arguing that a clean sea will boost property values by enormous margins. In healthier days, the sea was a playground for the cities of the coast. Restored, it could be so again—and thereby set in motion billions of dollars of economic growth.

Not all the region's dreams have been as grandiose as that. The natives of the region, especially the powerful Yuma-speaking tribes who farmed the floodplains of the Colorado River, coveted dreams of an entirely different order and built their lives around them. They set as much store by dreaming as perhaps any people who ever lived. Their story is a vital element in the archaeology of this place.

So is the story of legions of newer arrivals, the individuals and families who sank roots in the desert and made their home there. They are a rare breed. Americans are known better for restlessness than for sticking to a place, and sticking and building in the Salton Sink is no small endeavor. No landscape can have been more intimidating to settlers than the vast and empty bowl of heat and blown dirt that now contains the Salton Sea and the Imperial Valley.[4]

The ruins of the Salton Bay Yacht Club may stand for one approach to settlement. Valley towns like Brawley and El Centro stand for another. In the workaday world of the heart of the valley, you might, of a spring morning, step out and breathe your first lungfull of the outside day, tasting air redolent of fertilizer, heavy with the tang of ammonia, and think, this is the taste of home. You might go to the edge of town, while the sun is low and shadows fill the furrows between the crops, and rejoice in the stark beauty of wide green fields stretching horizon to horizon. For outsiders, the aesthetic of this place is an acquired taste: it takes time to appreciate a landscape that lacks a focal center and consists simply of rows of leafy plants spreading outward to a cloudless, boundless sky. Anyone raised in woodlands or city canyons tends to recoil from the valley's emptiness, to say nothing of the slow crescendo of heat that builds through its day. But wait a while: the place will work on you, you will make emptiness your friend, and you will listen with appreciation for the sounds of a rough, hardworking world.

Those sounds tell stories of private struggle and personal yearning—for mundane comforts, of course, but also for exalted ones—for freedom and for opportunity. More often than you might think, such stories reveal a quiet but very real heroism. It is the heroism of people who never sought to test their limits but who found themselves tested anyway, and who rose to the challenge.

The lands of the Salton trough bear an unhappy legacy of racism and labor turmoil. Today they struggle with the full suite of contemporary problems: embattled schools, drug use, severe unemployment, crowding, pollution; yet the place remains, gritty and persistent. It provides a home for about the same proportion of idiots and idealists, brutes and belles, saints and sickos as you find in other places. The valley may not be a garden of arts and humanism, but it has heart, and you note gratefully that, irrespective of ethnicity, its people retain a habit of courtesy and openness that has largely vanished from the cold-shoulder cities and that seems today a relic of rural lands.

THE DREAMS TO be found in this great desert bowl are many and varied, but one dream in particular stands out, for it has shaped the place and its landscape more than any other. It is the dream of reclamation.

At the turn of the last century, American westerners and westerners-to-be contemplated a half-continent of arid lands. The humid East had been settled. The

Pacific Northwest, rich in water and timber, was growing apace. San Francisco remained the greatest city in California, and the riches of the rain-blessed northern half of the state flowed through it. But the drylands of the West, stippled with dusty and forgettable oases like Albuquerque and Tucson, shared little in the growth of the American nation and still less in the realization of its dream. The region could promise little—*unless* its deserts might be made to bloom.

Tactics for changing dry to wet might vary. Details of ownership, payment, pace, and priority ran the political gamut and not infrequently touched lunacy, but in the West and in the halls of power attentive to the West, every economically active white man from John Wesley Powell and Theodore Roosevelt to the storekeepers of Provo and Yuma fundamentally agreed that western rivers should be harnessed and that as much of the desert as might be moistened should be made into farms.

The empire builders and visionaries of American expansion had never conceived their hopes in solely economic terms, nor did they now. They cast the business of controlling western rivers in a moral context and strove to weave the idea of desert reclamation into the fabric of the American Dream.

Consider the term on which their argument was built: *reclamation*.

A standard dictionary defines the word as "a restoration, as to productivity, usefulness, or morality." One might wonder what kind of restoration dams and irrigation accomplish. The arid lands of western American are deserts, dry by nature; they are not lapsed farms, although that is how westering Americans, looking through a biblical lens, came to see them.

The idea of reclamation—as opposed to the simple colonization or conversion of land—drew its persuasive power from an evangelical view of landscape: deserts were indeed conceived as fallen lands, requiring redemption. When the Colorado River overflowed its banks in 1891 and spilled northward into the Salton basin, the secretary of the Historical Society of Southern California wrote, "The Colorado River seems to have repented of its evil work, and is now seeking to atone for its great sin, in desolating so large a portion of the earth, by refilling the desert sea."[5] This was not an isolated point of view. Writer after writer described desert lands in the terms known best throughout the culture: it was a story of creation, a fall from grace, and redemption through God's favor. In many cases it was the only story the writers knew.

A desert—in this context—was the geographic equivalent of a soul in perdition. It was barren, useless, all but dead. Reclamation could save it. Put more exactly, reclamation was the *act* of saving it. The prior state to which it was to be *re*stored or *re*claimed was not a state in which it had ever existed; it was an idealized state that existed *in principle*. It was a state that might be made to exist in reality through the realization of a divine plan, executed by a chosen people. The Latter-Day Saints of Utah saw themselves as such a people, obedient to such a plan, and their labors in bringing water to the land they called Deseret helped chart the path that the later "reclamation movement" would follow.[6]

As the nineteenth century drew to a close and the twentieth began, Protestant Americans waved the reclamation banner as fervently as the saints from Great Salt Lake. They, too, possessed a desire to make the desert pleasing in God's sight. To this they added a sense of national mission, for many of them believed that reclamation logically extended the process of continental conquest that a previous generation had called Manifest Destiny. Reclamation would help people the West and bring it the blessings of civilization.

Continuities with the past may have given the nascent reclamation movement its fundamental strength, but its allure derived from how it addressed concerns about the future. As the century turned, Americans worried about the waste of natural resources, and reclamation fought waste by converting deserts to usefulness, capturing water that would otherwise flow uselessly into the sea. Americans wanted the growing promise of science and technology to be realized, and reclamation did that by charging the engineer, who embodied practical know-how, with the construction and operation of complex dams and waterworks. At times, it placed the engineer in charge of planning the new towns that reclamation made possible. Reclamation became a central element of the Progressive agenda, and it epitomized the Progressives' attitude toward the use of land: it was highly structured, required management by experts, and promised the delivery of benefits not just for the year ahead, or even the decade, but for generation after generation.[7]

Reclamation also answered the needs of the huddled masses of the East. The West had always been the nation's safety valve, a place where the restless could go who found no opportunity elsewhere. Jefferson himself had premised the survival of a democratic, agrarian republic on the availability of cheap new lands. Even as that supply seemed exhausted, with the proclamation by the 1890 census of the "closing of the frontier," reclamationists argued that large-scale irrigation would open vast territories for settlement, relieving pressure on the cities.

Reclamation, however, saved its greatest virtue for the nation's greatest challenge: the rescue of American democracy. In the telling of William Ellsworth Smythe, whose evangelism on behalf of reclamation led to deep involvement in the campaign to promote and settle the Imperial Valley, reclamation would cure the ailing body politic. "The essence of the industrial life which springs from irrigation is its democracy," he wrote. The "small farm blesses its proprietor with industrial independence and crowns him with social equality."[8] Smythe and others grounded their arguments in Jefferson's widely shared belief that the yeoman farmer and his family would accept nothing less than fair-minded, full-bore democracy. They maintained that desert farm families would embody Jefferson's model of yeomanry better than any predecessor. Despots and demagogues might mislead the urban masses, but never the freehold farmer.

The coming freeholders of the desert would bolster democracy because they would be more prosperous, satisfied, and cultivated than farmers of moister lands. Of poverty there would be none, because the endless desert growing season would guarantee protection from the vagaries of climate and rainfall. Isolation and backwardness, the dual curses of the rural poor, would be banished, because with intensive cultivation and multiple harvests, farms might be small and neighbors, therefore, close. In reclamation's world, no family need live so remotely as to miss the benefits of schools and town life, and none would lack opportunity for the cultivation of manners as well as the soil.

In 1899, Smythe gathered his arguments in favor of irrigation into a single book that he called *The Conquest of Arid America*. The book gave the reclamation movement the bible it needed, and it made Smythe famous. He wrote with millenarian fervor, as though his own and his readers' salvation were at stake. When he achieved the full stride of his rhetoric, it was as if he breathed not ordinary air but a mixture of nitrous oxide and other laughing gases. A desert stand of mesquite suggested the "good cultivated orchards" of coastal California. The silt-laden waters of the Colorado became "like a stream of golden dollars which spendthrift Nature pours into the sea," and the deltaic soils of the Colorado Desert represented deposits in a bank "where, when the hour should strike, the children of men might draw their checks against it and never see them dishonored."

Smythe lent his full energies to promoting the reclamation of the Colorado Desert. Before the effort was far advanced, he would suffer disillusionment and come to fight the developers he had earlier served. But at the beginning, while the dreamers of the desert still dreamed in harmony, Smythe promised that in the Colorado Desert, there would be no rape of virgin land. The uses to which the kiln-dried wastes and their great river would be harnessed were divinely ordained. Only consummation was lacking. Consummation, that is, preceded and blessed by the sacrament of matrimony:

> In no part of the wide world is there a place where Nature has provided so perfectly for a stupendous achievement by means of irrigation as in that place where the Colorado River flows uselessly past the international desert which Nature intended for its bride. Some time the wedding of the waters to the soil will be celebrated, and the child of that union will be a new civilization.[9]

Smythe was partly right. A new civilization was indeed born in the irrigated desert, although it turned out differently from what he had in mind. Its character and complexities tell us much about our society and ourselves, more perhaps than we might learn from any other place. It may be that the best vantage from which to view a land and its people is not from the eagle's perch gazing down, but looking upward from the bottom, where consequences collect.

PART ONE

ANTEDILUVIA

2 | DREAMS OF EARTH

Come over a little farther. Climb this gravel terrace and scan the open, naked desert. You must squint to see the few scrawny shrubs that stipple these badlands, and you can't escape the sense of being cut off and alone. Feel how hard the ground is underfoot. Hear how the gravel crunches like oyster shells. Mexicans call this kind of land *pedregal:* it is a pavement of cobbles and pebbles from which the wind has lifted every grain of soil. Thus armored by subtraction, the barren ground has lain like this a thousand years. Barring skid marks or spinning tires, it might so lie a thousand more.

This is a sacred place. The broad circle etched in these gravels is the outline of a shaman's hut, long vanished from the site. The curling path that doubles back on itself and comes twisting round again marks where the shaman's people danced, year after year the same pattern, until their feet hammered a discernible trail into the sun-varnished ground. They also sang here. And they recited tales that took a week of nights to tell—weird, convoluted tales that never really ended and wouldn't fit our idea of what a story is, tales with passages that were little more than recitations of the names of place after place: canyon, butte, hill, and plain, scattered over hundreds of desert miles.[1]

The tales' obsession with geography exceeds even the obsession of the tribes of Israel with genealogy, which produced the monotonous "begats" of the Old Testament. Clearly the stories recited on this barranca helped teach their listeners the people's mythology. It is tempting to believe they also helped teach geography, so that among a people who ranged far and wide across the deserts, the right song or story might hold clues enough to guide a person where he might never have been—all the way to the Hopi mesas, say, or westward to the sea.

Those who came here also came to dream, and they put great stock in dreams, as much as any people ever have. They believed that all things worth knowing were to be learned through dreams and that little of importance might be learned in any other way. The most powerful among them said they could dream their way back into the actual primordial enactment of the events of the mythic tales and songs they recited, back into the decisive moments in the lives of their gods, back even to the earliest moments of creation. At such a place as this gravel terrace, dreamers sought those dreams, or, sleeping elsewhere, they traveled here in dream, or thought they did, for purposes beyond the limits of our ken.

If ever you would contemplate the dreams of those who wished not to reclaim the earth, only to inhabit it, absorb this place in your mind. You may be sure *they* did. Look there: the outline of a lizard, ten yards long, lies embedded in the *pedregal*. They shaped the body by removing the desert gravel to bare the clay beneath, and they bermed the lifted gravel around the edges of the figure to emphasize its outline. The lizard's legs they made a different way, beating the gravels deeper in the clay, much like the path of the dance pattern.

And over here. Come on, though it's a fair walk. Now see what they scraped and tamped in the gravels. The bison on the walls of Lascaux are not more beautiful than this snorting, coiled-neck horse with flowing mane and a luxuriant fountain of a tail. This beast might haunt the dreams of anyone, including us. It is the soul of this place and the heart of our story, which is the story of the greatest encounter in the history of these lands—or of any lands on the continent. This story, repeated with variation everywhere in North, Central, and South America, is the bizarre and only half-articulated story of first encounters between the native people of this hemisphere and the emissaries of Europe.

We will soon speak more of the horse of the pedregal. But for now look south and see the patrol roads of the international border. If distances do not deter you and you've the constitution of a fit camel, you might walk from here to a place of comfort where cottonwoods will shade you. Allow several days for the trek, unless you undertake it unprepared in the full sun of high summer, in which case allow eternity.

We'll not further suggest the location of this place. We have no wish to attract the attention of barbarians.

Perhaps you hear them on the wind. The flatulent complaint of their motors breaks the silence of the desert: looking for the sound, you see a clot of them, in the padded suits and hard-shell headdresses of their own tribe, fleeing their dust cloud across a far barranca. Their boots, tires, and blindness could in moments destroy the gifts that lie embedded at our feet. These gifts are the property of no one. They belong to the memory of the continent.

Two trails that cross near here proceed to the cardinal directions. Westward lies the ocean, which drew many past this place. Some who passed by belonged to tribes, local or distant; others were Spanish, including, in 1776, a long caravan of colonists bound for the new settlement that became San Francisco. Later, and far more numerous, came an army of invading *norteamericanos,* soon followed by legions of gold-mad forty-niners, both gringo and Mexican. Outsiders had less interest in the trail that ran north and south. This trail was sacred to the people who used this place for dreaming. They said their ancestors followed it when they journeyed downriver from Avikwame, a mountain far to the north, where the gods had made them. Since then, the people of innumerable generations had used the trail for war and commerce and also to travel to Avikwame, in body or in dream, there to reenter the continuous and unending moment of original Time. They called the trail *xam kwacan* and took their name from the second of those strange words: Quechan (kweh-*tsan*), the people who came down.[2]

BY THE TIME Juan Bautista de Anza crossed the Colorado in 1774, searching out the route (along trails known to the Quechan) by which he would later lead colonists from Sonora to Alta California, Spaniards had learned a good deal about the Quechan and other Yuman-speaking tribes who dwelled beside the river. The honor of first contact on the European side belonged to Hernán de Alarcón, who commanded a fleet in support of Francisco Vásquez de Coronado's northbound land expedition of 1540. Alarcón had sailed to the head of the Gulf of California and anchored his ships, then ascended the Colorado in launches, perhaps as far as Yuma Crossing. Though he failed to rendezvous with Coronado, he found occasion to share the corn, beans, and squash of the river people, which they grew in the floodplain, and he tasted the bread they made from ground beans of the honey and screwbean mesquites.[3] He also drew from them all the news he could of Coronado's violent arrival at Zuni two months earlier, in July. The Indians were well informed on this, though Zuni, or Cíbola as the Spaniards then knew it, lay more than the breadth of Arizona away.[4] Alarcón quickly realized that the people of the river were not isolates. They traded far across the deserts that enfolded them, and their world was large.

In 1604, Juan de Oñate struck west from the Hopi pueblos and descended the Bill Williams Fork to the main stem of the Colorado, whence he continued downstream to the gulf, contacting numerous Yuman tribes, from the Mohave in the north to the Cocopa at the river's mouth. One village chief regaled the dour Spaniard with tales of other tribes, farther along, "with ears so large they dragged on the ground," and of a one-footed people who slept under water, and of still others who slept in trees or standing upright with burdens on their heads, and others who subsisted solely on the odor of their food or—most shocking, in the teller's view—who were entirely bereft of hair. Francisco de Escobar, Oñate's Franciscan chronicler, expressed skepticism that "there should be so many monstrosities in so short a distance" but, reflecting that the Almighty might produce freaks if He so chose, acknowledged that "since He is able to create them, He may have done so."[5]

It remained for the Italian Jesuit Eusebio Francisco Kino, in the course of three visits to the junction of the Gila and Colorado rivers, to fix the understandings of his adopted countrymen in tighter focus. When he approached the confluence in 1699, Kino encountered the Quechan people well settled there. Evidently the Quechan, whom Kino called Yumas, had migrated downriver and won control of the area since Oñate's time. Lieutenant Juan Mateo Manje, who accompanied Kino not so much to protect the padre as to serve as secular witness to his discoveries, described the Yumas as "a well-featured and large people," adding that "their women are pretty, and much whiter than those of any other nation of Indians known in New Spain."[6] Manje was able to observe both sexes directly and without the interference of clothing, for they went about entirely naked, save that the women wore skirts made of strands of willow bark, which rustled as they moved.

Kino recorded little of his first contact with the Quechan. What most fired his imagination was a gift the Indians made him of large blue abalone shells, which, so far as he understood the provenance of such things, seemed likely to have come from the Pacific. This was disconcerting, for Kino then believed, as all of Europe did, that California was an island that lay separated from the continent by the as yet unlocated and much sought after Straits of Anian, which were presumed to connect with the head of the Gulf of California. The blue shells, however, suggested that a land route to California might exist and that to reach it one need only cross the thousand-foot-wide, skin-brown waters of the Colorado, a feat which the Quechan, masterful swimmers, accomplished almost daily. Further assertions by the Quechan that clothed white men were known to dwell beyond the deserts to the west lent support for a new geography. Those whites might be the Spaniards of the coastal missions at San Diego and San Gabriel.

But what was one to make of such tales? The tellers, after all, were related to the same naked and painted people who had beguiled Oñate and Escobar with preposterous yarns and who even now spoke of a white woman, clothed head to foot in blue, who years ago had visited the river people bearing a cross and speaking unintelligibly. Kino and Manje were neither so gullible as to wholly believe such tales nor so rash as to dismiss them out of hand. In those days of early contact between whites and Indians, the hot sun of history may have been at the Europeans' backs, but myth still lay like dew on the new lands, and its vapors scented every breath that anyone, white or Indian, might draw.

For the Quechan, the physical world of known trails that stretched from the Pacific coast to (probably) the Rio Grande and from the Sea of Cortez to sacred Avikwame was not disjunct from the mythic, dreamtime plane where Kumastamxo, their foremost deity, and the first people forever dwelled in the moment of creation. The buttes, mountains, spires, and sentinels that marked their heat-soaked homeland were not separate from the monsters and heroes whom, according to myth, the landforms formerly had been. In the Quechan's view of the world, things mythological and empirical might exist together without conflict, with no epistemological Straits of Anian separating one class of knowing from another.[7]

So, too, for the creole Spaniard Manje, born in the New World, and the Italian Kino, who did God's work in Spain's name. For both, the Almighty was an immanence, not an idea. They traveled always in His presence and prayed daily for such guidance, blessing, and favorable intervention as He might bestow. Like Escobar, they acknowledged that for Him all things were possible, and so the mysteries of strange lands and stranger people like the Quechan, whose men wore sticks in their noses and whose women, like the men, painted their bodies most outlandishly, contained nothing that could not be contained by Him. As to the Lady in Blue, Manje and Kino allowed that, notwithstanding the greater likelihood that God would have given her the power of tongues in order to communicate with the natives, "perhaps the visitor was the Venerable María de Jesús de Agreda," who in the year 1630, through the medium of heavenly transport, "preached to the heathen Indians of this North America and the borders of New Mexico."[8]

Kino, true son of both his holy faith and his knowledge-hungry time, bridged the worlds of myth and history. In 1700 he returned to the confluence of the Colorado and the Gila, carrying with him a telescope. From a summit in the Gila Mountains east of present-day Yuma he descried the head of a large body of water on the southwest horizon. Another year passed, and he returned a third time, intent upon descending the Colorado to the bay he had seen and confirming that neither it nor any other barrier of water separated the deserts of familiar Sonora from the fabled coast of California. Manje was not with him; the lieutenant had been called away to chastise "sorcerers" elsewhere in Sonora. As Kino passed downstream on the east side of the river, having never crossed to the far bank, his small retinue of Indian servants was soon swallowed amid a gathering throng of several hundred Quechans and Pimas.

The tumult of so many savages—for surely from a Spanish point of view that is what they were—was too much for Kino's sole white companion, a servant whose name is lost to history. As the ragged procession departed Quechan land and arrived downstream at the first village of the Quíquima,[9] the Spanish servant contrived to fall behind. Seizing a moment when he was little noticed, he turned his horse and spurred for Sonora. Kino, hearing the news, dispatched two Indian boys on fast mounts to catch the deserter, but in vain. The panicked servant had too great a lead. Though worried by this abandonment, Kino pressed on. He paused to treat for peace between the Quechan and their Quíquima enemies (with temporary success), then continued southward, the river always on his right.

There followed, two days later, a spectacle that must have been one of the most extraordinary sights human eyes have witnessed in North America. The natives who had followed Kino, together with gathering bands of Quíquima, all of them turned out in their painted best—stripes, say, on the torso, dots across the face, an arm red, a leg black or white, and everyone different—crowded both banks of the river. The Quíquima cut a path through the jungled thickets to the water's edge for Kino and his horses, but the horses mired in the river mud and could not pass. Never mind. With Kino's encouragement the natives lashed cottonwood logs together to make a raft, and a great reed basket, waterproofed with pitch, was placed upon it. The black-robed Jesuit then climbed into the basket while crowds on either bank made "dances and entertainments after their fashion."[10] From our vantage, centuries later, we may forever wonder whether Kino next spread his hands and smiled, or prayed for strength and grimly eyed the turbid river with its dark relentless flow. We know only that at last, installed in a basket atop his tippy raft, the padre committed himself to the current and to the care of at least a score of Quíquima swimmers, who surrounded his unseaworthy craft and pushed it toward the farther shore. Thus was Eusebio Kino, with all the pomp and fanfare the Colorado delta in 1701 could muster, ferried to the land that by then he knew to be most surely California.

KINO'S GEOGRAPHICAL discoveries made possible, three-quarters of a century later, the pathfinding of Juan Bautista de Anza, from whom we begin to learn in detail of the Quechan, their neighbors, and the river world of delta and desert which they inhabited.

Had there been at crucial times a few more men like Anza, the history of northern New Spain and Mexico might have been different. Anza, his stern eyes glowing from a face masked by a thick, pointed beard, was a man of extreme discipline and unquestionable courage, who by one estimate may have traveled twenty thousand miles in the saddle.[11] No wonder he died a little short of reaching fifty. Like Kino, he lived almost always in motion, rarely resting, but unlike the padre, he was a native son of Sonora and a warrior. Born in 1736 at the presidio of Fronteras, which his father commanded, Anza enlisted at the age of sixteen and fought Apaches, Seris, and other desert tribes for most of the next thirty years. His greatest military achievement came in 1779 when he delivered the beleaguered province of New Mexico, which he served as governor, from the thrall of the Comanches. His victory in pitched battle over the Comanches' most redoubtable chief, Cuerno Verde, laid a foundation for the closest thing to peace New Mexico had known in a century.[12]

Anza, however, earned his greatest fame in two expeditions to Alta California a few years earlier. The first was a journey of exploration in 1774 to find a route across uncharted deserts and mountains; the second, in the winter of 1775–1776, constituted a veritable migration of 240 soldiers and colonists, plus a thousand head of livestock, along that route to San Francisco Bay. Crucial to both expeditions was Anza's cultivation of good relations with the Quechan, who controlled the vital crossing of the Colorado near its confluence with the Gila. From an encampment near the river, Anza wrote to Viceroy Bucareli in Mexico City:

> The people who live on this Colorado River are the tallest and the most robust that I have seen in all the provinces, and their nakedness the most complete; their weapons are few, as I have informed your Excellency on another occasion; their affability such as is never seen in an Indian toward a Spaniard; and their number must run into thousands, for in a league and a half which I have travelled on both rivers I have seen about two thousand persons, notwithstanding that they are afraid of us because of our color and our clothing, which seem more strange to them than their entire nakedness to us. As a result of the inundations, their lands are so fertile that they yield grain in the greatest abundance.[13]

Anza had discerned the key to the natives' tall physical stature and, indeed, to much of their culture: it was the generosity of the river. Nearly every year the Colorado and the Gila flooded in spring, depositing fresh silt on their alluvial plains, which the rivers also braided with sloughs and twisting ponds. In some areas the plains stretched a mile or more back from the river. In spring, when the floods receded and the drying silt began to crack, the Quechan, like the other river Yumans, moved down from their winter villages in the uplands at the desert's edge and spread out, family by family, in "rancherías" scattered along the floodplain. They planted maize, tepary beans, and various cucurbits (squashes and pumpkins) in the drying mud, and nature did almost all the rest. In most years no additional irrigation was necessary, for the water table was high and the soils held moisture well, but if the crops began to wilt, the Quechan filled clay pots from the brown river and watered them by hand. The river gave abundantly of fish, too—mainly the humpback sucker and several species of minnows, some up to three feet long. The natives shot them with arrows in the shallows or caught them any way they could: bare-handed, with baskets, nets, or weirs, or with thorn hooks baited with grasshoppers.[14]

They also planted a grass, known today as Sonoran panic grass (*Panicum sonorum*), that was like wild millet. They filled their mouths with seeds and blew them onto moist soil, mainly in places too boggy for corn.[15] And they hunted rabbits and other small game and gathered the beans of the mesquite, both honey and screwbean, which grew on the river terraces a level or two above the floodplain. When the river flooded too much or too little, the protein-rich mesquite beans helped carry them through times of want.

With food generally plentiful, the river Yumans may have been as well nourished as any native people on the continent. They lived, to be sure, at the mercy of their environment, but it was a reliable and predictable environment, which may partly explain their character. Compared with the Arizona Hopis and the Pueblos of New Mexico, who depended on rainfall that was famously capricious in timing and amount, the Yumans lived in a world of regularity and security.

One can argue that a variable environment helped engender the Puebloans' elaborate ritual, dance, and cosmography—much of which was directed toward guaranteeing a needed but too often interrupted orderliness within the cosmos, which in turn might assure abundant crops and game.[16] But for the river Yumans, such assurance need not have been so assiduously sought, for the Colorado and the Gila nearly always gave it freely. The point is not to argue for environmental determinism, only to say that the environment in which the Quechan dwelled helped sustain the outward simplicity of their tribal lives.

This simplicity pleased Spanish missionaries like the fussy Pedro Font, a Franciscan who accompanied Anza on his second expedition. With customary condescension Font noted in his diary the kinds of shamans who would have presided at the gravel terrace where the horse was etched: "There are some wizards, or humbugs, and doctors among them, who exercise their offices by yelling, blowing, and gestures."[17] But there were no full-fledged priests commanding the day-to-day obedience of the people. Nor was there much evidence of effigies, icons, sacrifices, or intricate ritual. For a man like Font, the scarcity of such things was reassuring: it meant less to erase, less to overcome in the course of bringing such heathens to the Christian fold. Still, he found much to dislike about them—their "disorderly and beastlike" ways of living, their body painting and piercings, and their many repugnant personal habits, including a penchant for dust-raising flatulence, which Font described in passages so detailed that a century and a half later his circumspect American translator elected to leave them, for propriety's sake, in the original Spanish.[18]

Font also detected among the Quechan what he felt to be a troubling combativeness. He knew little of their habits of warfare, but had he inquired, he might have learned that indeed they were fond of fighting and campaigned in alliance with the Mohave and other groups almost yearly against the downstream Cocopa and the Maricopa, who lived some distance up the Gila. The Quechan disdained attacking the enemy from a distance, Apache-style, with the kind of long-range bow Anza might have better respected. Their way, like that of all the river Yumans, was to close with the enemy in organized ranks and batter them hand to hand with maces and stabbing clubs. Indeed, as Font suspected, they were not a docile people, and their way of combat was no mere chest beating. (One reason the Quechan and Mohave did not more stoutly resist U.S. forces in the late 1850s and 1860s was that the ranks of their warriors had been thinned—to the point, for the Quechan, of virtual annihilation—by a disastrous battle with the Maricopa in September 1857, which left some 140 of their number dead.)[19]

But Font, while acknowledging the need for a large presidio to answer the Quechan's truculence, looked on the bright side: "Thus, it is seen that these people are greatly disposed to enter the Holy Church as soon as provision may be made for it, and that they are not repugnant to subjection to the law of God and of our sovereign, for they say they will be glad when Spaniards and fathers come to live with them. It seems to me that a great Christendom may be won among these tribes."[20]

IN 1779, Spaniards and fathers duly came to stay, and gradually, as more colonists continued to arrive, their numbers swelled, though the band of troops protecting them did not. The error was fatal. Too many times, Spanish soldiers molested Quechan women, and too many times their livestock trampled Quechan crops. By July 1781 the Quechan had had enough of Christendom. They rose up and destroyed the two small Spanish settlements at Yuma Crossing, killing four priests and fifty-five male settlers. They took the women and children captive, releasing many of them a few months later. In spite of repeated attempts to bring them to heel, never again would the Quechan submit to Spanish control.[21]

Thus, in anger, ended almost two and a half centuries of intermittent, friendly contact between the *adelanteros* of New Spain and the people of the river.

AMERICAN HISTORY was born in incompleteness. Its most profound moment—that of first contact between Europeans and Native Americans—we know almost exclusively from only one side of the encounter. The other left comparatively little record of what it felt, thought, and experienced.

And what a void this silence leaves. After hearing from Alarcón, Kino, Anza, and their like, we still know relatively little of those first cultural temblors along the Colorado—or anywhere else. In that moment of early contact, whole societies of native people, encompassing complete and unduplicated ways of seeing the world and being in it, apprehended the perfect foreignness of the European, his animals, plants, and weapons, and perforce thought long and hard about the obdurate strangeness of them. But sadly for posterity, those thinkers, for want of means if not desire, laid up no comments against the ages. A few legends have come down, a very few, and time has greatly blurred their images. The moment of contact, induplicable in every instance yet endlessly replicated across an entire continent, band by clan by tribe by nation, comes to us only by half.

No doubt in many early episodes of contact the actors scarcely knew the import of their actions. Natives may have thought that the newly arrived strangers would soon go away and never, if prayers be answered, come back. And for many Europeans, one native group seemed little different from any other. They often failed to grasp the extraordinary diversity of the societies through which they cut their path.

It is not just history we are missing, but also literature. European accounts of first contact, whether by friar, soldier, or journal-keeping wanderer, give us the author's view well enough, though usually tailored to please the viceroy or *custos* or commanding general to whom the writer reported. But while such narratives help establish who said what to whom and when, they seldom suggest what, from the natives' point of view, might have been the *feel* of events.[22]

We will never truly know the counsels of the tribe, the vying of factions within it, the calculations of trust and mistrust, the assessments of power, each side measuring the other, man by man, weapon by weapon; nor the shock of ideas or fear for the old order's crumbling, reactions that probably came later, with reflection. More immediately, there was the envy of possessions like knives and colored cloth; astonishment at the frightful energy of gunpowder; the exhilarating sensation of the first taste of wheat flour or dried apricot; the odd entanglement of curiosity and contempt as the natives pondered the Others' beards, clothes, and modesty or lack of it; and not least their amazement at the strangers' lack of women, soon followed, we may assume, by wariness and jealousy as the eyes of the new men took in the women of the tribe and as the women, their minds equally consumed with assessment, responded or shied away.

In the absent native literature of first contact there would have been as much marvel and mystery, as much horror and tragedy, as might be found in any of the literatures of the world. Even then, it would have been incomplete, for the truest first contacts were often not perceived as such. These would have been the terrible first epidemics of European diseases, which in many cases preceded the arrival of the Europeans themselves. For want of understanding the biological cause of these irruptions, blame would have been fixed on whatever fault of gods or ritual or human error was most readily observed. Perhaps such a dynamic explains the following dread recollections, by an old Mohave, who in turn heard the story from the Kamia, concerning the Alakwisa, a river tribe long presumed to be extinct when the story was recorded:

> There was a small pond from which the Alakwisa used to draw their drinking water, and which had never contained fish. Suddenly it swarmed with fish. Some dug wells to drink from, but these, too, were full of fish. They took them, and, although a few predicted disaster, ate the catch. Women began to fall over dead at the metate or while stirring fish mush, and men at their occupations. They were playing at hoop and darts, when eagles fought in the air, killed each other, and fell down. The Alakwisa clapped their hands, ran up, and gleefully divided the feathers, not knowing that deaths had already occurred in their homes. As they wrapped the eagle feathers, some of them fell over dead; others lived only long enough to put the feathers on.[23]

Such fragments are the little we have of the missing literature of America: a few tales warped by time, some scraps of codexes, a smattering here and there of petroglyphs and pictographs. Each relic sharpens our hunger for what was lost. In the land of the Colorado Desert, nothing more fiercely hones that appetite than the image in sun-darkened gravels of the snorting, coiled-neck horse.

THE PEDREGAL crunches underfoot as you approach. In this place, outside the hut whose outline is preserved here, Quechan elders in the time of Kino and Anza, as well as elders of the predecessor tribes who held this ground in the time of Oñate and Alarcón, would have pondered the identity and intentions of the bearded white-skins who arrived from afar.

This much was obvious: the foreigners knew hunger and thirst the same as the people of the tribe. They knew suspicions and fears. They had a language, albeit an unintelligible one. They had the same audacity to impose their will on others. And to do so, they had swords and guns, which were equivalent, if superior, to the weapons of the people. One thing which they had, however, was utterly incomparable. More than the strangers' weapons, their clothing, their odd foods, or their fixation with the symbol of the cross, more than the outlandish-looking strangers themselves, the beasts the Spaniards rode were of an order that the people of the river had never seen before on earth.

Those who squatted on the gravel barranca and surveyed the trails the Spaniards traveled must have long pondered such matters. Among them, perhaps foremost among them at some indeterminate time, was the artist who executed, alone or with others, what we shall call the horse intaglio.

He had no doubt seen other four-legged creatures: deer in the bosques by the river, desert bighorn in the canyons toward the Mohave country. If he had ever joined a trading expedition to Hopi, where the river people bartered for cotton blankets, he might have seen elk among the forests and prairies of what is now Arizona. And although he might not have seen the beasts for himself, he had surely spoken with others who had seen bison on the plains far to the east of Zuni; the Indians of the delta described such creatures to Alarcón in 1540 and showed him war shields made of their nearly impenetrable hide.[24]

But none of those creatures possessed the feature that so distinguished the horse. None of them had a long, flourishing, abundant tail, which is the most pronounced characteristic of the horse intaglio. The artist rendered it with evident fascination. The narrow stump of the earthen tail bursts into a cascade of hair that falls nearly to the horse's heels. Not that the tail is excessive. Like every other detail of the intaglio, it is properly proportioned to the whole.

The image exceeds twenty feet in length from the rearmost curve of the tail to the tip of the creature's nose, and its features are true: the bend of the rear hocks, the arch of the neck, the poise and carriage of the head. The creature captured here is an intelligent and vital horse, rendered by methodical removal of desert-varnished cobble and pebble. The artist scraped out the stones to a depth of one or two inches, piling most of the material around the outline of the image. He etched ears where ears should be and an eye where an eye should be. He made four long, strong legs by wetting the ground and tamping the gravels into separate relief from the gravels around them. No doubt when the intaglio was fresh, one could tell the beast had hooves—solid hooves, not split like those of a deer. Strangely, though, the legs are straight and static, while the rest of the animal seems enlivened with motion.

There are other uncommon touches: at the withers and continuing forward to the throat, the artist removed less of the cobble, so that an indistinct dusky area remained. And in the space between the front knees and the muzzle—the domain of breath and vocal sound—the reverse exists: here the artist made a shallow, separate excavation, not quite as deep as the basin of the body. Was water offered here, or grain? Or something that surpasses facile logic? It is easy to imagine, perhaps too easy, that through use of this minute depression the artist and shaman (it is impossible to believe they were not the same person) attempted connection to, if not mastery of, the spirit of the powerful, enviable, and quintessential horse.

IT IS USELESS to attempt saying what the horse intaglio "means." One can no more say what is meant by the cave paintings at Lascaux, or by Botticelli's Venus or Andy Warhol's soup cans. Although a legion of art historians may bend their efforts to the task, things ineffable cannot fully be explained.

What little we know of the intaglio comes from understanding the context of the work, rather than the work itself. First, it is unique—at least now. Cocopa lore includes mention of a second horse intaglio paired with this one. The presumed site of the second, however, is now a dusty void—part of the pit of a gravel quarry.[25]

Second, intaglios and other forms of "earth art" lie scattered far and wide across the deserts of the lower Colorado. Traditions of sand painting (known to the Quechan, Cocopa, Navajo, and others) probably attach to the same deep cultural root that produced these extraordinary forms. Many of the desert earth figures—"geoglyphs" in the parlance of archaeologists—depict abstract shapes or stylized humans. Some are hundreds of feet long; only a few are as realistic as the horse. Although dating is difficult, most geoglyphs substantially predate European contact, some possibly reaching back thousands of years to the arrival of the area's earliest people.[26] Virtually all are at risk to vandals.

Third, the river Yumans would have first encountered horses in the winter of 1540–1541, when Melchior Díaz, in command of a third division of Coronado's expedition, journeyed to the Colorado delta in search of news either of his general or of Alarcón, who had departed the area a few months earlier. Horses next arrived in the possession of Oñate in 1604, but there is no evidence that the Quechan kept horses of their own until they picked up strays from Kino—and most of these they appear to have eaten or returned.[27] In subsequent years, through intermittent contact with Sonora, they acquired other horses which they refrained from eating, so that when Anza passed through in 1774 he could observe, with just a hint of jealous contempt, that "the horses kept by the Yumas, however it may be, are seen to be fat in the extreme."[28] Anza's hard-used mounts, we may assume, showed plenty of rib, while the Quechan horses did little but loaf and graze.

Fourth, in keeping with the themes of their cosmology, the river Yumans undoubtedly sought to come to terms with the horse through dreaming. They believed deeply in the power of dream. In his encyclopedic and authoritative *Handbook of the Indians of California,* the usually restrained Alfred Kroeber made this categorical observation: "There is no people whose activities are more shaped by this psychic state, or what they believe to be such, and none whose civilization is so completely, so deliberately, reflected in their myths."[29]

The Yuman people believed that all the knowledge and power, skills and understanding a person might possess were acquired through an extraordinary process of prenatal dreaming. While still in the womb, the spirit of the individual dreamed its way back to Avikwame, where in the presence of Kumastamxo such gifts were conferred as the unborn being might in life enjoy. At birth, this dream was forgotten, but in adolescence the individual might begin, through dreaming, to reenter the earlier dream and to recapture the gifts with which he or she had been endowed.[30] Approaching the puzzle of life from this direction, the river Yumans saw learning not as a process of acquisition, let alone of study, but of recovery—a return to an earlier, more complete state. It was a kind of *reclamation* of the spirit and psyche. The means for this return was "proper" dreaming, supplemented and reinforced by the telling of dreams, which was a tribal obsession. Elders listened carefully to the dream stories of those who aspired to possess influence within the tribe, and the seasoned old ones would judge whether the dreams of the young might be classed as merely ordinary or as "great" dreams, which brought power. They further judged whether in life the dreamer's actions validated the claims he made for his dreams.[31]

Such a dynamic of knowledge would not easily have accommodated new information. Dreams, passed down through generations, became myths. Those myths and dreams coalesced into song cycles, which were the core of the tribe's shared knowledge and religious understanding. The telling of any one of them might last all night, night after night for a week, and still not be exhausted. How puzzling the horse must have been. How difficult to find a place in a universe, already deemed complete, for a creature that had never before existed.

It is not hard to imagine that the artist and shaman who created the horse intaglio would have strived to dream a dream by which to integrate the horse with the store of tribal knowledge. One wishes that all the problems resulting from contact between Europeans and Native Americans had been as incorporeal.

Except in their taste for war, the river Yumans contrasted as much with Europeans as any people on the continent. Their idea of the character of knowledge and how it might be transmitted was utterly ascientific and nonrational—a barbarous worldview from the perspective of the heirs of Descartes and Galileo. Differences ran no less deeply in the approach of natives and Europeans to tangible things. Though the Quechan and other river Yumans possessed the means, through the generosity of the Colorado, to amass material wealth, they disdained accumulation. When a member of the tribe died, they burned not just the body but all the decedent's possessions, including his house.

It comes as no surprise, then, that the first Europeans who found their way to Yuma Crossing considered the natives so variously amusing, appalling, and astounding. It should come as still less a surprise that the litany of damages ensuing from contact with Europeans included for the Quechan and other Yumans the same sequence of disease, dispossession, and dependency that has afflicted virtually every other tribe and band on the continent. The dawn of history wakened them, as it wakened natives throughout the Americas, to conditions more nightmarish than they had ever dreamt.

It may be useless to wish that things had been different, yet to wish is as natural as dreaming. Imagine, for a moment, that by dream or any other means we might reenter those earlier times. Imagine going back to those moments of first contact, armed with what we now know of the promises and threats they embodied. Imagine discovering the means to steer a different course across the sea of intervening years and to land here on the shore of the present with a full continental cargo, still intact, of people, ideas, and understandings.

Side Trip: Jacumba Pass

If so many of the dream-based myths of the natives of these lands had not been lost, there would likely still be a sacred epic, a solemn tale requiring a week of nights to recite, that would guide the pilgrim up from the ocean and over the brute geography of the Peninsular Range, through and past the dangers of the route.

Not to demean the chronicles of bands known variously as Tipai, Diegueño, Kamia, or Kumeyaay, who eked out existence from these stingy mountains and stingier deserts, nor to mock the annals of the Yuman tribes of the Colorado River who plied this route in pursuit of abalone, but in contemporary terms, can one not transpose the realities of travel into a kind of metaphysics? Who among us, if not with corn pollen then with fuzzy dice or membership in AAA, fails to propitiate the evil twins, Breakdown and Blowout? Foremost among the presences in this landscape is the powerful repulsion of the Border, which hurls the highway back in strangled loops each time it seeks the ease and logic of a more southerly passage. Confined to rough country, the highway hews toward Jacumba Pass, the descent of which is marked in one direction by run-outs for brakeless trucks and in the other by shrinelike cairns promising water for foaming radiators. The closer one approaches the inland desert, the more mythic the landscape becomes. Tight canyons provide scant shade. Mile after mile reveals little but uplifted stone, stone that has been tilted, cracked, and shattered. One descends whole mountains of orange rockfall, a jumble of giant boulders. Erect, leaning, reclining, jutting into space, the rounded shapes give the appearance of a population, and so they have been described. "They are the souls of those who died in Imperial Valley," said one veteran of the farm labor crews, a soul long since departed to boulder land. And why are they here? "It's too damn hot in the valley," he is remembered to have said. "Soon as they died, they came up here to get cool."[1]

Onward one descends into the Land of Heat, the Land of Below, the land that for so many has been a land of Dreams.

3 | DEAD MULES AND NIGHTMARES

Mythic California, land of fair-weather ease and bounty, depends for its reputation not just on a balmy coast and golden, oak-stippled hills but on the deserts that surround it like a moat. Lasting images require strong contrasts. The orange grove demands the salt flat. The bubbling fountain needs the sun-seared gravel wash. The luster of gold shines the brighter when polished with suffering, and California polished it well. As America is alone among nations, California is alone among states: it has a dream of its own. Today we remember the preeminent California dreamers—the argonauts of 1849—as much for what they endured as for what they found.

The Quechan dwelling by the Colorado, in their way, were right. Dreams are the source of all power, and the degree of power a dream confers depends on the dream and the dreamer. But nothing the Quechan had dreamt, within the womb or out of it, could have prepared them to comprehend the magnitude of collective yearning that infected the thousands of over-equipped and ill-prepared whites who appeared at Yuma Crossing in 1849. Nor did most of the argonauts understand the fix they had gotten into. Wrote one,

> I deem it a matter of importance, for there has been so much suffering and loss of life not only on this but on other routes, that if I can be the means of saving one victim to the California fever, I shall be amply repaid. Hence my journal is for the public use in the shortest possible time.[1]

Journals there were aplenty. Then as now, education was no stranger to money, and the forty-niners as a group were not poor. It cost a fair bundle to abandon family and friends and wend one's way across a roadless continent. Some eschewed the continent entirely, booking expensive passage on crowded sailing ships hastily pressed into service. They rounded windy Cape Horn, then beat their way up the west coast of South America to the doldrums where the sun blazed and their gums bled from rank food and scurvy. Others sailed to Panama, trekked across the isthmus through leechy swamps and clouds of mosquitoes, then languished on the Pacific side in camps rife with dysentery and malaria. The infrequent northbound ships that delivered them from misery relieved them also of the weight of their pocketbooks.

Overland travel was hardly cheaper.[2] One had to purchase draft animals, harness, wagons, and camp outfits. Only the cholera that plagued emigrants on the northern trail up the Platte River came without charge. Possibly thirty thousand people, a third of the total migration, took that route in 1849, including most of the thousand or so women and children in the otherwise masculine stream.

Another nine or ten thousand chose the southern route—or routes, for many trails braided toward Yuma Crossing. They came down the Santa Fe Trail from Missouri, or across Texas to El Paso, or sailed the Gulf of Mexico and then hopscotched through the pueblos and placitas of northern Mexico into Sonora, ultimately to converge on the Pima villages along the Gila River, which (prior to the Gadsden Purchase) marked the international border, north and west of Tucson. From there they descended the Gila to the Colorado and the stares of the puzzled Quechan.

Many took the southern route for the simple sake of speed, hoping to steal a march on those who waited in Missouri for grass to green on the northern plains. Some, veterans of desert campaigns in the Mexican War, felt they knew the country and accepted its arduous *jornadas*—passages without water—as a calculated risk. Like all of the forty-niners, they were afflicted with the fervor of the times, which was pecuniary at the core but connected to much else. Their material hunger conjoined with a sense of hurtling destiny—a destiny that they, more than any generation before or since, believed to be divinely sanctioned and, for all with eyes to see and ears to hear, manifest.

Three years earlier, they and their countrymen had warred with Mexico to assert the right of Americans to as much of America as they wished to seize. Their sense of the rightness of the enterprise was embedded in their appropriation of the name of the continent for their national identity: they were *Americans*. The gold rush seemed to vindicate the war by suggesting that Americans were meant to find wealth where Spain and Mexico had dawdled. The gold seekers thought they saw a new day dawning, with themselves at the edge of the light. They thought they detected in their adventure the elements of classical literature, for they were after not just gold but new lands and new selves. They quested for a kind of Golden Fleece, and so, like Jason and his shipmates, they called themselves Argonauts. Through different lenses their intrepidity might appear noble or foolhardy, but the forty-niners believed they were rushing into a future of limitless promise, which had become incarnate in a place called California.

Predictably, experience tarnished the glitter of their dreams. On the southern trail, thirst, hunger, and exhaustion stalked every group. Draft animals weakened and died along the grassless *jornadas*. Pacts and partnerships that had formed farther east on full stomachs grew rancorous as belts tightened. Some argonauts consoled themselves that conditions could not worsen. Then they reached the Colorado Desert.

The passage from Yuma Crossing westward to the next reliable water on Carrizo Creek at the foot of the Vallecitos mountains was the severest trial of the so-called Southern Emigrant Route. It led across a hundred miles of harsh desert, skirting the southern rim of the Salton Sink. Anza made the crossing twice, not without difficulty. He had the advantage of military discipline and the good sense, both times, to cross in late fall or winter, when the weather was most forgiving.[3] Later, General Stephen Watts Kearny, at the head of an American army of conquest in the Mexican War, struggled across in the relatively cool month of November, 1846. The next major expedition, consisting of the Mormon Battalion under Captain Philip St. George Cooke, crossed in January, suffering by turns from frost and heat in six days of torment—mules failing and dying, wagons abandoned, and many of the men "so nearly used up from thirst, hunger, and fatigue, that they were unable to speak until they reached the water [of Carrizo Creek] or had it brought to them."[4] All this in midwinter. The passage won, Cooke looked back and wrote, "The extreme cold braced all and postponed the torture of thirst."[5]

But every forty-niner knew that in a race to riches, laggards were losers. And so in June, July, August, and September of 1849 they plunged into the summer furnace of what is now Imperial Valley, and paid dearly for their haste.[6]

ON JUNE 19, 1849, the rigors of the Colorado Desert lay not far ahead for newspaperman John E. Durivage, formerly of New Orleans, but on that day, as Durivage contemplated the Colorado River from its eastern bank, the concern foremost on his mind was the possibility of death by drowning. A few days earlier, three Americans had lost their lives when their raft struck a rock and broke apart as they attempted to cross the swollen river. Now Durivage and his companions contrived to fashion a boat from "a willow framework, covered with tents and India-rubber blankets."

No stranger to adventure, Durivage had accompanied General John E. Wool in his invasion of Coahuila during the Mexican War. In that capacity he posted regular dispatches to the *New Orleans Daily Picayune,* a practice he attempted to continue as he pressed toward "El Dorado."[7] A New Orleans friend had recorded the onset of Durivage's gold fever to the day: February 3, 1849. By March 4 he was shipboard and bound for the mouth of the Rio Grande. Six days later he landed. Then came passage upriver in a shallow-draft steamboat to Rio Grande City, a meager and dusty town rendered all but deserted by an outbreak of cholera. By land, Durivage continued upriver through other, similarly afflicted towns, then struck west to the dry Mexican interior. He soon overtook the rest of his party, which had left the river ahead of him. Although the expedition had organized under the high-sounding name of the Hampden Mining and Trading Company, for the present it traded only in broken-down mules and nags, the good stock of the region having been swept up by argonauts ahead of them.

After three months' hard travel across northern Mexico and down the Gila, Durivage and his party finally attained the Colorado: "The stream at this point is over one hundred and fifty yards wide, about twenty feet deep in the channel, and runs at the rate of seven miles an hour. It was a booming stream, sure enough."

They tested their boat, and it served well. Fourteen feet long, it could carry eight hundred pounds in relative safety. That took care of ferrying their men and gear. For their stock, they turned to the Quechan Indians: "They are splendid looking men. . . . An Indian charges a shirt to swim across with an animal. I have seen many expert swimmers, but none that can equal these—men, women, and children all swim well."[8]

As custodians of livestock, however, the Quechan suffered a conflict of interest—they relished the taste of muleflesh and satisfied it, justifiably enough, by dining on animals that strayed into their floodplain fields to browse upon their crops. Notwithstanding their ability as swimmers, they also compiled a less than perfect record guiding animals across the river, lead rope in their teeth. As fate and the Quechan had it, a conspicuous number of mule drownings occurred close to the far shore where it was not too difficult to get the animals landed. Durivage observed one such event, his wryness undamaged by rigors of the trail:

> In swimming over a friend's mule, he was drowned; but the Indian who had him in charge got him ashore, and in five minutes afterward a dozen Indians were seated near him, close to a blazing fire, and feasting from his carcass, which they roasted in the ashes. I was offered a very large chunk of underdone mule, but declined the *bonne bouche* on the plea of having just dined, notwithstanding the pressing of my mahogany-colored friend.[9]

By June 1849, relations between Quechan and forty-niners were in decline; as the summer progressed, they approached open hostility. Wave upon wave of argonauts scoured the land for what sustenance it provided, including those resources on which the Quechan depended. As Durivage observed,

> Our animals were fed on what few mesquite beans we could find and a very small quantity of cane. The numerous parties which had preceded us had almost stripped the trees of leaves and cut nearly all the cane, and it is my private opinion that those who follow will have very short commons.[10]

Some of those who followed broke into Quechan granaries and stole what they needed, taxing Quechan subsistence to the utmost and using their firearms when the Indians protested or retaliated.[11] Durivage, however, would not be present for the exchange of murders in the weeks ahead. California and the scalded desert beckoned. He was already down to two mules: one to ride and another shared for packing. Compared with their freshly drowned brethren, these last beleaguered beasts may have been unluckier still.

First, Durivage and the co-owner of his pack mule, a Dr. Brent, lightened their load: "We discarded everything that was not absolutely necessary for our trip, and doing so, scattered a good many tricks to the four kind winds of heaven."[12] Then they started downriver to the lower ford near present-day Algodones, a short distance below the international border. There they turned their backs to the Colorado and followed a well-trodden trail into the desert.

Their route roughly paralleled the channel of the Alamo River, skirting the southern limit of the Algodones sand dunes. The Alamo, unfortunately, was hardly a river. A discharge channel for the Colorado, it carried water only when the big river flooded. What the Alamo chiefly offered were several widely separated wells yielding small amounts of somewhat brackish, frequently contaminated water—but water it was. The first well lay about sixteen miles from the Colorado and afforded a spare campsite used by nearly every party attempting the *jornada* of the Colorado Desert. Durivage lost the trail as he approached the place, but his anxiety eased as he realized he need only follow his nose: "A mile or two before we reached this camp we smelled an intolerable stench of dead animals, which we afterward ascertained arose from beasts that had fallen into the first well."[13]

Some of these life-saving, if repugnant, water holes may have been the work of Kearny's or Cooke's men.[14] But not all. Most were uncovered by desert-dwelling Kamia Indians,[15] neighbors and allies of the Quechan, who wrung a tenuous existence from the Imperial Valley and farmed, when water was available, in overflow sloughs like the Alamo. Their wells might reach depths of twenty-five feet, the bottom being accessed by a sloping trench up to seventy-five feet long. With sides funneled out to prevent slumping, these excavations represented prodigious undertakings for people equipped only with crude shovels made of mesquite wood and woven baskets with which they hauled out the loosened earth.[16]

The second well lay twenty-six miles farther along, the route leading to it being especially hard going, for the trail cut across the dunes and required uphill pulls in loose sand. Wagons sank to their hubs and animals to their knees as men and beasts labored in the heat, expending precious energy and fluids. John W. Audubon, son of the famous painter, compared the dunes to those he'd seen on the Outer Banks of North Carolina. Another traveler wrote that "they resembled nothing so much as the crested billows of the gulf stream when lashed into fury by a violent North Wester."[17]

Durivage stood little chance of getting lost along this portion of the route because the trail, increasingly, was flanked by goods jettisoned by travelers:

> On every side the eye encountered objects attesting the distresses and misfortunes of parties who had preceded us. Every few yards on each side of the road, marking the track plainly, were dead mules by scores, saddles, bridles, blankets, broken trunks, bags, pantaloons, *cantinas* (hide panniers), and all sorts of articles.[18]

The scatter was even greater as Durivage neared the second well:

> Literally covering the ground were dead mules, fragments of harness, gun barrels, trunks, wearing apparel, barrels, casks, saws, bottles, and quantities of articles too numerous to mention. The hot air was laden with the fetid smell of dead mules and horses, and on all sides misery and death seemed to prevail.[19]

This place was called Alamo Mocho, a name describing a cottonwood tree that has been topped or dismasted. The presence of cottonwoods gave the Alamo River its name, and indeed, it seems that there had been a pretty grove at Alamo Mocho until forty-niners came through and chopped down every tree—to feed the foliage to their famished animals and for poles and stakes with which to shore up the well and keep the ever-sloughing sand from reburying the water.

Durivage found the site entirely barren, affording nothing for his mules to eat, and so, like most travelers, he and his small company of five or six (months having passed since the last mention in his journal of the Hampden Trading and Mining Company) pressed on as soon as they had watered, in spite of their exhaustion and the approaching darkness. The slough of the Alamo River here bent to the north, and the travelers departed it to the west, first to a copse of mesquite which they ransacked for beans to feed their animals, and thence onto a vast and ashen plain. Before striking out, Durivage and Brent sought to lighten their load further, lest their faltering pack mule give out:

> Notwithstanding we had left every article we thought we could possibly dispense with at the Colorado, we deemed it necessary to make still further sacrifices. Away went a bag of beans; out tumbled a suit of clothes; Major Emory's report and a canister of powder followed suit; a case of surgical instruments followed; and a jar containing five pounds of quicksilver with a small bag of bullets brought up the rear. To lighten my riding animal, I threw away a double-barreled fowling piece I had procured in Chihuahua, my heavy holster pistols, and a powder flask.[20]

Onward they crept across a monotone landscape, the ground so hard it bore no tracks, and now they risked becoming fatally lost. By this point, every party had been compelled to throw away all it could, and there was no longer any jettisoned baggage to mark the trail. The next well was twenty-eight brutal miles away, and its water was reputed to be unfit for man or beast. This was the *pozo hondo*, "deep well" (near present-day Seeley), also known as Indian Well. It lay in the trough of a second discharge slough of the Colorado, similar to the Alamo. This slough, in the course of the summer of 1849, would earn the name New River. As Durivage and his companions approached it on June 25, "a hot and disagreeable wind" blew against them. "Charged with the most intense heat, it came across the plain with the greatest violence, and it was with difficulty that the pedestrian could breast it." Durivage's party, now reduced to four white men and a black slave, dissolved into a line of fainting stragglers.

> A young friend of mine told me that he felt he must give out, and begged me for God's sake to bring him water if he did. The froth stood on his lips, and he could hardly articulate, while his blanched cheek denoted the dreadful thirst and exhaustion under which he was laboring. I cheered him up as well as I could, and as he was riding a fine animal, compelled him to put spurs to him and make for the well as fast as he could.[21]

Two and a half years earlier and blessed with cooler conditions, Cooke described this portion of the southern Imperial Valley as "a great flat of baked clay over which a sheet of water had evidently stood. . . . I saw large quantities of sea shells, some perfect. The ground has evidently been the bottom of the gulf, which has now receded a hundred miles. The salt on this plain confirms the idea." Soon enough, others with similar insights would begin to guess the true slope and elevation of the desert and to imagine the possibility of bringing water there, but in the summer of '49, the otherwise observant and wryly cheerful Durivage could think of nothing but his swelling tongue and failing strength. At last, the flats rose slightly and gave a farther view. In the distance, Durivage discerned "a few solitary, withered tree tops. . . . I knew it was the well." He signaled to his companions behind him, "and fainting, panting, and staggering from exhaustion, they rushed on to the well."

A note tacked to a tree said a previous party had used the water without ill effect. Durivage and the others greedily fell to:

> The water was detestable and at any other time would have proved a powerful emetic, but now it was *aqua dulce*. A tincture of bluelick, iodides of sulphur, Epsom salts, and a strong decoction of decomposed mule flesh were the component parts of this delectable compound.[22]

They waited until the sun was low, then set off, resting again after the moon had set and resuming their march at four in the morning as the eastern sky offered a promise of dawn. They headed for Carrizo Creek, an intermittent trickle that descended between two small ranges of mountains but lost its water for good where it met the valley floor.[23] By mid-morning the vanguard of their group had reached the first waters of the creek—another scene abundantly decorated with desiccated mule carcasses—and there they drank the salty liquid, carried water back to stragglers, and, when all were in, slept hot and dreamlessly in the meager shade of such bushes as they could crawl beneath. Durivage and his companions had come nearly a hundred miles from the Colorado River in four days of almost constant movement, drinking little, eating less, and sleeping only a few hours at a time. Hard miles still lay ahead. As evening came on they started again, and Durivage's mule, despite a long drink from the Carrizo, began to fail. While the others went ahead, Durivage stayed with the beast, but in vain. At last he abandoned the dying creature and staggered alone up the darkening valley of the Carrizo, burdened now with a pack that contained his last possessions. He had separated from his party and was using the last of his strength, but he was in watered country now, and for the first time in many days he felt sure he would survive.

ONE SUNNY July morning I visited the stretch of Carrizo Wash where Durivage lost his mule but celebrated the saving of his life. *Sunny* may understate matters. I climbed a sandy ridge to gain a better view of the forty-niners' route and checked the air temperature with a pocket thermometer: 98 degrees and the day still warming. I lay the thermometer in the sand at my feet and unsnapped my camera from its cover. Quickly, in less time than it takes you to read this, I shot two views: the first, down the wash whence the emigrants had come. Through the viewfinder I saw a near-infinity of creosote all the way to a shimmering horizon, which was bounded on the right by the scalded hulk of the Coyote Mountains. I pressed the button. Click. Then I turned and shot a second picture facing up the wash toward Vallecitos. In this direction the emigrants had departed, plunging into yet more creosote, mesquite, and burnt-orange salt grass and bound for a wall of purplish, unpromising mountains. Click again. Then I snapped the cover of the camera closed and picked up the thermometer. The mercury had swelled past the top gradation of 120 degrees Fahrenheit and had disappeared into the burning aluminum head, which I could not bear to hold.

I walked back to the car. With almost every step I felt the desert crust break underfoot, and several times I stepped through, shin deep, into cavities beneath. In these lands, as the rodents well knew, it was better to live beneath the ground than on it. Alarmed jackrabbits zigged and zagged ahead of me, seeming almost to fly, their feet scarcely touching the sand. Hardly a square inch of the sunbleached ground was untracked by one rodent or another. Tiny footprints with wispy lines traced between them I took to be the tracks of tail-dragging kangaroo rats. I marveled at what a busy, ground-crawling place this must be at night and wondered how well I'd sleep amid so much scurrying. In Durivage's condition, on the other hand, having come from even hotter desert where few creatures of any kind could live, I might have welcomed the company.

A few doves hurtled overhead, wind mewing through their wingtips, and flurries of tiny dun and green dickybirds feuded in and out of thickets of mesquite. I felt the soles of my shoes warming like the burners of an electric stove, and I sensed that if I overstayed my visit I would become a kind of human double-boiler, my steaming blood leaving my brains well poached. The day is July 27, 1993. Durivage passed through here in 1849 exactly one month earlier. He was in a condition that nowadays would earn him helicopter transport to a hospital and immediate intravenous fluids, but as he passed through this place, he was light of heart. He'd outlasted his mule and made it to California.

I admit to being less a desert rat than a sort of semiarid mouse. My body is not conditioned to hoard its water and salt the way a desert dweller should. Air temperature has now climbed past 105 degrees, and in this stretch of desert, which lies within California's magnificent Anza-Borrego State Park, I must make a conscious effort, before I surrender to the heat, to take note of my surroundings. In the journey here from the salt pan of the valley floor, the desert enriched itself with every foot of elevation. First, ocotillos rose like sentinels, gray and dead-looking at this time of year but sure to swell with winter rain, turn green, and erupt in scarlet flowers. Next came chollas furred with thorns, and the green, crenelated stumps of hedgehog cacti. The clustered daggers of several species of yucca also appeared, joined by an even more diverse array of leather-leaved and smoke-green shrubs. Still higher, as the land began to buckle into ridges, every hillside and sloping plain, every bend of wash and splay of alluvium showed a different pointillism of vegetation, patterns endlessly variable and the palette always softly undertoned with pink, gray-blue, and purple. Smoke trees lifted pale green arms from the washes, and the ever-present ocotillos, twenty feet high and treelike on the rubble slopes, stood like visitors from another world.

I asked my friends in the Imperial Valley to tell me how it is they know when it is really hot. "When the old folks who've lived here all their lives start dropping like flies," answered one. "When you don't have to tell the kids to put on their shoes," said another. "When the air you breathe burns the insides of your nostrils," offered a third. And then this from the shores of Salton Sea: "When you get up at dawn in your air-conditioned house and putting the dog out the door is like opening the oven to check on a Thanksgiving turkey."

But when you are really hot, you do not think about what it is like to be hot. You do not even imagine being cool. You just shrink away like an injured spider, withdrawing into smaller and smaller parts of yourself, until the last of you is gone.

I RETREAT to the car and drive a mile or so by dirt road to Palm Spring, an oasis that succored many forty-niners. Not to be confused with the city of the same but plural name, this waterhole never supported more in the way of human habitation than a primitive stage station—and that only briefly. A grove of palms thirty feet high once graced the place, but forty-niners and later emigrants cut down every one, as they had the cottonwoods at Alamo Mocho, to feed the foliage to their stock.[24] Such destruction must have placed hardship on the Kamia, who exploited every advantage in food and shelter that such a place might give. Along the emigrant trail, the Kamia's wells, mesquite groves, and other useful places were hard hit by the westering tide, and the Indians were soon displaced. For compensation, they received a year's bounty of dead and dying mules, a sudden potlatch of protein which they did not disdain to use; for a short while they may even have counted themselves fortunate.[25]

Today the Palm Spring feeds a cement basin where desert pupfish and tadpoles circle in algae-thick soup. From the thicket behind the spring, which includes a few young palms planted in recent years by park rangers from Anza-Borrego, you hear the cries and occasional explosive take-offs of Gambel's quail, which abound here by the water, closely watched, you may be sure, by coyotes and a Cooper's hawk whose discharges have whitened the top of a dead mesquite.

From such a perch on this day in 1849, a hawk would have seen a succession of slat-ribbed mules stumbling toward water, and behind them, exhausted men alone or in small groups, their clothes begrimed and skin darkened by exposure, staggering through the scrub, giddy at the prospect of survival.

What Durivage and others endured to arrive at this place is part of the raw experience that metamorphosed into a dream of California. His experience—or something akin to it—was shared by many of the ninety thousand others in that first gold rush year, and men like him soon dominated the society and spirit of the territory. Before James Marshall spotted yellow nuggets in the race of John Sutter's sawmill in January 1848, only about fifteen thousand non-Indians claimed California as their dwelling place. By the close of 1852 the number of non-Indians gathered there had risen to nearly a quarter million.[26]

Few brought wives or children, and so the beginnings of American California acquired a more individual, less familial character than that of other territories where families shared in the earliest waves of white settlement. California became a collection of single, independent operators, each a kind of soldier of fortune hunting not work or a new home but riches, and each having endured a rite of literal passage, a harrowing initiation, just to earn admission to the fray. To the extent that passage to California was dire and trying, it was also transforming. One left the past behind and sloughed off his old self like a snake skin, leaving it beside the trail among the dead mules and discarded baggage. In reaching California a man learned to jettison his former life—possessions, mementos, attachments. There was no looking back; survival prohibited nostalgia. One stripped to bare essentials and faced only the future.[27]

Some prospered by the transformation; some were broken. Among the latter was John W. Audubon, whose party, one of the largest and best capitalized to attempt the southern trail, languished in Rio Grande City while Durivage passed through. Cholera struck their group, killing at least four. Soon after, they were robbed of much of their cash, and their eighty-man expedition began to break up.[28] Audubon straggled on with a handful of others, reaching the Colorado three months behind Durivage. Like so many others, he kept a journal, and its bleakest passage describes the *jornada* of the Colorado Desert. Here is the crescendo of what he recorded:

> Truly here was a scene of desolation. Broken wagons, dead shrivelled-up cattle, horses and mules as well, lay baking in the sun, around the dried-up wells that had been opened, in the hopes of getting water. Not a blade of grass or green thing of any kind relieved the monotony of the parched, ash-colored earth, and the most melancholy scene presented itself that I have seen.[29]

The desert crossing broke both Audubon's spirit and his health. His daughter later wrote that when he died, short of his fiftieth birthday, "especially was the California trip present in his fevered mind and scenes were once more vividly before him."[30]

The desert passage had an opposite effect on other argonauts. An emigrant who called himself "Rambler" camped within a few miles of the horrific scene Audubon described and confided the following in a letter in which he referred to himself always in third person:

> The long time since "Rambler" has been heard from . . . has led to the conclusion he is no longer in the land of the living. The contrary is the fact. Privation and exposure have to a great extent made a new man of him, and he is now in the enjoyment of better and more robust health than he has felt for a number of years; and ere his return he anticipates such a change as will be, as it were, a transformation from sickly middle age to the vigor and energy of youth.[31]

Perhaps Rambler returned, perhaps not. Many like him failed to carry out the promises they made; many never had the chance. Rambler may have been murdered in a mining camp or spent the rest of his days like a California grandee. We may conclude only that he went on from the desert, imbued with a new spirit and accompanied by others more or less equally transformed. The same, and no more, might be said of Durivage, whose trail grows faint and then disappears as he nears the coast. Like so many of their fellow travelers of 1849, he and Rambler probably slipped anonymously into the ranks of a more or less new cultural species: the Anglo Californian. They were adventurous, individual, content to be greedy and greedy for contentment. Over the mountains they went to Warner's Ranch, thence to the coast. After they recouped their strength, they headed for the gold fields, intent on making not a living but a killing.

The forty-niners set the pattern for every future California boom and every migration to the Golden State—in agriculture, in films and entertainment, in defense contracting, in real estate. The message and style have always been the same. Just as America, alone among nations, stands for a dream, so does California, alone among states. The California version ratchets the American Dream to a higher, more compulsively happy, selfish, and sybaritic level.[32] In California the gamble is headier, and the money, when it comes, comes bigger and faster. The idea is to score big, to strike pay dirt, and then to savor and flaunt it as though it had been stolen from the jaws of death, which in the forty-niners' case, it had.

Side Trip: Yuma Crossing

At mid-century, even after the passage of Kearny's small but potent army in 1846 and Cooke's Mormon Battalion in 1847, and in spite of the presence of Lieutenant Cave Johnson Couts and his small patrol during the gold rush, the lands of the Quechan and Kamia belonged to the United States only in theory, and a singular theory it was. Under the 1848 Treaty of Guadalupe Hidalgo, Mexico had ceded to the United States a vast portion of the continent which included those lands. Neither party inquired how Mexico might cede what it did not truly possess. Mexico had exercised neither political nor military control of Yuma Crossing since the Quechan uprising of 1781, and the Quechan, to be sure, acknowledged no impairment of their sovereignty.

It remained for a bloody chain of cause and effect, catalyzed by possibly the worst cutthroats in the history of the American West, to seal the fate of the Quechan people and to assure inclusion of their lands within the growing empire of the United States. The chain begins with a Doctor Lincoln, said by some to have been a distant relative of the future president. In 1849 Dr. Lincoln organized a ferry service at Yuma Crossing with at least the tacit approval of the Quechan. He did not operate it for long.

In early 1850, out of the deserts of Sonora appeared John Glanton and thirteen bloodstained and conscienceless trailmates. Taking note of Lincoln's steady earnings, Glanton and his men made Lincoln an offer which the ferryman, covetous of life and limb, could not refuse.

Lincoln was right to fear Glanton and his gang. Some years earlier, a series of murders in San Antonio had rendered Glanton persona non grata in Texas. He recruited a small band and pushed south. He and his men soon entered the service of the governor of Chihuahua, having offered to exterminate Apaches in the greatest number possible. Compensation was fixed at so many pesos per delivered scalp. The hair of actual Apaches proving difficult to obtain, the Glanton outfit availed themselves of the hair of a substantial number of ordinary Mexicans abiding in villages scattered through the territories of Apachería. The villagers' hair, on shrivelled patches of skin, was as black and straight as any Apache's, and many

loyal citizens of the Republic of Mexico perished before the government understood the counterfeit it had engendered. At long last, *federales* set out on Glanton's trail, and a price of $8,000 was put upon his head. Glanton and his clutch of devils thereupon decamped for Sonora and, finding no welcome there, fetched up at Yuma Crossing, where, in spite of their pennilessness, they successfully negotiated a controlling interest in Dr. Lincoln's ferry operation.

Glanton's business practices, which featured murder (of a rival ferryman named Callahan), extortion of exorbitant fees, rape of *mexicana* passengers, kidnap and repeated rape of Quechan women, and frequent, indiscriminate threats of bodily harm, proved unpopular with all users of the crossing and residents of the area. A consensus soon emerged that the perpetrators of this reign of terror stood in need of killing. On April 21 or 23, 1850 (accounts vary), Quechan warriors performed the necessary service by clubbing, shooting with arrows, hacking, and stabbing to death eleven members of the Glanton gang, including Glanton himself, whose face was smashed in with a large rock. Quechan efforts were incomplete, however, and three of the scalp-hunters escaped by scrambling across the desert to San Diego, where their distorted report inspired formation of a militia that intended to teach the Injuns a lesson.

The mobilization was hardly speedy. It was late summer before a 125-man unit, under command of Quartermaster General J. C. Morehead, reached the Colorado. By then, peace had once more broken out, and a new ferry was operating without hindrance. But Morehead had come a long way for justice and glory, and the quiet along the river did not restrain him and his men from laying waste to Quechan fields and provoking a fight that produced casualties on both sides.

Recognizing the potential for calamity, higher authorities dispatched three companies of U.S. infantry under Major Samuel P. Heintzelman to subdue the Quechan and establish a permanent military post at Yuma Crossing. During their march to the river, Heintzelman's column passed the returning Morehead militia, in whose claims of victory one of Heintzelman's lieutenants detected seeds of doubt: "As we were going out we met the general and his army on their triumphal return, but the speed with which he traveled seemed hardly consistent with the dignified march of a conqueror. . . . If what the ferry company told us when we struck the Colorado be true, the alleged victor was only too glad to get back with a whole scalp to the settlements again."[1]

Heintzelman established Fort Yuma on a bluff above the river's west bank, and though an uneasy peace persisted through most of 1851, in the following year Heintzelman waged a systematic campaign of crop destruction against the Quechan and starved them into submission. His task was rendered easier by the heavy casualties the Quechan had suffered in their nearly continuous warfare against the Cocopa and Maricopa.

Heintzelman's greater problem turned out to be logistical. Even with a skeleton garrison, Fort Yuma required steadier and more constant supply than desert mule trains could provide, a need that, in turn, seeded development of a vigorous steamboating business on the lower Colorado. Ocean-going vessels sailed to a miserable anchorage at the head of the Gulf of California, where some of the world's most extreme tides mediated between saltwater and mudflat. There the ships transferred freight to sternwheelers that carried it upriver. After several days of incessant mosquito slapping, snag dodging, and winching across sandbars, the steamers tied up opposite the fort at a village that sprang up on the east bank of the river. Originally called Colorado City, the settlement soon became known as Yuma. Considering the chain of events that brought it into existence, the town has every right to claim John Glanton as its founding father, and every reason not to.

4 | MEMORIES OF SEAS

Fate played roughly with Durivage and charged him a toll it waived for later emigrants. Had Durivage delayed his desert crossing by two or three weeks and attempted the *jornada* in the very maw of July heat, he might have had an easier time, not because of any mercy from the weather but because of an extraordinary gift from the Colorado River itself. The river, however, first masked its gift by presenting a more daunting challenge than ever at Yuma Crossing.

When Durivage crossed the Colorado in the third week of June, he estimated its width at Yuma Crossing to be about 150 yards. Roughly four weeks later, a Texan forty-niner, name unknown, described the same stretch of river as "five or six hundred yards wide, with a very deep and rapid current."[1] When a member of the Texan's party attempted to emulate the Quechan, who routinely swam the river even in high flood, "the task proved too great for him, and he was drowned."

Having crossed to the California side, the Texan and his fellows struck out across the desert following the same route as Durivage and having much the same experience. At their first camp the water "was shockingly bad on account of a dead mule having been in the well; still we were compelled to use it." Next day they made the twenty-six-mile push to Alamo Mocho, where they rested, then by five in the afternoon got under way again, across country that "reminds one of descriptions given by travelers of the great Sahara desert."

Twenty-two miles farther the Texan and his party came to something entirely unexpected. Midway across the fearsome *jornada,* instead of desolation, he and his fellows encountered "a large, beautiful lake of water." The water was fresh and "excellent." They camped there, "in a pleasant grove of mesquite trees."[2]

The source of the unforeseen liquid lay downstream of Yuma Crossing. Along a section of channel unvisited and unobserved by forty-niners, the flooding Colorado had topped its banks, spilling water into a labyrinth of flats and sloughs. The water spread into tangles of willow and arrowweed, clumps of mesquite, and beyond, into pure desert bleakness. It flowed as though onto a teeter-totter, for stretches of the land were almost perfectly flat and tipped the flow of water now to the south, now to the north. The imperceptible divide between the two drifted across the drenched plain according to the logic of accident and chaos. A raft of snagged vegetation, an accumulation of sand or silt, a toppled cottonwood, a sudden pulse of current speeding up or slowing down, any of these changes might cause the flow on a given acre to veer along a new path. As acres added to acres, patterns of drainage might shift and rearrange. In such a way did the restless wilderness of the delta divide southbound waters destined for the Sea of Cortez from those headed northward to the still unnamed Salton Sink.

Some of the northbound waters collected in a discharge slough that ran through the present site of Mexicali and into the path of the emigrants. Where the slough widened, the waters spread out to form small lakes or lagoons.

The Quechan and their desert neighbors the Kamia were intimately familiar with this phenomenon. It occurred not every year but often enough that exploitation of overflow lands was a central strategy of Kamia survival.[3] Put simply, they planted fast-growing crops in moistened soils as waters receded. But there was no predicting precisely when or where such bounty might occur. Year by year, the amount of water involved, its distribution between sloughs, and the timing of flows varied hugely.

But for Americans on the gold trail, and even for experienced Mexicans who might have made the trip several times, the sight of a significant body of open water in midsummer in the midst of the continent's most barren desert seemed as unprecedented as it was astonishing. Travelers in that busy summer of 1849 drew on no great store of insight to call this visitation the *New River.*[4]

The river no doubt saved a share of lives. When Audubon came through in October, the waters had shrunk to a string of small lagoons with no surface flow between them, but where the waters had evaporated, another boon appeared: fresh grasses flourished in the moistened soil. The new forage was abundant enough that Rambler's party, in November, camped by the meadows for "two days and a half making hay," which they packed upon their animals to sustain them through the hard stretch ahead.

Lieutenant Cave Johnson Couts, who commanded a patrol that summer in support of surveyors marking the international boundary, deemed the sudden materialization of a river in the desert proof of divine providence: "This particular place of the river, favored with such luxuriant grass, can only be the work of an Invisible Hand to aid the thousands of distressed emigrants."[5] Couts established a base near the New River, which he named Camp Salvation, and from it he and later others dispensed relief supplies to needy travelers.[6]

William Hunter, a forty-niner from Missouri, reached the New River on November 25. By then the remnants of the summer overflow were

> contained in holes varying from a few yards to half a mile in length, and from a few feet to a quarter mile in breadth. The bed is at present shallow, and if the water ever runs at all, from the observation I have been able to make, I cannot say which course they take. I should, however, imagine that they must find an exit to the South East.[7]

Hunter's observations were accurate, but his conclusion was wrong. The New River flowed north. Those who later divined the source of the providential river would unlock the central mystery of the region and lay a foundation for the transformation of the Colorado Desert into the Imperial Valley. Many travelers and observers understood pieces of the puzzle, but no one yet grasped the pattern of the whole. The first to do so would be a young geologist in the employ of the army. His name was William Phipps Blake.

A NATION claims new lands twice. First, in military terms, it must hold them against the interests of rivals. Later, the ties of ownership deepen; the nation begins to claim its land by right of occupation and understanding, believing that it knows, better than its rivals, how to use and settle the new land, how to capture and multiply its riches. Understanding the Colorado Desert, however, was no easy matter. Before Americans could grasp the character of the newly won territory, a fresh chain of events, arising from the nation's severest tensions, had to unfold.

As mid-century approached, divisions over slavery and other matters widened between North and South, threatening to break the United States apart. But there was also a second great fracture in the frame of the nation. For all practical purposes the nation's far western holdings in Oregon and California lay as distant from eastern seats of power as had the thirteen original colonies from England. Without strong ties to bind them to the whole, ties strong enough to reach across a vast and thinly populated land, the western empire might break away.

The renowned explorer John James Abert, commander of the Army Corps of Topographical Engineers and namesake of squirrels and towhees, understood the problem better, perhaps, than anyone. In May 1849, the urgency of strengthening links to the west increased daily as reports on the magnitude of Sutter's gold strike trickled back from California. Legions of aspiring tycoons were packing their kits, kissing their wives, mothers, and sweethearts goodbye, and setting out by every conceivable path and conveyance for the land of golden opportunity. Abert wrote: "Unless some easy, cheap, and rapid means of communicating with these distant provinces be accomplished, there is danger, great danger, that they will not constitute parts of our Union. Then what will become of our great moral power, our great commerce, our infinite resources . . . ? We shall sink into two second rate governments if we are even able to maintain as good a position as that of second rate."[8]

Abert well knew that the steam locomotive should provide the answer. Steel rails could stitch the overgrown nation together, establishing bonds of economic interdependence and quick communication. Building such a railroad, however, lay beyond the capability of private capital. Leaders throughout the nation assented that government should undertake the effort, but the problem of selecting a transcontinental route posed enormous political difficulty. The eastern terminus would bring incalculable economic benefit to the surrounding region, and so northerners insisted it be in the North, southerners in the South. Compromise on this question, as on the extension of slavery into new territories, proved elusive.

At last Congress tried to place the choice outside politics. Dispassionate science, it decreed, should determine the decision. By an act of March 3, 1853, Congress instructed the War Department, and specifically its Corps of Topographical Engineers, to "ascertain the most practicable and economical route for a railroad from the Mississippi river to the Pacific ocean." The hope was to identify a single route that, on its merits, might command acquiescence, if not support, from competing sectional interests. In the end, what Congress got was far from definitive: several good routes existed, each with advantages and liabilities. Ultimately, the nation would not decide how to bind itself together until it had first come apart: civil war would rage and pass before the United States built rails to the far sea. In the meantime, the Pacific Railroad Surveys produced a flood of information about the western half of the continent and heightened immeasurably the eastern half's estimation of the value of its new empire.[9]

IN 1853 Congress was in a hurry. It gave the new secretary of war, Jefferson Davis, already a national figure and second to none in the fervency of his sectionalism, ten months to field the necessary survey parties, conduct reconnaissance of all possible railroad routes across a universe of unmapped plains, mountains, and deserts, and compile the findings.[10]

Davis dispatched four parties to the field, three to operate from east to west and a fourth to scour California's Sierra Nevada and Peninsular Range for mountain passes, low and even-graded, that might allow an iron horse to reach the coast. He placed the California reconnaissance under the command of Lieutenant R. S. Williamson, with Lieutenant John G. Parke assisting. Davis empowered the officers to requisition packers and livestock; he provided them an escort of twenty-eight mounted troops—"picked men and horses only"—to protect the surveyors from Indians and outlaws; and he appointed to the expedition a small array of civilian specialists: "one mineralogist and geologist; one physician and naturalist; two civil engineers; a draughtsman: who, in addition to their stipulated compensation, will be allowed the actual cost of their transportation to and from California."[11]

The last position to be filled in Williamson's party was that of mineralogist and geologist, which fell to William Phipps Blake, a recent graduate of Yale's new School of Applied Chemistry (soon renamed the Sheffield Scientific School) and the scion of a distinguished Yankee family. It was the chance of a lifetime for the promising young scientist, who became one of the first professional geologists in the United States.[12]

Within days of receiving his appointment, Blake embarked on the steamer *Georgia,* bound for Panama. Less than seven weeks later he landed in the Sacramento delta port of Benecia, where Williamson was equipping his party. Only twenty-seven years old, Blake cut a fine figure. He was six feet tall and possessed the strong, angular features and flowing mane that portraitists sometimes give their subjects whether they have them or not. The day after he arrived, he drew equipment for the trail and put his things in order. Next morning, July 10, amid the braying of mules and the groaning of wagon timbers, he and the rest of the survey party, complete with sabred escorts, head-tossing remounts, and canvas-topped supply train, set out for terra incognita.

For the rest of the summer and straight through to fall they crawled along the western face of the Sierra Nevada, examining every gap in its upthrust wall. Six passes in particular drew their attention, but one needed four miles of tunnel, others were too steep, and none, in Blake's words, "offered an especially favorable and easy route or inviting grades."[13] With time running out and nothing favorable to report concerning the southern route that Secretary Davis fervently desired, Williamson in early November split his command and sent Parke and Blake south to investigate the mountains east of the Mormon settlement of San Bernardino.

After resupplying, Parke's train of men, horses, and wagons began a gradual ascent through a sparsely settled land of large cattle ranches, toward the saddle in the mountain divide separating Mount San Bernardino on the north from Mount San Gorgonio (known today as San Jacinto) on the south. The surveyors' path rose smoothly and gradually all the day of November 12 and all the day following, at the end of which they found themselves, to their intense delight, "on a broad and gently sloping plain," which proved to be the summit of the pass.[14] Decades later, at the age of eighty-four, Blake described the party's discovery this way:

> Imagine, then, the enthusiasm with which the unknown great break in the mountain range . . . was approached by the members of the party as we made our way eastward from the region, then practically unoccupied but now including the towns of Colton and Redlands, and found an easy grade and open country for our train of wagons to the summit, only 2,580 feet above the sea. Here, at last, was discovered the greatest break through the western cordillera, leading from the slopes of Los Angeles and the Pacific into the interior wilderness. It had no place upon the maps and had not been traversed by surveying parties or wagons. From the summit, we could look eastward and southward into a deep and apparently interminable valley stretching off in the direction of the Gulf of California.[15]

On November 14, the party began descending the eastern slope of the pass through country so vast and devoid of vegetation (where now the hundred golf courses of greater Palm Springs humidly abound) that Blake, upon later viewing an illustration prepared for his report, apologized for the minimal liberties the artist had taken. It seems that the eastern engraver, working from Blake's quite competent sketch, failed to grasp the true and, for the artist, unimaginable austerity of this most austere of deserts. The "brown and barren-looking mountains," Blake wrote, "are represented in the engraving, but appear too near, and the plain too narrow. Grass and vegetation among the rocks, and clouds in the sky, have also been added by mistake." He forgave as he corrected: "It is seldom that an artist is called upon to picture a scene so barren and so desolate."[16]

A factor in that desolation, and the first of Blake's notable discoveries, was the erosive power of wind-driven sand. San Gorgonio Pass today is home to electricity-producing wind farms, great orchards of windmills, row on row, topped with propellers big enough for a B-29. It was Blake who first called attention to the winds of the pass. He deduced the constancy of their direction and the magnitude of their power from the evident weathering of the landscape's naked rock. Not even resistant materials were immune: "Long parallel grooves, deep enough to receive a lead-pencil, were cut on the surface of the hard and homogenous granite." Such observations of "the cutting power of drifting sand" might seem trivial today, but in 1853 they were at the leading edge of science.

Charles Lyell had published his three-volume *Principles of Geology* between 1830 and 1833, and at Yale Blake no doubt studied it minutely. In that work Lyell advanced the concept that came to be known as uniformitarianism—the idea that the earth attained its present form as a result of processes of sedimentation, erosion, and volcanism that continue to operate and to be observable today. Lands and seas formed, according to Lyell, not in a single day or week of Genesis but over vast stretches of what, thanks largely to him, we are accustomed to calling "geologic time." Analyzing the landscapes of the American West in uniformitarian terms became a chief preoccupation for Blake and his professional contemporaries, although the spotlight of scientific discovery soon focused elsewhere. In 1859, while Blake struggled to make a living as a consulting geologist, Charles Darwin, another Lyell disciple, published a treatise in which he applied Lyell's notion of continuous change to the process for which he named his book, *The Origin of Species.*

BLAKE'S HARVEST of discovery had only begun. Onward the surveyors descended into the bleak but chromatic bowl of distance and void. Despite the harshness and sterility of the land, its topography posed no obstacles. The eastern slope of the pass was as gradual and even-graded as the western had been. They followed no road, for none existed, yet they easily drove their wagons over the wind-hardened clay. For all they knew, they were the first white men ever to penetrate these unmapped precincts, and they were not far wrong. History records only an obscure few who preceded them.[17] Anza and his colonists, Kearny and the Army of the West, Cooke and the Mormon Battalion, and the legions of forty-niners had all crossed the mountains farther to the south.

Near present-day Palm Springs the surveyors stopped at a lone thicket of willow and mesquite concealing a hot spring. A band of Cahuilla Indians were encamped there—"many Indian boys and girls were bathing in the warm spring and a group of squaws were engaged in cooking a meal." Blake was pleased to find, amid the thicket, "a young palm tree spreading its broad, fan-like leaves." Like many of his future readers, Blake had formed most of his ideas about deserts from the tales of Ali Baba. He soon drafted the Cahuilla into a romantic vision of New World Bedouins: "The surrounding desert, and this palm tree, gave the scene an Oriental aspect; and the similarity was made still more striking by the groups of Arab-like Indians."[18]

By the standards of the day, imagining Indians as Arabs, bizarre though it may now seem, showed commendable open-mindedness. Blake's willingness to view Indians in a favorable light was hardly universal among early Anglos in California. Far commoner was the idea that desert bands deserved "to be regarded as the lowest scale of humanity."[19] And thus easily dismissed and dispossessed.

Two days later, on November 17, Blake and his trailmates encountered an astonishing sight. The party rounded a projecting ridge of the mountain and beheld "an unlimited view" down the long axis of the Salton trough, south and east toward the head of the Gulf of California. Immediately Blake noticed "a discoloration of the rocks extending for a long distance in a horizontal line on the side of the mountains," which he identified as a waterline. It followed "all the angles and sinuosities of the ridges for many miles—always reserving its horizontality." Soon, as they descended below the waterline, they discovered an abundance of shells littering the desert floor, which Blake rightly identified as having belonged to freshwater snails and mussels.

This was geology at its most exciting: in the same moment that he stood on the previously submerged floor of a now vanished lake or sea, Blake also stood on the verge of identifying and describing a vast, dramatic transformation of the surface of the earth. That night the survey party encamped near present-day Martinez, close to the largest Cahuilla village they had yet encountered. Other village clusters of Cahuilla dwellings, low domes thatched with arrowweed, lay scattered and concealed among the mesquite thickets of the valley floor. "Our coming having been duly heralded by *smokes* and excited express riders from one village to the other, the whole population had collected to gaze on us in wonder."[20]

The wonder was mutual. In correspondence dispatched from Yuma a few weeks later, Blake wrote,

> The squaws were horribly tattooed as well as painted, and manifested much interest in the advent of our wagons and such a number of "pale faces" to their hitherto almost unvisited retreat. Many of the men had often been into the settlements, and had learned some Spanish and accumulated quite a stock of second-hand clothing. . . . Old hats and odd buttons seemed to have been in especial demand, and were worn with much dignity on the occasion of our arrival.[21]

Throughout the evening's councils, Parke and Blake questioned the "principal men" about "the shore-line and water marks of the ancient lake." By way of answer,

> the chief gave an account of a tradition they have of a *great water (agua grande)* which covered the whole valley and was filled with fine fish. There was also plenty of geese and ducks. Their fathers lived in the mountains and used to come down to the lake to fish and hunt. The water gradually subsided "*poco,*" "*poco,*" (little by little,) and their villages were moved down from the mountains, into the valley it had left. They also said that the waters once returned very suddenly and overwhelmed many of their people and drove the rest back to the mountains. . . .
>
> The Indians had a grand feast and dance during the night, keeping us awake by their strange songs and indescribable noises.[22]

By the time sleep came, Blake was far along in deducing the history of the landscape he had entered. Although the barometer the party carried was fragile and imprecise (and had not been freshly calibrated), Blake knew that they had passed below sea level somewhere close to the mark of the old waterline and that they were still descending.

Next day, the surveyors tried to recruit a Cahuilla guide to go with them as they struck southward along the ancient lake bed toward the emigrant trail linking Yuma and Carrizo Creek (which in their reports they promoted to the status of a "wagon road"). "Mucho malo," however, "was the only information that could be extracted from him regarding the section we were to go over."[23] The reluctant guide

and every other Cahuilla whom the surveyors importuned knew a bad gamble when they saw one and warned the whites that their route afforded neither water nor grass and that they would not get their wagons through. Nevertheless, Parke, Blake, and the rest of the party set forth.

Toward sundown they came to what today is known as Travertine Point, a sharp, granitic projection of the Santa Rosa Mountains. The formation fascinated Blake. All the way up to the ancient waterline, which stood out brilliantly, the crags were covered by a calcareous deposit, in places two feet thick. This calcium-based deposit, variously known as travertine, tufa, or sinter, had formed as freshwater evaporated, leaving crystallized minerals behind. In lakes it forms at waterline, where wind and wave action bathe the shore with moisture, which then evaporates. In large lakes, which Blake's inland sea certainly had been, the effect is magnified by *seiches*—wind-driven lake tides. A visitor today to Parker Dam, which impounds Colorado River water to form Lake Havasu, will find similar deposits on the concrete wings of the dam.

Although he was embarked on a dire *jornada,* Blake lavished much time and energy on the exploration of Travertine Point. He clambered over its white crust, collected samples, took barometric readings, and, to judge by the detail of his final report, made voluminous notes. His romantic sensibility came to the fore. Laying science momentarily to the side, Blake deepened the recesses of Travertine Point into darker grottoes and shadier caverns than actually are there; then he mused upon the forces that shaped them:

> When wandering over these great masses of rock, and standing in the once subaqueous galleries and passages, with their walls and ceilings of the coral-like crust, the surfaces looked so new and fresh that it was not difficult to imagine that I heard the measured swell of the waves resounding in the dim caverns, and it was impossible to resist a feeling of dread that the great waters might suddenly return and claim their former sway over the deserted halls.[24]

Blake observed much at Travertine Point, but he made no note of the petroglyphs the Cahuilla and their ancestors had chipped into the tufa. Nor did he notice the stone fish traps they had built on the basin slopes north of the point. The boulder-rimmed traps dot the hills along a narrow band of elevation. They functioned as shallow corrals into which the natives harried the mullet and humpback suckers of the prehistoric lake as its wind tides receded.[25]

As for returning seas, Blake might better have saved his dread for the fearsome aridity of the land ahead. He and all his party soon recalled the warnings of the Cahuilla as they struggled to find passage for their wagons across the steep-banked arroyos that descended, west to east, from foothill badlands to the still unvisited bottom of the desert bowl. They clung to the belief that the spur of hills they presently faced was the last that separated them from the emigrant trail and its guarantee of water at Carrizo Creek. Again and again the land dashed their hopes. For two days and two nights they traveled continuously, their canteens soon as empty as the landscape. They were fortunate to attempt the trek in the relative cool of November, and without winter rains turning the clay over which they traveled into glue. Nevertheless, on the second day they faltered. "The poor mules began to fail, and cried out in their peculiar, plaintive manner."

They pushed on through a hellish night, yet did not reach water. As dawn approached, they were on the point of abandoning "the wagons and all heavy articles" when they arrived at the brink of yet another daunting ravine. This time, however, a trickle of wetness snaked down the bottom: "The cry of 'Water!' arose from those who first reached it, and it was repeated with loud shouts of joy from one end of the train to the other."[26]

The salty but life-saving waters were probably San Felipe Creek. The party rested a day in baking sun, then pressed the last twenty miles to the Carrizo and the well-used emigrant trail. From there they turned west and ascended the mountains to Warner's Ranch, where they rendezvoused with Williamson and the rest of the command. Williamson then sent Parke south to evaluate Jacumba Pass (which Parke rejected as a feasible railroad route), while he, with Blake, marched back into the valley and continued east along the wagon road to Yuma Crossing.

THE TRIP across the desert, past New River, Alamo Mocho, and the sand dunes, afforded Blake the opportunity to assemble in his mind still more of the region's geologic puzzle. Others had already suspected its character—Cooke in 1847, for instance, who theorized that the desert had once been submerged, and Heintzelman a few years later, who reached the same conclusion and with Quechan or Kamia guides rode almost to the bottom of the basin to inspect the mysterious mud volcanoes that boiled and bubbled there with infernal gases.[27] But it was Blake who finally joined the available fragments of information into one logical and explanatory picture. While many details have since been added to Blake's portrait of the region, its main elements remain unchanged. One may say today, as was said in 1909: "The work, though done in the briefest time and under great physical difficulties, has borne well the tests of all later investigations."[28]

Blake, however, did not reach the right conclusions all at once. In his ride across the desert to Yuma he correctly reasoned that the desert basin had once been an arm of the Gulf of California and that the emigrant trail followed the crest of an earthen barrier that had formed across a narrow waist of the sea. He further theorized that with the barrier established, the Colorado might at times have flowed into the northern half of the divided basin, filling that section with a freshwater lake which then overflowed the barrier southward, possibly discharging along the upper channel of the New.

What Blake did not immediately divine was the cause of the barrier's formation. In a letter to the *San Francisco Commercial Advertiser* written from Yuma, he ascribed the development of the barrier to "submarine elevation"—the same set of forces, he said, that had raised the continents from the oceans. As for the ancient lake, he vastly overestimated its size: "Its extent and boundaries cannot be precisely determined until the maps of the region are completed, but it is probable that its area will not be less than 7,500 square miles"—an area the size of Connecticut.[29]

Perhaps while at Fort Yuma Blake examined the various escarpments and terraces of the desert and refined his earlier conclusions. He realized that the eastern border of the ancient lake had stopped well short of the Colorado. He may also have found time to draw a volume of Lyell's *Principles* from his saddlebags and reread the section on river deltas: "Sir Charles Lyell gives a statement, made by Colonel Rawlinson, that the delta of [the Tigris and Euphrates] has advanced two miles in the last sixty years, and is supposed to have encroached about forty miles upon the Gulf of Persia in the course of the last twenty-five centuries."[30]

With Lyell's help, he discovered the central dynamic of the basin's topographic evolution, which he subsequently explained in his report. The northern arm of the Gulf of California had formerly reached far up the Salton trough "to the base of the San Bernardino Mountains." The mouth of the Colorado River would then have been located near Fort Yuma, discharging both water and sediments under the shadow of Pilot Knob. The river's sediments, dropped from suspension, accumulated to form a delta, and "the encroachment of this delta, and its final extension to the opposite shore, was sufficient to shut off the waters of the upper end of the gulf, leaving them in the condition of a lake, connected with the river and Gulf by a narrow channel or slough."[31] Blake named the postulated lake for the Indians whose tradition of a great water abounding in geese, ducks, and fish lent further evidence of its previous existence. "The Great Salt Lake of Utah is the residual lake of Lake Bonneville much as the Salton Sea is the residual lake of Lake Cahuilla."[32]

DELTAS ARE the most inconstant of landforms. A river with an active delta may raise itself through deposition above the level of the surrounding land, building a perch that becomes increasingly precarious. In flood the river may deposit a new plug of sediment or vegetation, or rip through an old plug with new strength, so that it breaches its banks and spills on one occasion one way, and on the next, another. In such a way the Colorado may have repeatedly formed Lake Cahuilla over previous millennia and just as repeatedly abandoned it to evaporation. When the lake was full, it presented a bounty to the Cahuilla on its north shore and, presumably, to Yumans or others on the south. The most recent recession of the lake probably occurred between 1400 and 1500 A.D., only a few generations before Alarcón sailed into the mouth of the Colorado with the royal banner of Spain flapping from his mast.[33] The desiccation of the lake may have forced lake-dependent bands to compete for territory on the already occupied banks of the Colorado, which might partly account for the enduring hostility among river tribes that Alarcón and every European who followed him observed.

Spaniards bestowed the names of colors on both the Vermillion Sea and the Colorado River because of the heavy load of sediments that reddened them. Said one geographer in 1932, before Hoover Dam was built: "As a carrier of silt the Colorado is probably without a peer among the greater streams of the world."[34] Other researchers have estimated that the Colorado's burden of sediment approaches seventeen times that of the Mississippi and ten times that of the Nile.

And so when one gazes into the Grand Canyon and wonders where the content of such a mighty rent in the earth's crust might have gone, the answer, simply, is *downstream*. The soils of Imperial Valley, the clays beneath Salton Sea, the plain on which Mexicali sprawls, and all the salinizing fields southward, down to the mudflats at tidewater and the sea bottom off San Felipe—all these are the Grand anti-Canyon of North America. The miracle is not the quantity of material but the speed with which it accumulated. The formation of the modern Colorado and its canyons remains imperfectly understood, but the river probably did not commence its great work of erosion until about sixty-five million years ago, when the Rocky Mountains began to take their present shape. In the brief span of geological time since then, the river and its tributaries have abraded the intermountain West and deposited the sloughings in the Salton trough and the upper Gulf of California. Without dams like Hoover, Glen Canyon, and Imperial, it would continue to do so now. Instead, it deposits its million tons per day of sand and silt in lakes like Mead and Powell, thus guaranteeing their eventual obsolescence.[35]

WHEN BLAKE wrote to the *Commercial Advertiser* from Fort Yuma, the agricultural potential of the desert was already on his mind, and he promised that he would include his observations on the subject in his official report. Chief among these was that wherever water was applied to the desert soils—as at the hot springs of the Cahuilla villages or among the ephemeral hay meadows of the New River bottoms—trees and grasses sprang forth luxuriantly. "From the preceding facts," wrote Blake, "it becomes evident that the alluvial soil of the Desert is capable of sustaining a vigorous vegetation. The only apparent reason for its sterility is the absence of water." He further mused that "by deepening the channel of New River, or cutting a canal so low that the water of the Colorado would enter at all seasons of the year, a constant supply could be furnished to the interior portions of the Desert."

Blake was neither a farmer nor a developer. When he looked at the desert, he did not think economically. He did not imagine the fortunes that might be made by selling land or raising crops. He was a scientist, not yet thirty. When he looked at the desert, he thought of questions, each of which possessed interest, including the question of whether the desert might again be inundated. Prophetically, he noted that bringing irrigation waters to the desert would not be without danger: "It is, indeed, a serious question whether a canal would not cause the overflow of a vast surface, and refill, to a certain extent, the dry valley of the Ancient Lake."[36]

PART TWO

THE GREAT DIVERSION

5 | LOOMINGS

In the twilight of the Ming Dynasty, soldiers under the order of General Gao Mingheng breached the dikes of the Yellow River where it snaked across the plains of Kaifeng. The general wished to drown a peasant revolt and so employed the great, soup-thick river as a weapon of war. For vast distances, the Yellow flowed on a bed that its own depositions had elevated above the surrounding landscape, with the result that it had acquired the character of an aqueduct. By cutting the dikes, General Gao allowed the river to pour out and submerge the plain, painting a sea of crops and untold corpses with a coat of silt. Alas for the Mings, the general's ruthlessness did not long preserve them. Two years later, their dynasty collapsed.[1]

As General Gao loosed a flood from a perched river in 1642, so, too, did the California Development Company loose a flood in 1905 from a similarly perched stream—but not out of anger or tactical intent. In the case of the CDC, bungling, negligence, and greed sufficed to generate disaster. But nature assisted. The Colorado was then entering a period of natural restlessness, when its main channel, having moved to the eastern limit of its delta, should begin new movement to the west. This predisposition, combined with monumental volumes of snow and rain far upstream, made the months leading up to the 1905 flood a particularly dangerous period for tinkering with the river. Indeed, from the present day back as far as the time of General Gao, the CDC could hardly have selected a worse time to demonstrate its ineptitude.[2]

The great flood of the Colorado Desert defied efforts to contain it for two years—from February and March 1905 until February 1907. The flood accomplished a number of things. It created the modern Salton Sea, which, while not the largest man-made lake in existence, surely qualifies as the most inadvertent. It reworked the topography of the Colorado Desert, carving canyons where only sloughs had formerly existed and wiping out the chain of shallow lakes that on occasion had nourished forty-niners and other weary travelers. Most significantly, the flood—or more accurately, *floods*—contributed mightily, and probably decisively, to redefining the respective roles of the United States government and private capital in the reclamation of arid lands.

The floods were amply foreshadowed. A summer overflow in 1849 left pools of water in the channel of the New River and saved the skin of scores of argonauts. A similar overflow in the winter of 1891 stimulated considerable curiosity about the water's source and produced, as we shall see, one of the more remarkable boating excursions in desert exploration.

In the years between these two widely noted overflows, prospectors probed the desert's slopes and canyons and scored paying strikes along the Colorado River in Mohave and Chemehuevi country, well upstream of Yuma. A regular road soon linked the mines to the coast. It descended San Gorgonio Pass and touched the northern rim of the Salton Sink before slipping north of the Chocolate Mountains and traversing the Mohave Desert to the river. Along this route traveled a steady flow of men, trade goods, and, to the everlasting misfortune of the Cahuilla, smallpox. Within a few years, stage lines began to serve the mines and forged new links southward to Yuma and the old emigrant trail, where, for a time, the coaches of the Butterfield Overland Mail Line rumbled, supported by relay stations at every waterhole.[3] Stockmen, pushing their herds into every corner of the West, did not neglect the sink. In winters when grass was exhausted in the mountains, they drove their herds into the basin to graze the less and less luxuriant grasses in the sloughs of the New and the Alamo, a practice that consumed resources essential to the Kamia.[4]

In 1877 the Southern Pacific Railroad finally penetrated the low hot lands of the Colorado Desert. It pushed its tracks across the salty, sub-sea-level flats at the foot of the Chocolate Mountains. The sidings it established for watering, loading, and maintenance gave the desert a new crop of names: Glamis, Acolita, Mammoth, Flowing Well, Volcano, Frink, Durmid, and Salton, the last of which acknowledged the salt beds of remarkable purity that lay at the bottom of the sink.[5]

Industry, in the form of the New Liverpool Salt Company, soon followed the railroad to the basin. The salt that the company plowed up, sacked, and shipped out on the Southern Pacific came not from deposits of ancient oceans or residues of river floods but from saline hot springs that emptied in the basin. The evaporation of their bitter waters left behind deposits of sodium chloride more than 95 percent pure.[6]

The salt beds were less a mine than a white-fielded farm whose sterile crop renewed itself, harvest after harvest. A desert rat by the name of Arthur Burdick called it "the most remarkable harvest-field in the United States, if not the whole world."[7] The beds lay some 265 feet below sea level, perhaps 10 feet higher than the absolute bottom of the basin. Stationary steam engines winched gang-plows back and forth across the beds to break the crust of salt. Each plow could loosen seven hundred tons of salt a day, and each was guided by one or two Cahuilla Indians. All the field hands of this salt farm were either Indian or Japanese, for, as Burdick noted, "in the summer season the temperature reaches 130 to 140 degrees at Salton, and white men are unable to endure the work." Laborers followed behind the rust-ravaged plows and raked the salt into neat cones. Their feet were likely never dry of the corrosive brine that leaked beneath the crust and, like their rough hands, must have been marvels of callus—as well as deeply cracked by sores that could not heal. The laborers loaded the salt onto wagons and then onto a tram that carried it to a drying house for grinding, sorting, and sacking. More workers ranged the sacks in boxcars for shipment to the California coast, where the salt sold for six to thirty-six dollars a ton, rated by quality. The salt beds, meanwhile, were impervious to exploitation. The mushy, caustic furrows torn by the plows soon filled again with brine and scabbed new crusts. Thus did the salty farm grow crop after crop, unbidden by its farmer and tended, as nearly all the crops of the region were destined to be, by brown hands.

IN 1891 Cahuilla laborers at the salt works spotted a sheen of water in the lowest part of the basin. The sheen grew; its edge moved closer. Without delay, the Cahuillas collected their few things and left for higher ground, "muttering statements of impending disaster."[8] As Blake had noted, Cahuilla oral tradition included tales about the sudden and dangerous appearance of a lake in the desert. Warned by legend, the workers at the salt beds knew what to do.

From somewhere southeast of Salton that February, the passengers and crew of a Southern Pacific train beheld the same unanticipated sight. At first they believed it to be a mirage, for the refraction of light on the featureless desert often gave the impression of a flat, reflective sea. But the water they thought they saw was actual water: a lake had formed in the bottom of the basin. The train carried word of this to Yuma, where it provoked puzzlement. Floods had ravaged Yuma only days before, the result of heavy rains in Arizona that had caused collapse of the Walnut

Grove Dam, killing hundreds. The Gila River surged to a calamitously high stage and, joining with the already hefty Colorado just north of town, swamped most of Yuma. An observer, Godfrey Sykes, described the scene:

> The main and only street was a muddy gully, adobe and willow-pole houses, stores and saloons were mostly heaps of ruin and the inhabitants were either grubbing about in the piles of debris in search of treasured belongings or sitting perched upon the higher heaps, considering matters. . . .
>
> Structures of the types of which pre-diluvian Yuma consisted readily dissolve under such circumstances, and the heavy, heat-resisting earth roofs which were almost universal had merely added to the volume of the mud-heaps.[9]

Sykes had arrived in Yuma by a most roundabout route. Although he never settled in the region, in the decades ahead he proved to be a sympathetic witness to its troubles and transformation. By his own account, Sykes suffered from a rare affliction. He called it the "wanderlust bacillus," a restlessness that forced him at the age of eighteen to leave his native England in search of horizons less dreary and gray. A steamer carried him to New York in 1879, when the harbor was still a forest of sailing masts and the cables of Brooklyn Bridge had yet to be hung. By the following spring he was in the heart of Texas, living by his wits, which included extraordinary abilities as a tinker and mechanic. Years later he would supervise construction of a dome for the Lowell Observatory near Flagstaff, but in the Abilene of 1880, he was content to undertake the repair of a barrel of condemned army pistols, a labor that earned him the ownership of what he thought was indispensable equipment for the time and place—a serviceable six-gun.

Sykes's happiness grew when he was hired to replace an ailing cowpuncher in a trail crew bound with longhorns for Dodge City. The owner of the herd solemnly conferred on him the departed man's ponies and bedroll. Thus equipped, Sykes officially became a cowboy.[10]

Sykes rode the long trails for several years, sometimes as far north as Ogallala or Cheyenne. With irregular success he also took time for other ventures. He caught wild horses in west Kansas—or tried to. He became a mule skinner hauling freight to Fort Sill, and he prospected for gold in Indian Territory, as Oklahoma was then known. When some of the translocated tenants of that vast reservation took him prisoner, an army detachment rescued him. Somewhat later, he prematurely ended a game of pool in a Dodge City saloon by riding his horse on top of the pool table, a tactic that irritated his competitors but allowed him a rapid departure when gunfire erupted.

Sykes wandered west to San Francisco, then east to the 1884 Republican Convention in Chicago, where he hired out as a paid shouter, west again to Japan for a commission on which his autobiography does not elaborate, then to the South

China Sea, where he bummed passage on tramp steamers in the same waters where Joseph Conrad was sailing. Next he fetched up in Australia, where he found brief employment in "water development."

Eventually he returned to the small property near Flagstaff, Arizona, that had become a base for him and his brother. By dabbling in cattle and logging, they subsidized their other interests, and Sykes, still feverish with wanderlust, resolved to explore a place described to him by the master of the ship that had taken him to Japan. In earlier years the officer had served on a coastal vessel delivering freight to sternwheelers on the lower Colorado. He spoke of "uncharted dangers, strong currents, and the enormous tides that rose over the mud-flats surrounding the mouth of the Colorado," and he described "river navigation as an advanced form of lunacy well worthy of investigation by anyone as crazy as I appeared to be."[11]

Crazy indeed. Sykes's neighbors in Flagstaff had no doubt of it as they watched him build a twenty-two-foot sailing vessel in the shadow of the San Francisco Peaks, hundreds of miles from the nearest glint of open water. Sykes completed the fitting-out of his craft and cadged free transport on an Atchison, Topeka and Santa Fe flatcar. The railroad delivered the boat to Needles, California, on the Colorado River, where it aroused such interest that the townsmen who passed for musicians broke out their instruments. To the rat-a-tat of drums and bleating of horns, the boat slid into the river.

With a friend to help with oars and sails, Sykes drifted down to saltwater at the head of the Gulf of California and there survived the springtime's tidal bore, which in the days before the Colorado was dammed sometimes generated a wave six or seven feet high that roared up the river faster than any boat could flee.[12]

Safely under sail in open water, Sykes and his partner cruised the gulf for several weeks, exploring both shores of the narrow sea, until disaster struck. They had camped some distance north of Punta San Felipe, today a Baja California beach resort and fishing village but then a deserted stretch of sand. The wind blew hard. Sykes went back to the boat to light a lantern under the shelter of the bow decking. He used a number of matches. One must have smoldered on a oily rag, for minutes after he returned to the campfire, he and his partner were startled "by a muffled explosion and red glare showing over the shoulder of our protective dune."[13] Their store of kerosene had burst; the boat was engulfed in flame. And the powerful tide was out: no water, only mud, as far as the eye could see.

The fire utterly destroyed the boat and most of their provisions. Nor was there another human, nor wood for a raft, nor road or trail, nor even freshwater for many a mile. Nothing but sand and mud and saltwater tides. Sykes had managed to recover his tinker's tools, however, and he fashioned a pair of large canteens from some galvanized sheet metal he salvaged from the wreckage. Then he and his understandably disconsolate shipmate set out for Punta San Felipe, where they hoped to find a spring. They dined on the gristle of a very lean coyote along the way.

Resupplied with water, they reversed their path to the north and then northeast, struggling across the soft-surfaced brush-jungles of the Colorado delta.

They swam the deep water of the Río Hardy and the main-stem Colorado and waded the bayous and backwaters in between. They slogged across mudflats and through thickets. They failed to starve, thanks to a small supply of hardtack they saved from the boat and also to a fortuitous encounter with a large mullet, which they cooked over a salt-grass fire. At last they came upon a human track. It led to a camp of hunters, whose success in pursuing the delta's feral hogs allowed Sykes and friend to feast on pork and bacon. The last fifty miles to Yuma passed quickly, and there Sykes beheld the flooded wreckage of the town. While his erstwhile co-voyager elected to retire from exploration, Sykes, typically, felt the call of a new adventure.

THE APPEARANCE of a lake in the Salton Sink, within sight of the railroad, had alarmed the businessmen of Yuma perhaps even more than the flooding of their town a few days before. They feared lest the growing lake submerge the tracks and isolate Yuma from commerce. The source of the new lake was a mystery. Some said rains in the San Jacinto Mountains must have caused it. Some hypothesized the existence of subterranean rivers or other notions equally strange.[14] Sounder minds postulated a connection between the new lake and the flooding Colorado, but the path of that connection was unknown. Sykes, being freshly arrived from the delta, was consulted, and he parlayed the consultation into a commission: he would drift downstream on the Colorado and find the channel to the sink. His commissioners provided him a stack of salvaged boards with which to build a skiff, and a similarly salvaged supply of tinned provisions. The canned food promised to augment the sense of mystery attending the expedition, for the flood had washed off the labels. At any meal Sykes might open a can to discover hash or apricots, or beans, beans, beans.

Sykes quickly built his boat and recruited a new and amiable companion known as "Beer-Keg Tex." The explorers shoved off from Yuma into the muddy current and spun away downstream.

They drifted a day or two, probing the right bank of the river until they found a gap through which water flowed. They turned their craft through the gap and soon found themselves in a maze of brushy channels. Mudbanks had begun to emerge as the water level fell, but the land was still awash. Sykes had planned to draft a map of their route, but he soon abandoned the task. The amphibious plain afforded no landmarks, no view longer than a stone's throw, no means to count the twists and turns of the current that bore them onward.

Eventually their course verged northward, and the vegetation that confined them thinned, then fell away entirely. They floated north through the barren gravels that had so disheartened the forty-niners. Mile by mile the water shallowed, then gave out. Their boat grounded in a drying mudflat. Only emptiness lay between them and the distant mountains. At length, they descried a plume of smoke inching past the foot of a purple range. A locomotive, they judged, and started walking. Sykes later calculated that they abandoned their boat "a few miles northwest of the present town of Holtville." From there they trekked to the railroad siding at Volcano, a distance of thirty-five miles.[15]

SYKES WAS almost certainly the first to travel by water to the arid heart of the Salton Sink, but he was not the last. The Colorado topped its banks again the following summer, and H. W. Patton, a newspaper reporter representing both the *San Francisco Examiner* and the *Banning Herald,* duplicated Sykes's feat with somewhat greater fanfare—his account of the dangers he faced in the badlands through which he floated was widely published. It bears on future events to note the pattern of the floods: high water from winter rains followed by high water again in late spring and summer. The logic of such a pattern is quickly evident. Winter weather systems that pour rain on Arizona and prompt the Gila to surge also cause heavy snowfall in the southern Rockies, especially on the western slopes that feed the Colorado. Deep snowpacks produce high runoff when warm weather arrives. Augmented by spring rains, meltwater floods can reach heroic size.

It might have been wise for the developers and early settlers of the Imperial Valley to reflect on such patterns and to ponder the substantial variability of the Colorado's flows. Instead, they elevated wishful thinking to the status of knowledge, assuring themselves that the river "rises and falls with almost clock-work regularity."[16] They might also have remembered that the break in the bank of the Colorado through which Patton drifted was over a mile wide and that he floated through the delta on north-flowing sloughs that converged into rivers of considerable size. The desert floods were large and anomalous enough to attract the attention of the American Geographical Society, which summarized the factual material in Patton's account:

> Turning into the main channel [of the Alamo River], which is about 200 yards wide and 18 feet deep, with a current of 4 miles per hour, [Patton's party] floated along over the old overland stage route. Marks on the trees showed that the water had been 4 feet higher.
>
> At Alamo Mocho, an old station 52 miles southwest from Yuma, there was water all over the country as far as they could see.[17]

Ten miles farther west, the stream they rode converged with the New River. The combined flows, according to the geographical society, made a river "300 yards wide and 14 feet deep with a current of 6 miles per hour." When they passed the site of modern Mexicali, their channel "was 200 yards wide and 7 feet deep, and the current was very swift. After going through this break they saw 10 large streams join their channel on the west, showing that there are other breaks in that direction."

The author of the summary advised that events "will eventually turn the whole volume of the river into the desert," a dire prediction but not a new one. A similar thought had occurred to William Phipps Blake nearly forty years before.[18]

PORTFOLIO I

The Colorado Desert

Seventeen Palms, an oasis in the Anza-Borrego badlands

Petroglyphs in Quechan territory

Geoglyph

View near Travertine Point. The line of whitened rock marks a shoreline of ancient Lake Cahuilla

Petroglyphs incised into travertine

Algodones sand dunes

Horse intaglio

Mud volcanoes in the Colorado River delta

Mud volcanoes
on the east side
of the Salton Sea

Yuha Well, which furnished water for Anza's expeditions in 1774 and 1775-1776

Box Canyon, on the Southern Emigrant Trail

Palm Spring

Carrizo Creek

The Algodones dunes posed a formidable obstacle to early travelers, which construction of a plank road in 1916-1917 somewhat overcame.

Beall's Well in the Chocolate Mountains.

Port Isabel, a steamboat landing and dry-dock in the lowermost Colorado River delta

Salton Sink, near Niland

Alamo River

All-American Canal

Headgate machinery at Hanlon Heading

Yuma Crossing of the Colorado River

Flooded shoreline of Salton Sea, late 1980s

New River canyon at Calexico, carved by the Great Diversion

25
M.P.H.

Near the headquarters of the Salton Sea National Wildlife Refuge

6 | NATURE REDREAMT AND REDRAWN

Picture in your mind a cinematic landscape: a sun-bronzed, Stetsoned man stands beside a beautiful young woman in a windblown waste. She is chaste, well-spoken, and in love, in love with the desert. She is also capable and resolute. She has just put away the pistol she drew when the handsome man surprised her in her place of solitude—a sandy hilltop overlooking a vast desert basin.

"Do you come here often?" he asked curiously.

"Yes, often," she answered. "I could not get along without my Desert and this is the finest place to see it. The Seer always comes out here with me when he can. Do you think that land will ever be reclaimed?" She faced him with the question.

"Why, no one can say about that, you know," he answered slowly. "There has never been a survey."

"Well," she declared emphatically, "I know. It will be. Listen! Don't you hear it calling? I think it's for that it has been waiting all these ages."[1]

THE YOUNG woman is Barbara Worth, heroine of Harold Bell Wright's melodramatic novel about the settling of the Imperial Valley, *The Winning of Barbara Worth* (1911). It is a preachy and sentimental tale, but it struck a nerve with the reading public, eventually selling over a million and a half copies.[2] When the *Winning of Barbara Worth* became a (silent) film in 1926, none other than Gary Cooper, in his first screen role, played the strong but self-effacing Abe Lee, who portentously confirms to Barbara, "There has never been a survey."[3]

A survey does several things. It converts wild land into property susceptible of ownership and available to markets. More than the removal of native people or the eradication of predators, a survey confirms the taming and domestication of space. A reclamation survey does even more: it measures the fall and drop of potential waterways, the quantity of land that might be placed "under ditch," the spacing and alignment of canals, the location of diversions, headgates, and overflow channels. It provides a blueprint for a closely controlled physical world.

In the universe of Wright's surrogate Salton Sink—which his Anglo characters called "King's Basin" and his Mexicans, always more poetic, termed "La Palma de la Mano de Díos," the palm of God's hand—civilization advanced inexorably as white settlers moved west, and good consistently won out. The moral hierarchy governing human relations, though it might shake, never toppled. At the top of Wright's moral ladder stood a noble character called the Seer, whose destiny it was to release the wealth of the land, wealth that Providence had hidden "until Time should make the giants that could take it."[4]

These giants were surveyors and engineers, men (always men) who possessed the knowledge and daring to reshape the world. The Seer, their natural chief, was a visionary high priest of reclamation. His grasp of the possibilities of the desert, combined with uncommon perseverance, would transform the region, create homes for thousands, and advance the progress of humankind.

Wright based the unflawed character of the Seer on the altogether fallible Charles Robinson Rockwood, born in Flint, Michigan, in 1860, who came to the Colorado River in September 1892. Rockwood studied engineering at the University of Michigan in the late 1870s but left the school without earning a degree—he studied so hard, he said, that "his eyes failed him."[5] Rockwood struck out for the open spaces of the West, serving as a construction engineer for the Denver and Rio Grande Railway and the Southern Pacific and, after 1889, as an irrigation engineer for the U.S. Geological Survey and the Northern Pacific Railroad. Then came an opportunity for fame and fortune.

A group of investors hired him to evaluate the possibility of irrigating a million-and-a-half-acre tract in Sonora, on the east side of the Colorado River, just below the international line. This he found infeasible, reporting that the Sonora land was "not worth two cents an acre and never could be made worth any more."[6] But he soon chanced to examine maps a railroad engineer had made of the previous year's overflow on the California side of the river. He read William Blake's report for the railroad survey, and with wagon and team he reconnoitered the lands Blake described. By the time he returned to Yuma, Rockwood had committed himself to the idea of transforming the Colorado Desert. All the land needed was a survey.

ROCKWOOD CONTACTED his employer, the high-sounding Arizona and Sonora Land and Irrigation Company, and requested clearance to undertake a survey of lands lying west of Yuma, on both sides of the international border. He explained that he believed, "without doubt, one of the most meritorious irrigation projects in the country would be bringing together the land of the Colorado Desert and the water of the Colorado River."[7] Rockwood was then in his early thirties, and his life to that point had not presented opportunities commensurate with his considerable ambition. He wanted money, and he had none. He wanted responsibility and the power that came with it. Judging from his future behavior, he particularly wanted recognition from the world in general. He needed a break, a big one. The authorization he soon received to survey the Colorado Desert seemed proof that the opportunity for which he had prayed was at last at hand.

Rockwood presented preliminary results of his survey to the company's directors early in 1893. Years later, he crowed, "They were so well assured . . . that the Colorado Desert project was a meritorious one that they immediately took steps to change the name of their company from the Arizona and Sonora Land and Irrigation Company to that of the Colorado River Irrigation Company."[8]

No sooner had the name been changed than the financial panic of 1893 wiped out the fortunes of the company's putative investors. Rockwood soon learned that the company consisted of little but a sheaf of papers, a set of closely protected and rarely seen financial books, and the rather too fecund imagination and manic energies of its principal, John C. Beatty.

These were a meager store of assets with which to launch a complex international project, for international it would have to be. The Algodones Dunes blocked the westward movement of water, as well as people. Rockwood's canal, like the forty-niners, would have to circle the dunes to the south. His plans called for a diversion at a place called the Pot Holes, twelve miles above Yuma. A canal roughly parallel to the river would carry the water into Mexico and connect with the usually dry overflow channel of the Alamo River, which then would convey the water to the irrigable desert lands on the United States side of the border.

Acquiring ownership or rights-of-way across the land to be traversed was an obvious requisite, but Beatty soon offended the owner of the Mexican land, General Guillermo Andrade, who was then Mexican consul-general in Los Angeles. Unable to strike a quick deal, Beatty allied himself with a rival claimant to Andrade's ownership, thereby foreclosing all possibility of obtaining the general's cooperation.[9]

Meanwhile, Rockwood was soon broke. In March 1894, he threatened suit if he

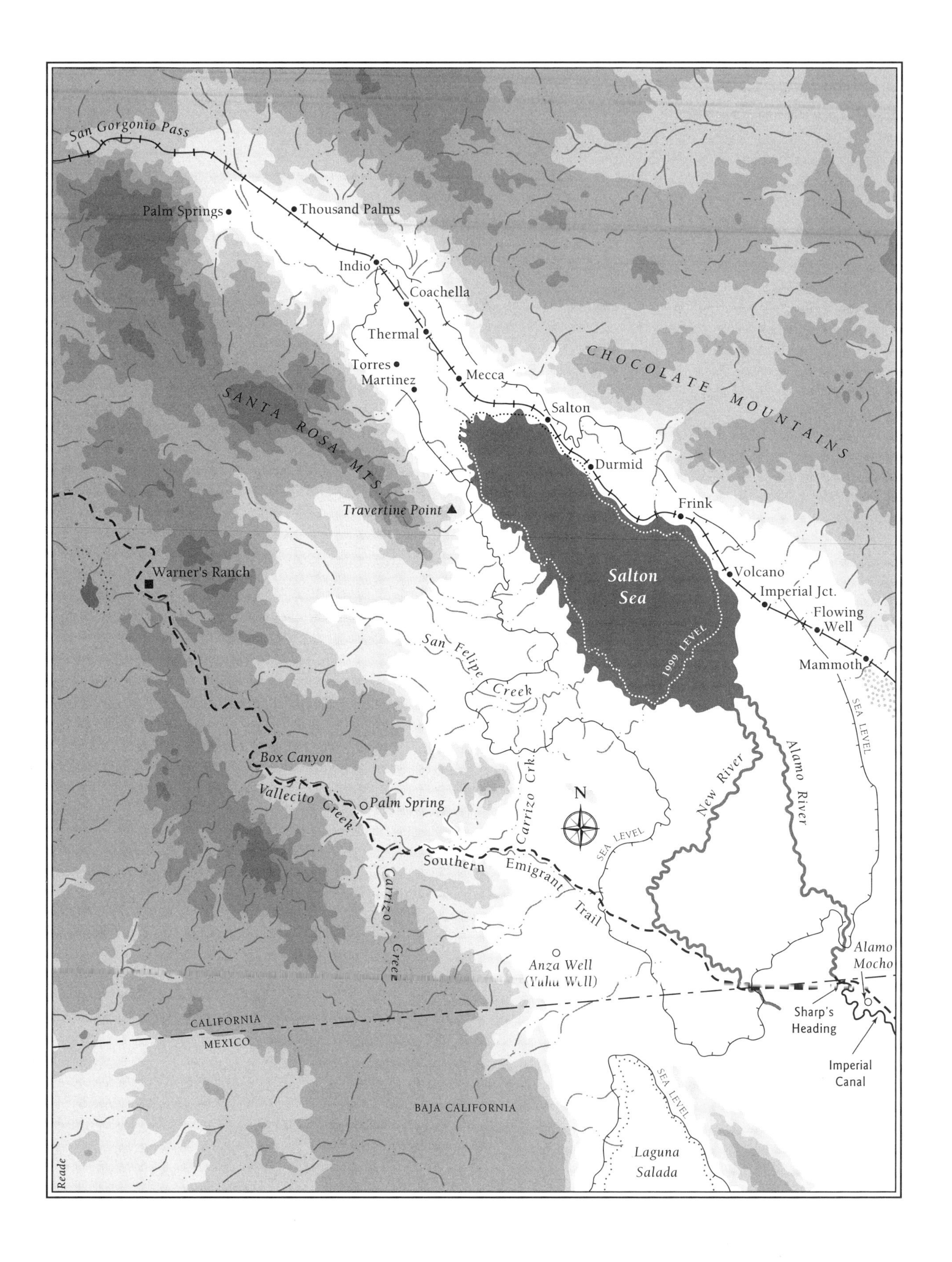

San Gorgonio Pass
Palm Springs
Thousand Palms
Indio
Coachella
Thermal
Torres Martinez
Mecca
Salton
CHOCOLATE MOUNTAINS
SANTA ROSA MTS.
Durmid
Frink
Travertine Point
Salton Sea
1999 LEVEL
Volcano
Imperial Jct.
Flowing Well
Mammoth
Warner's Ranch
San Felipe Creek
SEA LEVEL
Box Canyon
Vallecito Creek
Palm Spring
Carrizo Crk.
N
New River
Alamo River
SEA LEVEL
Southern Emigrant Trail
Carrizo Creek
Anza Well (Yuha Well)
Alamo Mocho
Sharp's Heading
Imperial Canal
CALIFORNIA
MEXICO
BAJA CALIFORNIA
SEA LEVEL
Laguna Salada
Reade

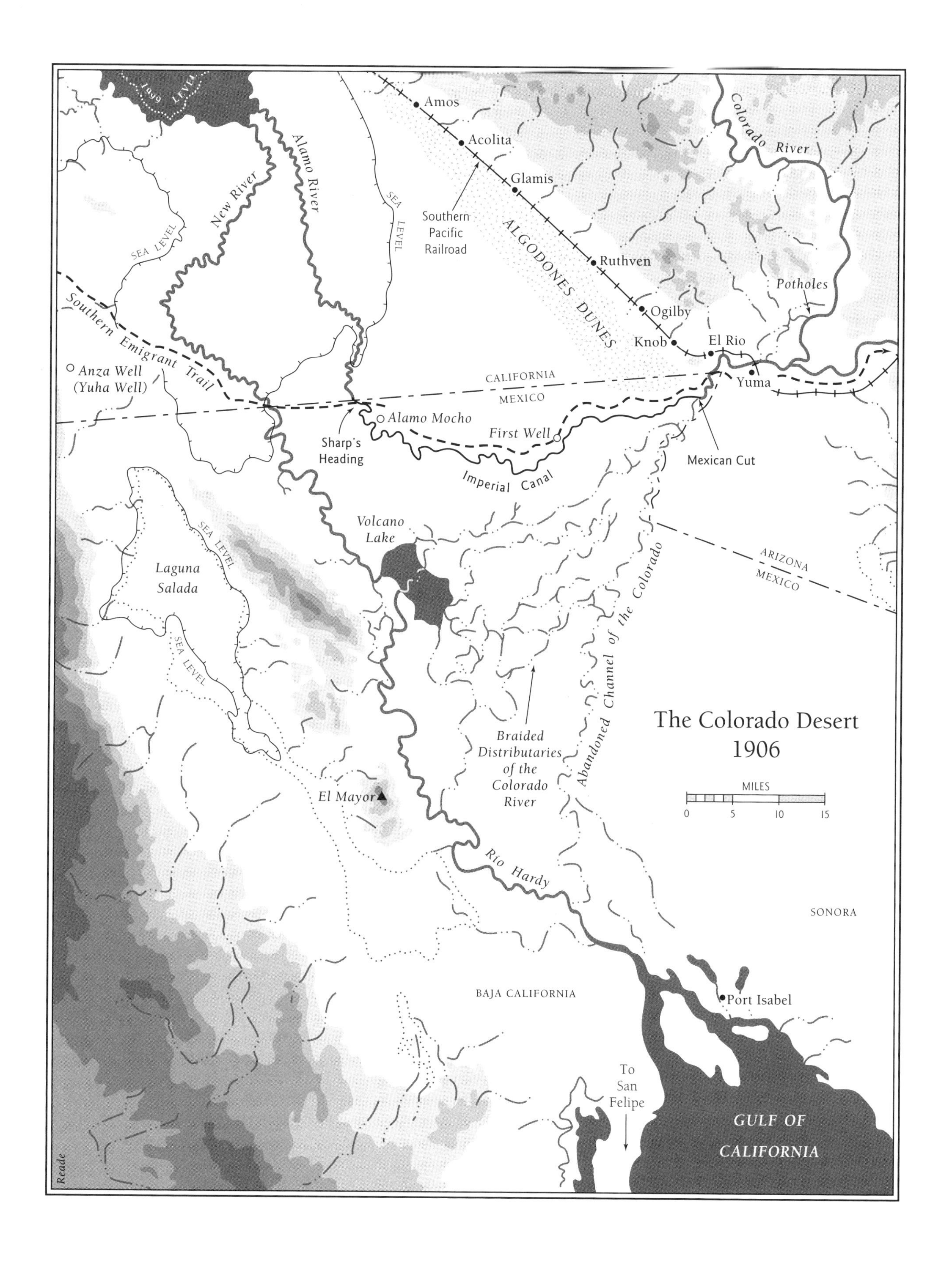

1999 LEVEL
Amos
Acolita
Glamis
Colorado River
Alamo River
New River
SEA LEVEL
Southern Pacific Railroad
ALGODONES DUNES
Ruthven
Potholes
Ogilby
Knob
El Rio
Southern Emigrant Trail
Anza Well (Yuha Well)
CALIFORNIA
MEXICO
Yuma
Alamo Mocho
First Well
Sharp's Heading
Mexican Cut
Imperial Canal
Volcano Lake
ARIZONA
MEXICO
Laguna Salada
Abandoned Channel of the Colorado
Braided Distributaries of the Colorado River
The Colorado Desert 1906
MILES
0 5 10 15
El Mayor
Rio Hardy
SONORA
BAJA CALIFORNIA
Port Isabel
To San Felipe
GULF OF CALIFORNIA
Reade

were not paid the wages owed him and reimbursed for money he had advanced his field crew. Beatty came to Los Angeles to placate him, and his powers in that line must have been impressive, for he succeeded. He even cadged ten dollars from Rockwood before setting off by train for Mexico City, where he hope to persuade the government to overrule Andrade's opposition. Predictably, he failed, and he managed to buy passage back to the States only after borrowing the money from an acquaintance whom he chanced to meet.[10]

Despite continuing problems of "temporary liquidity," Beatty soon ensconced himself in a suite of offices in the best building in Providence, Rhode Island. When Rockwood visited him, he was astonished to see that the suite featured a set of display cases, arranged for the benefit of potential investors, that showed off the fruits of the Colorado Desert: "oranges, lemons, bananas, figs, apricots"—although, as Rockwood well knew, the desert was then "producing nothing but a few horned toads and once in a while a coyote."[11]

In his memoirs, Rockwood wrote of these revelations with a mounting sense of horror, yet he was also learning the tricks of the developer's trade. Before the decade was out, Rockwood was doing what Harold Bell Wright's noble Seer would never have done: he was begging loans from his friends and relations and courting investors with hollow, Beatty-like promises. He was also issuing stock in only nominally existent land and water companies and selling it at a fraction of its face value in such a way as to bury prospects for future solvency under a snowfall of cheap paper. The fictional Seer, of course, would never have resorted to such stratagems. Poor Rockwood was in the real world.

ROCKWOOD WAS neither the first to dream of reclaiming the Colorado Desert nor the first to suffer frustration in bringing such a dream to life. In 1860, Oliver Wozencraft, a forty-niner from Ohio, petitioned Congress to grant him title to the Salton basin in return for his undertaking its reclamation. He pressed his case for nearly three decades, but Congress never granted him the fiefdom he desired. He died in Washington, still in active pursuit of his dream, only a few years before the Colorado River Irrigation Company was born.[12]

Wozencraft had predicated his plan on receiving a valuable grant of land from the government. Rockwood, by contrast, would depend entirely on private capital. By 1895 he had completed his split with Beatty while retaining control of the remnants of the company. In 1896, Rockwood and several others reorganized the enterprise as the California Development Company and started afresh.

They wooed every prospect who might help or hurt them: the government of Mexico, General Andrade, the Mennonite Church, the president of Bell Telephone, New York banks, and wealthy Europeans. But luck refused to smile. Their overtures to the Mexican government elicited no cooperation. General Andrade happily sold them successive options—which they paid for with borrowed cash at prices they could not afford. The Mennonites, whom they expected to colonize their desert, had second thoughts and never arrived. After a long courtship, the man from Bell withdrew, convinced that "the country was so hot that no white man could possibly live in it." Of the Europeans, one, a Scot, drowned in a stream, and another, a Swiss, dropped dead before a deal could be consummated. At last, on February 14, 1898, the Hamilton Trust Company of New York promised support. Lawyers drew up the necessary papers; formal ratification by the bank's directors was to follow within days. That night, Rockwood and Anthony Heber, a small, viperish man who had become the project's second most ardent advocate, celebrated with a modest meal, spending their last dollars. The next morning, they awoke to hear newsboys shouting that the U.S.S. *Maine* had been blown up and sunk in Havana harbor. War with Spain was certain. Equally certain was that the bankers at Hamilton Trust were spooked: the deal was off.[13]

Rockwood persevered. He trimmed engineering plans to lower project costs and continued to pursue investors. Finally, in 1899, his luck turned. George Chaffey, a pioneer of irrigation projects on two continents, expressed interest in the Colorado Desert. His meeting with Rockwood led to plans for a detailed reconnaissance of the project area in December, the last month of the century.

IN 1882, George Chaffey, father of irrigation colonies in the Los Angeles basin at Etiwanda and Ontario (named for Chaffey's native Canadian home), had discussed the Colorado Desert with an aging Oliver Wozencraft. Chaffey, then in his thirties, had declined involvement in Wozencraft's plan. Like the president of Bell Telephone, he believed the Colorado Desert was simply too hot: white men, that is to say, the kinds of men who had money enough to make a reclamation venture profitable, could not and would not work in temperatures that might consistently attain 110 degrees, day after day, weeks at a time.[14]

A few years later, the Australian government recruited Chaffey to develop an irrigation colony at Mildura in the torrid interior northeast of Adelaide. There Chaffey learned firsthand that whites could farm land as hot and forbidding as the Colorado Desert. Just as important, in Australia Chaffey suffered defeat. Both Mildura and a second irrigation colony foundered. When Chaffey returned to California in 1898, he hungered for redemption.[15]

And so, in the company of Charles Rockwood, Chaffey toured the desert in the last month of the old century and the first of the new one. They were joined by William Ellsworth Smythe, who came along, probably at Rockwood's behest, to publicize the expedition. Before the new year was out, Smythe published his high-gloss version of the trip in *Sunset* magazine. In his telling, the trip was no mere reconnaissance; it was a triumphal inspection of future domain. The "chieftain" of the group was "the indomitable man who sees in the transformation of this mighty desert the crowning work of his life." This was George Chaffey. Rockwood was presented as "the efficient engineer upon whose surveys and technical studies the physical plan must be built." The land was as good as conquered when two such Seer-like men conceived of conquering it.[16]

In actuality, the trip little resembled Smythe's account. Chaffey returned to Yuma skeptical of the venture. He doubted that colonists would take up land fast enough to make the enterprise viable, let alone profitable. And he thought Rockwood's plans for delivering irrigation water to the valley too expensive.

It is an understatement to say that Chaffey and Rockwood did not "hit it off." To a large degree, Chaffey was the man Rockwood wanted to be, and Rockwood may have sensed he had little chance of attaining Chaffey's stature. Rockwood had become a wheedler and a plotter, a self-justifying advocate of the grandiose. His soft ego sought continuous praise and approval, especially from Chaffey.

Chaffey, for his part, was supremely self-confident, and his confidence, as Rockwood surely noted, was both his greatest strength and his greatest weakness. He handled nearly all his business affairs himself, attending personally to legal and financial matters as well as to the exhausting work of directing projects in the field. Chaffey took a scornful view of Rockwood's capabilities as an engineer and businessman, and Rockwood became resentful. From the start, friendship was out of the question, cordiality probably a strain.[17]

After the reconnaissance, Chaffey's rejection of the project seemed complete, and so Rockwood returned to New York in pursuit of financing. But Chaffey had merely withdrawn for a time to ponder. Very soon, he went back to the desert. Accompanied only by an Indian guide, the fifty-two-year-old embarked on an exhausting second reconnaissance of potential canal routes and the lands to be served.[18]

By March he was ready to deal. Chaffey offered to build a system of canals meeting his own specifications and ignoring Rockwood's design. The canals, he pledged, would cost no more than $150,000, all of which he would raise. The system would deliver no less than 400,000 acre-feet of water annually to desert lands on the U.S. side of the border. In return, Chaffey would receive one-fourth of the capital stock, "fully paid up and non-assessable," of the California Development Company, plus $60,000 in cash or equivalent securities, payable after the company had sufficient cash flow. Additionally, he was to receive proxies for a majority of the company's remaining stock, giving him control of the company for five years. With acceptance of the offer, Chaffey would become the company's president, chief financial officer, and chief engineer.

Rockwood and his cohorts considered the proposal. It was a humiliating bid to take over their company, but they were on their financial knees. A year earlier, Anthony Heber, newly returned to the venture, had pawned his wife's jewelry to meet expenses. Now they owed the Mexican government several hundred dollars, which they did not have, and the state of New Jersey, where the CDC was headquartered, was about to revoke their corporate charter for failure to pay $1,000 in fees and taxes. In bitterness, Rockwood and the others signed Chaffey's contract on April 3, 1900.[19] Preparations for construction began immediately.

Chaffey had driven a hard bargain, but not hard enough. He soon realized that he had committed the most fundamental error in business: he had failed to understand what he was buying. He had never actually seen the books of the company, only unaudited representations of them. During negotiations, Rockwood and his allies had convinced Chaffey that the company's official records, kept at its New Jersey headquarters, were temporarily unavailable. What they did not say was that the company's agent refused to release them because his bills had gone too long unpaid. This situation, however, suited Rockwood and the others very well, for the books contained evidence of misrepresentation that would have caused Chaffey to turn his back on the enterprise.

Chaffey believed that by securing control of the CDC he had secured control of three assets essential to bringing water to the valley. The first was the Mexican company, La Sociedad de Yrrigation y Terrenos de la Baja California, S.A., that CDC had created to do business in Mexico. The second consisted of the land and rights-of-way the Mexican company was represented to possess, and the third was an option to buy land on the west bank of the Colorado just north of the international line. It was there that the intake for the canal system was to be built. When at last Chaffey was able to examine the books, he was horrified to learn that the company possessed none of these things. The capital stock of the Mexican company was not in CDC's treasury; Andrade was not under contract; the option on the land where the heading was to be made had not been paid in full and was not in force. Instead, the books revealed a miscellany of debts and liens, which now Chaffey, if he were to continue with the project, would have to pay.

Perhaps he should have withdrawn, cutting his losses. Instead Chaffey waged a board-room battle. He created a new company, Delta Investment, to hold controlling stock in the CDC and made it CDC's financial agent, a position from which it might drain CDC of cash, when and if any came in. He demoted Rockwood to the mortifying position of assistant engineer for one of the new water companies building lateral canals on the California side of the line, and he bullied the other directors into accepting revisions in his contract.

Chaffey and the original directors of CDC were not the only speculators haggling over the future of the region. The land that Chaffey soon ordered to be renamed Imperial Valley was attracting the dollars of investors on both coasts and in Canada.[20] In the dry Southwest it generated something close to frenzy. Rockwood, his coterie, and countless others quarreled over water stock and land scrip, jostling with each other to build alliances that would control future townsites. All this was happening in a valley that still possessed no more permanent residents than it had when Durivage, with blackened tongue, staggered across it in 1849.

But earth was beginning to move. Chaffey built a temporary headgate on American soil at the present-day Algodones border crossing on property he had secured at great cost from a man named Hall Hanlon. His canal then stretched across the border and paralleled the river for four miles until it reached the bed of the Alamo River, which his crews graded and reshaped. Forty miles west, at a place called Sharp's Heading, the canal broke from the Alamo and cut toward the location of a tent encampment Chaffey had named Calexico, on the international border. On May 14, 1901, six and a half months after breaking ground, Chaffey ordered the gate at Hanlon Heading opened. A tongue of red, silty liquid shot through and flowed down the canal, over the border, and out of sight into Mexico.

Chaffey tersely telegraphed his son in Los Angeles: "Water turned through gate at 11 A.M. Everything all right." Years later, his biographer described the moment in terms Harold Bell Wright might have used: "The news behind that bare statement of fact was that man had at last obtained dominion over the Sahara of the West."[21]

IN THE VIEW of many observers, Chaffey was indeed a noble conqueror of nature. Rockwood, similarly if undeservedly, received near canonization at the hands of Wright and a separate group of partisans. As much as the two men disliked each other—and their enmity continued to deepen—they were united in a singular way. They sought economic gain by means of appropriating as their private property the waters of a free-flowing river. Such an action was then common enough, if rare on a scale as large as theirs. Equally, their shared objective required that they deliberately and systematically undermine the land laws of the United States, which was another common practice, although again their project surpassed all others in audacity.

Wozencraft had pursued the wrong strategy. He had based his efforts on gaining ownership of land. Chaffey and Rockwood were more cunning. They staked their claim of ownership on water, which they did not have to buy, only distribute. They would sell the water to settlers under terms that, in the event a settler defaulted, guaranteed reversion of his land to them. One way or the other, their treasury should swell with cash or deeds that might be resold for cash.

Here, in outline, is how the system worked. First, CDC filed for rights to divert ten thousand cubic feet of water per second from the Colorado River, a claim it soon doubled. Next it asserted that the water became the property of the Mexican company as soon as it entered Mexico and remained Mexican property as it reentered the United States, thus earning exemption from regulation by either the state of California or the federal government.[22] The Mexican company sold the water for a half dollar an acre-foot to a baker's dozen of mutual water companies that CDC organized to distribute water within the Imperial Valley. The mutual water companies then resold the water to settlers—but not nearly so directly as one might assume.

In order to buy water, a settler navigated a labyrinthine process. First, with the guidance of the Imperial Land Company (a subsidiary of CDC), he filed claim on 320 acres of public domain under the Desert Land Act or 160 acres under the Homestead Act, the former offering a simpler though more costly path that most settlers in the Imperial Valley preferred. Under the Desert Land Act a settler paid the government 25¢ per acre at the time of filing. Then he "proved up" his land, which meant making investments of at least $1 per acre per year for three years, at the end of which, for an additional $1 per acre, the government would issue him title.

In the Imperial Valley the improvement requirement was instantly fulfilled when the settler bought stock in the mutual water company that served his land. The stock, however, was not cheap. The California Development Company required the settler to buy "water stock" at a rate of $25 per acre. Owning the stock was equivalent to owning a water right, but the settler still had to buy water—at an annual cost of no less than $1.75 per acre.[23]

Unfortunately for the settlers, they were not the only investors in water stock. Long before Chaffey came on the scene, and years before water actually reached the valley, the CDC had sold various forms of water stock to its directors and other backers, often discounting it steeply. Such sales had been one of the company's few strategies for raising money and keeping its creditors at bay. Now, with water available and settlers arriving, those paper rights, sold for no more than $2.50 to the acre, controlled large areas and had real value. They might be assembled into a townsite; they might be resold for farms. Along with other holders of water stock, the company, which also acquired land through foreclosure, became an active reseller of land. After a branch rail line reached the valley in 1902, special trains brought buyers to the company's tent in Imperial for regular auctions of land and water stock.

This was not what the government had had in mind. The Desert Land Act, like the Homestead Act, was the legislative expression of a dream of widespread, small-scale landownership. The authors of the land laws of the United States, subscribing to the arguments of men like William Ellsworth Smythe, had intended to produce a landscape of farms large enough to be self-sufficient and prosperous yet small enough to be worked by a household. They believed that independent, landholding families were the essential unit of a democratic society. But the land laws were rarely a match for private enterprise. It wasn't that people like Chaffey and Rockwood set out to circumvent them; they set out to get rich, and circumvention followed naturally. When it suited them, they operated under the letter of the law; when it did not, they sidestepped it. In the process, they found a way to appropriate for private gain the waters of a great river and to cause that water to metamorphose into title to an inland empire. Theirs was a brilliant scheme and might have won them the fortune they craved, but for their own weakness. These self-believing Seers and master men failed to keep in check their impatience for wealth and their animosity toward each other.

The government also posed difficulty. When W. F. Holt, the founder of Holtville and a model, like Rockwood, for one of Harold Bell Wright's heroes, parlayed a $50,000 investment in discounted water stock into control of close to 18,000 acres, and when another speculator assembled ownership of a block that would reach 10,000 acres by 1904, government agents became concerned over the scale of abuse. In an attempt to slow the wheeling and dealing, they delayed the surveys necessary for land filings and patent approvals, prompting CDC and the Imperial Land Company to caterwaul in protest. The companies howled even louder when the government prosecuted some of the land company's directors for hiring shills to file falsely on desert land for which they held water stock. In 1908 a federal district court handed down several convictions.[24]

Under the direction of George Chaffey, water stock had become the currency of the valley, far dearer than the land itself. Ironically, it was a stock manipulation that led to Chaffey's undoing. From the moment Chaffey replaced Rockwood and set Delta Investments over the CDC, Rockwood had struggled to void Chaffey's contract and remove him. The contract gave Chaffey control of the CDC for a period of five years. To assure this, Chaffey obtained proxies controlling the voting rights of CDC's main investors. Busy with myriad details, however, he had failed to have the shares of stock underlying the proxies placed in trust and removed from the market. This was the weakness Rockwood exploited. As water reached the valley and the value of CDC stock rose, many early investors cashed in their profits. Rockwood and his allies, using (probably) money gained from sales of water stock, became steady buyers of stock in CDC. The proxies that had bound the original owners now became meaningless. Chaffey, who had been too confident to delegate his legal affairs to competent counsel, realized too late that Rockwood was assembling a controlling interest in the company. Soon it was Rockwood's turn to dish out the bitter medicine he had earlier been served. He insisted that Chaffey resign as president and withdraw from involvement with CDC, save for management of the main canal.

Chaffey had no taste for partial measures. In April 1902 he conceded Rockwood's victory and traded all his interest in CDC for $300,000 in securities. Even then his pain continued. Had he pressed for the full cash value of the paper he held, he would have bankrupted the company. Out of a sense of obligation to the valley's settlers, he instead sold his securities for a mere $100,000. Having built four hundred miles of canals and laterals in only twenty-two months, he departed Imperial Valley forever.[25]

Rockwood again was appointed the company's chief engineer, and his old comrade-in-flimflam, Anthony Heber, replaced Chaffey as president. The entire enterprise was again theirs, all theirs. Now Rockwood could again be the Seer, and Heber, restless and aggressive, could push and shove more powerfully than he ever had before, or would again.

Side Trip: The Shimmering Desert

The largest part of the Holly Sugar plant in the town of Imperial is a tall white tower. Part way up it, a hundred feet or so, some wag painted a blue equator around the tower's waist. Above the line, he added the words *sea level.*

It was somewhere near this beet refinery cum depth-gauge that Jose (*sic*, probably pronounced "Josie") Huddleston, a rare woman among the first white settlers of Imperial Valley, rented a frail tent in December 1901. Under it, in spite of sandstorm, heat, and greasy barrel water, she baked twenty-one loaves of bread a day in a gasoline oven and opened what she liked to call a restaurant. Like other whites who came to the valley, the bouyant Mrs. Huddleston hailed from greener vistas, and she relished any relief from the unending monotone of the "great sand waste." She was cheered by the verdure of a small plot of sorghum planted near the tents, and she cherished the mirages that hovered on the horizon, early and late in the day when the light was flat. Looking south, she was accustomed to seeing, besides inverted mountains, "a full-rigged battleship . . . so plainly that even the portholes were visible. Again we have seen the ocean and watched the breakers sweeping over the sands, and could see the spray from the rolling waves."[1]

Jose Huddleston let her imagination run, and in extracting images from the refractions of the desert air, she no doubt surpassed many other inhabitants of the tent cluster that bravely called itself Imperial. Eastward, she said she often saw "an immense castle with beautiful turrets with iron bars at the windows," or "a hole through the mountain . . . showing beautiful green on the other side," or "an immense bird feeding, a crane perhaps, with a beak a foot and a half long." In the days before the dust and humidity of settlement changed the very air of the valley, the desert and its shimmering sky became for Jose Huddleston a dreamscape of imaginings that she could not have conceived in any other place.

7 | LAND OF HEART'S DESIRE

In 1898 a lone horseman rode down from San Gorgonio Pass into the Salton Sink, a small terrier trotting behind him. Day by day they descended into what the rider called "the bottom of the bowl." Each day, when the heat rose and the gravels and sands of the desert grew hot, the terrier whimpered, and the rider stopped and lifted the dog atop the horse's rump. There it rode, leaning against the rider for balance, and the solitary rider took comfort from the trusting pressure.

A distinguishing feature of this rider, among all the eccentric wanderers who drifted down the trails of Cahuilla country, was his devotion to the desert. He loved it with the intensity of a Barbara Worth, but without her gushy sentiment. Dry-eyed and realistic, John C. Van Dyke praised the purity and ferocity of the land and all its qualities—its dazzling light, its brutal heat, and especially the purity of its color. In a manner remarkable for his society and time, he valued the desert not for what it could be but *as it was;* he did not want to change it.

Van Dyke was a severe asthmatic, and he had come to the desert to die or get well. He did neither. "I was just ill enough," he later wrote, "not to care about perils and morbid enough to prefer dying in the sand, alone, to passing out in a hotel with a roommaid weeping at the foot of the bed."[1] He set out without map or plan and wandered for three years, ranging far south into Mexico, east to Tucson, and north into Oregon. At one point, deep in Sonora, he traveled for six weeks without glimpsing another human. He lived off the land, subsisting on antelope, quail, and whatever else fell within his gunsights. He grew expert at finding water, digging for it, or stealing it from desert plants. Even his horse and dog accustomed themselves to drinking from gutted cacti. At the end of his peregrination, weakened by a mysterious fever he contracted in Mexico, Van Dyke returned to his professorship in art history at Rutgers University and resumed his prior life as one of the nation's leading aestheticians. He also completed the manuscript he had drafted during his travels. Scribners brought out *The Desert* in 1901, at about the time the first waters of Chaffey's irrigation system reached the Imperial Valley. The book has since become a classic of American nature writing; even today, no treatment of the desert surpasses it.

Van Dyke's long and solitary journey was a dual quest. He sought to flirt with death, to learn, in a sense, why he should remain alive. Undoubtedly, he encountered danger in many locations, though he rarely mentioned it in his writing. He also hungered for a purity of aesthetic experience. He found this purity nearly everywhere in the arid lands, but nowhere did his finely tuned, modernist sensibility rejoice more than in the shimmering, hallucinatory sea of light that filled the "the bottom of the bowl." It was, he said,

> the most decorative landscape in the world, a landscape all color, a dream landscape. Painters for years have been trying to put it upon a canvas—this landscape of color, light and air, with form almost obliterated, merely suggested, given only as a hint of the mysterious. Men like Corot and Monet have told us, again and again, that in painting, clearly delineated forms of mountains, valleys, trees, and rivers, kill the fine color-sentiment of the picture. The great struggle of the modern landscapist is to get on with the least possible form and to suggest everything by tones of color, shades of light, drifts of air. Why? Because these are the most sensuous qualities in nature and in art. The landscape that is the simplest in form and the finest in color is by all odds the most beautiful. It is owing to just these features that this Bowl of the desert is a thing of beauty instead of a dreary hollow in the hills.[2]

Nowhere else could eyes behold anything like it, with sunset approaching or sunrise just past: a gauzy sky; amethyst mountains; a cyan shimmer of horizon; and at the foot of the dark mountains, an unsettled pool of color, glinting now turquoise, now mirror-silver. As quickly as one might swallow deeply from the canteen, the angle of sunlight changed and recast the tableau as completely as ever the facade of Rouen's cathedral metamorphosed for Monet. Even the near-at-hand, ash-white desert floor revealed hidden nuances of brown and gray.

Such things Van Dyke rode into the desert wilderness to see, and he never saw them more superbly than in the sink. Regret sharpened his pleasure, for he knew that such sights would soon become extinct—the purity of the air would not survive the humidity and dust of reclamation, "not even the spot deserted by reptiles shall escape the industry or the avarice (as you please) of man." He lamented the California Development Company's widely publicized plans to bring the river to the basin, and his objections were the more vehement for being, as he knew, ineffectual:

> The "practical men," who seem forever on the throne, know very well that beauty is only meant for lovers and young persons—stuff to suckle fools withal. The main affair of life is to get the dollar, and if there is any money in cutting the throat of Beauty, why, by all means, cut her throat.[3]

The deserts, said Van Dyke, were the breathing spaces of the West and should never be reclaimed. They were to the air what the forested watersheds of the mountains, then receiving protection as forest reserves and national parks, were to the region's rivers. But before the ink of his book was dry, the "practical men," Chaffey and Rockwood, had changed the fortunes of the desert forever.

EVEN AS Van Dyke's *The Desert* reached its first readers, the dust of human activity was rising above the valley. Teams of men fanned out to drive the stakes that marked the boundaries of surveys; others operated the land dredges excavating canals; many more drove teams of horses pulling fresno scrapers, a kind of wheeled drag scoop with a dumping mechanism that was the basic ditch digging and field leveling machine of the day. The first settlers to arrive in the valley hailed from the Salt River country of Arizona, where they had tried to farm and failed. In the last days of 1900, the Van Horns and Giletts, three men, two women, and seventeen ceaselessly hungry children, rafted their five cows and a bull, their crates of chickens, and three wagons filled with their entire stock of worldly goods across the Colorado. They soon found Chaffey's crew out in the desert, where the men hired on for wages, and the women, besides tending their swarm of children, cooked for a camp of thirty-two men.[4]

As water trickled into the valley, so did more people. Some came the hard way, overland, as the Van Horns and Giletts had done, but most rode the train to Flowing Wells, then took a stage south to Imperial or one of the other new tent towns pegged to the desert floor. Until the middle of 1903 the stage still took the better part of two days to get all the way to Calexico, but for many it was worth it. Calexico's new adobe hotel featured a marvel of the desert—an actual "shower-bath." Although not a drop of drinkable alcohol was to be had in the valley, a man with a thirst water could not quench could cross the line to Mexicali, which consisted of "a single row of thatched huts and adobes strung along beside the canal." According to one observer, nearly every one of those unprepossessing buildings was "a saloon, or gambling den, or both."[5] Another visitor, not much later, noted the presence among the shacks of "three painted girls, or rather children, in dirty pink, who now and then ceased their crude blandishments of the men near them to shout the words of a ribald song."[6] Clearly, Mexicali's oldest service industries were born alongside the Imperial Valley.

The arduous first years of valley settlement demanded constant labor, and even more by way of enduring heat, wind, sun, and privation. For most, the struggle to gain a toehold in the desert carried the taste of sand between the teeth, which the silty water of the new canal hardly washed away. As elsewhere in the West, the harsh conditions weighed heavily on the few women who accompanied the early waves of settlement.[7] Some families arranged for their women and children to move to the mountains or coast during the summer months, while the men bacheIored as best they could and slept on cots outdoors. The effect of this familial strain and disorganization, in the opinion of one observer, was to give the farms and settlements "a slipshod appearance" lacking in settled domesticity.[8]

The settlers celebrated every milestone in the growth of their communities and heralded each marker of "civilized" life: the famous shower bath was one, but more important were the first newborn child (Cameron Beach in 1901), the first tree planting (at the Beach home in Imperial), the opening of the first barber shop, the first post office, the first store. Foremost of all achievements was the first school, which convened for classes on September 8, 1901, under an arrowweed ramada three miles northwest of Calexico, Professor J. E. Carr presiding. Many took note of the first Farmers' Institutes (which became the Farm Bureau), the first piano (available, in and around Calexico, for every social occasion), the first church (claimed by both Methodists and Congregationalists), and the first two-story building (Calexico, 1904). Certainly no achievement was more significant than the first harvest of marketable crops—mainly wheat and barley from almost twenty-five thousand acres, in 1903.[9]

The printed word came early to Imperial Valley. Even as sandstorms collapsed the tents of the earliest pioneers, the Imperial Valley Press began cranking out the news, asserting (to the dismay of other sub-sea-level blusterers near Indio) that it was the "lowdownest" paper in the country. The ingenious expedient of placing an evaporative bath of sawdust under the press kept the type from melting, but the Imperial Valley Press enjoyed its position of primacy only briefly. Where there was business for one, there was business for two, and soon the press had to cede its lowdown title to a lowdowner upstart in Brawley.[10]

In 1902, the Southern Pacific Railroad Company began construction of a spur line from Old Beach (now Niland) southward. Its completion as far as Imperial and the arrival of the first train in May 1903 occasioned much rejoicing, but not half as much as the arrival, soon after, of the first cargo of ice. The city fathers declared a holiday to celebrate the event, and the entire town devoted itself to an ice-cream-making party that lasted the day.[11] Other communities, such as El Centro, Silsbee, and Holtville, similarly celebrated as often as they had something to cheer about. For amusement, people organized horse races and greased-pole contests—with the pole cantilevered across an irrigation canal and a five- or ten-dollar bill at the end of it. Such an event might have stood for life itself. Everyone, in a sense, was shinnying a greased pole of risk, hoping to make it to the money. The settlers were tenacious and made much of nothing, when nothing was all they had. They swallowed their doubts along with the chewy canal water, and they crowed about the future of their towns, notwithstanding that what passed for a town in the early years of Imperial Valley "consisted principally of surveyor's pegs, magnificent distances, and the rose-tinted hopes of its projectors."[12]

In the early years, most valley events had Indian witnesses, although no one, least of all the Indians, troubled to record their observations. The Cocopa who drifted north out of the delta and the remnant Kamia of the valley won little sympathy from whites. One band, said to be "a large number," lived across the New River from Calexico at the time of the great floods. But they did not live there long. Within a decade most of them, according to a white physician, had "succumbed to the ravages of tuberculosis and venereal diseases."[13] "Old Borego," whom the people of Calexico came to view as the town's "Indian mascot" may have been a survivor of that band:

> He was past eighty, had no money and needed none. He lived on what he could pick up, an odd job for a meal here, another there. Often meals were given him. He slept anywhere. Why should he care where? His queerest trait was that of wearing everything he had. People were generous in gifts of discarded clothing and he wore them all at one time! Perhaps he would have three or four vests and as many coats on when the thermometer was over 100. When questioned as to why he wore them all, he would always reply with the question, "What else shall I do with them?"[14]

Most other portraits of Indians in that time and place were still less sympathetic. A 1918 history of Imperial County by Judge Finis C. Farr, the registrar of the United States Land Office in El Centro, described the region's previous inhabitants with unrestrained venom:

> That they were inhabiting such a desolate country of their own volition is hardly possible, and it has therefore been surmised that they were driven out of some

> more favored region by more powerful tribes, and then sought refuge among the vast wastes of this peninsula. They seemed devoid of all knowledge or even native intuition. They thought California was the entire world, visited no other people and had no visitors, cared mainly for filling their stomachs and toasting their shins in idleness. Even the native hunting instinct, so common with other Indians, seemed to be dormant in their minds if they had any minds at all.

Charity suffers in this treatment; so does accuracy—the assertion that the Indians of the quarter visited no one is patently false. Irony is lacking too: if the Indians abided there because they could not "make it" elsewhere, what did that say about the whites? Offensive as the passage may be, one need not look further to illustrate the chasm separating white from Indian. A man like Eusebio Kino might describe the natives in more compassionate terms, yet the naming of qualities is always a dangerous game, compromised by point of view and imbued with values that, in the fullness of time, appear no more fixed than a skiff on tidal water. If we say that the white settlers were materialistic, bigoted, pietistic, and bullying, then we might evenhandedly mention that the Indians they supplanted were, from a similar vantage, superstitious, improvident, and complacent.

On the other hand, if we say that the Indians were finely tuned to their environment—or at least to their former environment, when it was devoid of whites—and that they embodied the trial-and-error intelligence of centuries of interplay between natural selection and cultural evolution, that they had been tested under the most trying geographical circumstances this side of the Arctic, and that they were unpossessive, generous to a fault, and capable of inconceivable stamina—if we say all this, then we may also say that the whites were indomitable, energetic, adaptive, and inventive. With awe we may observe that the Indians knew uses for every plant in the desert and understood the habits of its creatures. We might note that they alone comprehended the unstable intricacies of the Delta—all this, yet they'd never made a wheel. And although the whites in a few years could increase the carrying capacity of the desert ten-thousandfold, when they were stripped of tools and gear they not infrequently perished beside the same desert trails that any Kamia adolescent negotiated with ease, or knew to avoid.

The tangle of the actual past never easily combs out. Looking back, we cannot be so bloodless as to abandon all judgment about people and cultures, but neither should we fall for simple categories.

AT THE START of 1902, the canals of the CDC brimmed with water, and the influx of settlers and speculators pushed the population of the valley toward two thousand. Land filings and sales of water stock proceeded briskly. But the valley's business soon suffered the first of a series of shocks. In January the Department of Agriculture issued *Bureau of Soils Circular No. 9,* in which soil analysts Garret

Holmes and Thomas Means asserted that a large portion of the 125,000 acres already filed on was too alkaline to support agriculture: "The land may produce a crop for a year, or even two years, and then, having become thoroughly saturated, the alkali will rise and kill the crops."[15] The hundreds of farmers sowing and irrigating their fields, the hundreds in the process of buying and preparing land, and the hundreds more considering investment in the valley drew back in anger and alarm. Banks cut off credit to the California Development Company, leaving it hard pressed to maintain its canals, let alone extend them for new settlers.

Anthony Heber, the newly named successor to George Chaffey as company president, rushed to Washington to seek remedy. Heber lacked his predecessor's dignity, but his tenacity and combativeness served well enough as he harried congressmen and bureaucrats, finally obtaining from the Department of Agriculture a milder (and more realistic) assessment of the alkalinity of valley soils. But the company had been damaged. The soils report put the company on the defensive, financially and politically, and there it stayed as other blows fell.[16]

The unexpected salvo from the government guaranteed continued tension between the company and federal officials, and it also sowed dissatisfaction among the settlers. If the government doubted CDC's integrity, so did they. The literature the company published to draw settlers and investors to the desert ran heavy with Old Testament imagery: the waters of the Colorado were as fertile as the Nile's, and the lands of Imperial Valley, like those of the Egyptian delta, would therefore birth a great civilization. While these images stopped short of mentioning Pharaoh, the people of the valley soon learned to complete the metaphor. Pharaoh was embodied in the company that meted out water to their fields, homes, and towns. Like the Pharaoh of Moses' time, this one proved by turns neglectful, fallible, ruthless, and avaricious.[17]

Pharaoh soon faced a new adversary. On June 17, 1902, as hubbub over the soils report subsided, President Roosevelt signed the Newlands Reclamation Act, and the national context for irrigation agriculture forever changed. The act placed the federal government squarely in the business of bringing irrigation water to western deserts, and the agent of this labor was to be a new division of the U.S. Geological Survey called the Reclamation Service.

Reclamation immediately began developing its own plan for irrigating the valleys of the lower Colorado River. It announced its intention to build dams, reservoirs, hydroelectric generators, and canals sufficient to irrigate ninety thousand acres downstream of Needles, most of it near Yuma.[18] The new agency invited comparison with the strained CDC. Reclamation would build its main diversion at Laguna Weir, near the Pot Holes, where Rockwood had originally intended his diversion to be. Laguna Dam would be stronger, safer, and more reliable than the intake and delivery system the CDC had installed. Once built, the Yuma project would be managed for the public good, not private gain. Land laws would be honored, settlers protected. When full details of the Yuma Project were unveiled in August 1903, Reclamation recommended that it take charge of delivering water to the Imperial Valley and that it integrate the valley's water infrastructure into the rest of the Yuma project. The project would thereby obtain economies of scale in the construction of dams and diversions, and the valley's settlers would gain a degree of security and safety which they lacked under the thrall of the CDC.

The intemperate Heber reacted with scorn. In February 1904 he boasted that the CDC's claim to a flow of twenty thousand cubic feet per second from the Colorado assured it permanent control of the river.[19] His audacity spurred the Reclamation Service to press arguments that it had been developing in concert with the Department of Justice. The two agencies argued that since the Colorado was a navigable stream, it was subject to regulation by the War Department, and no diversion of its waters might be made without the department's approval. Since the CDC had obtained no such approval, it had no right to take so much as a drop of the river's water. Again, the valley's farmers and investors grew panicky, and again Heber rushed to Washington to counterattack.

He persuaded members of California's congressional delegation to introduce emergency legislation declaring the Colorado to be more useful for irrigation than navigation—a determination that would exempt it from War Department regulation. A hearing on the bill commenced late in March, with Heber providing most of the favorable testimony. He found himself opposed by irrigation crusader William Ellsworth Smythe, who stepped forward as the opposition's star witness. Smythe condemned the CDC in the harshest terms, decrying its subversion of the nation's land laws, its profiteering at the expense of honest settlers, and its appropriation of public water for private gain. "These gentlemen," he said, "have gone in there and claimed the melting snows of the Rocky mountains as their property. . . . Should the Congress remove the government's only claim to the river by declaring it to be no longer a Navigable Stream, they would be giving it away forever to the California Development Company."[20]

Heber soon took his turn as witness, but before he finished testifying, he knew the bill was doomed. He did not accept defeat gracefully. He became peremptory and ill-mannered, lashing out at the committee: "It is my earnest desire to worship at our own altar and to receive the blessing from the shrine of our own government, but if such permission is not given, of necessity I will be compelled to worship elsewhere."[21]

In other words, if the United States did not give him what he wanted, he would strike a deal with Mexico.

This threat had as much to do with engineering and canal maintenance as it did with regulatory jurisdiction. In late 1903 and continuing well into 1904, the intake and first four miles of the main canal had silted up, denying water to the valley during the critical winter growing season. Farmers deluged the company with lawsuits arguing that the CDC had failed to meet its water delivery obligations under the water stock agreements. Their damage claims ran to $500,000—more than enough to break the company.[22] Dissatisfaction was rife, and the company's enemies, allied with the Reclamation Service, were quick to exploit it. Again William Smythe joined the fray at the head of the opposition. Allied with Paul Van Dimas, a leader among the valley's farmers, Smythe came to the valley to form a cooperative, valley-wide water users' association, the purpose of which would be to buy CDC's water delivery system and then to affiliate with the Yuma Project, as Reclamation desired.

Four years earlier, haste and cost-cutting had caused Chaffey to build his main intake five feet higher than his own engineering studies called for and to install only a temporary headgate.[23] The height of the intake meant that at low flows, the velocity of water in the canal was insufficient to prevent silt from settling out. In 1902 and 1903, again because of haste and cost-cutting, Rockwood had refrained from deepening the cut and building a concrete and steel headgate, as the departing Chaffey had specified should be done.[24] At every turn, haste and cost-cutting had caused CDC to skimp on investments in its own project: strapped for cash, it scrambled to stay a step ahead of its customers, its creditors, and the government. Ultimately, haste and cost-cutting caused Heber, on June 10, to sign a contract with the government of Mexico to open a new intake south of the border.[25]

Rockwood had been unable to solve the silting of the main intake. He cut bypasses around it in order to get water into the canal when the river was low, but these, too, silted up. He prepared a "waste gate" through which he attempted to sluice the canal clean during high water in the summer of 1904, but although the high flows carried water to the valley for a time, they did not scour out the sediment. When the floodwater subsided, Rockwood was dismayed to find that it had deposited yet another foot of silt. Rockwood tried every expedient: he kept his dredges running nonstop, but they were too few and too small to master their task—and the company was too broke to secure equipment equal to the challenge. He set a small river steamer to work trolling logs up and down the canal in an effort to loosen the deposits so that they might wash away. The effort was futile, if not pathetic. The vaunted canal, which Chaffey had completed too cheaply and too quickly and which Rockwood irresponsibly refrained from improving, had simply failed. It could not deliver water to the Imperial Valley in anything close to the needed or promised quantity.

Rockwood defended his subsequent actions in an apologia written in 1909:

> We were then confronted with the proposition of doing one of two things, either cutting a new heading from the canal to the river below the silted four miles [*sic*] section of the canal, or else allowing the Valley to pass through another winter with an insufficient water supply. The latter proposition we could not face for the

reason that the people of the Valley had an absolute right to demand that water should be furnished them, and it was questionable in our minds as to whether we would be able to keep out of bankruptcy if we were to be confronted by another period of shortage in this coming season of 1904–1905.

The cutting of the lower intake, after mature deliberation and upon the insistence of several of the leading men of the Valley, was decided upon.[26]

The passive voice provides refuge for cowards of all kinds, but here it affords Rockwood little protection. Although he lights candles at the shrines of "mature deliberation" and "leading men," the decision wasn't merely "decided upon." He and Heber made it, and Rockwood alone was responsible for its execution. In September 1904, his crews began excavating a new intake in the west bank of river approximately four miles below the border.

Heber, meanwhile, continued his frenetic and disingenuous mischief. Whether he meant only to mollify public opinion or hoped, through the stupidity of others, to come into a windfall, he offered to sell the company to the newly formed Water Users' Association. His initial price was $5 million, later reduced to $3 million. For that considerable sum the association might acquire title to a jerry-built irrigation infrastructure incapable of discharging its function, an empty corporate treasury, and a welter of unmet obligations. The association nevertheless pursued the deal. And in the tradition of desperate organizations chasing an "opportunity" on which no sane investor would risk funds, it appealed to the federal government to make the purchase.

Before an answer could be forthcoming, Rockwood opened a channel fifty feet wide and six to seven feet deep from the river to a point on the Imperial canal below the silted stretch. The new channel was only thirty-three hundred feet long and lacked water control structures of any kind. It was a ditch, nothing more, connecting one of the most powerful arid-land rivers in the world with a stretch of desert where by now—in the fall of 1904—some seven thousand too-trusting settlers had staked their fortunes.

As soon as water flowed freely again, Heber thumbed his nose at the Water Users' Association. In October, he appeared at a public hall in Brawley to debate whether CDC or Reclamation should run the valley's water system. Paul Van Dimas took the side of Reclamation. Heber stole advantage at the outset by announcing to a stunned and then cheering audience that the company was no longer for sale. Van Dimas tried to make his case, but Heber shouted him down. He painted the government as a demon, saying it would close the canal if it could, and recited horror after imaginary horror. By prearrangement, the crowd contained a fair salting of thugs who rushed forward and dragged Van Dimas from the speaker's platform, pulling him outside. Also by prearrangement, a goodly supply of tar and feathers had been delivered to that very spot. The torture of Van Dimas now began in earnest. The mob pasted him with both vile substances, sat him cruelly upon a rail, and carried him south through the towns of Imperial, Cameron Lake, and Calexico, at last flinging the miserable man across the international border—throwing him not just out of town, but out of the country. Their barbaric labors finished, these worthy exemplars of agrarian virtue turned to recreation, devoting the remainder of the day to a lavish barbecue, free and open to the public, which was followed as night fell by a dazzling display of fireworks, all courtesy of the Colorado Development Company.[27]

As the ruffians celebrated, water flowed from river to valley. The flow continued steadily for months more, but not a hand was lifted to build a headgate. Rockwood and his cronies later justified their dereliction on grounds that they did not wish to offend the Mexican government by commencing construction of a control structure without official approval. They claimed that Mexican officials were slow, criminally slow, in completing their review of the company's plans. Lamely they argued that the great floods of the next two years were, in effect, the Mexicans' fault. Moreover, they argued that in twenty-seven years of "rod-readings" at Yuma, the river had flooded only three times during the winter season, and never more than once in a winter.[28] Their decision to leave the new intake unprotected, they said, was therefore both defensible and prudent. These "practical men," to use Van Dyke's term, did not mention that they lacked funds to build even a temporary headgate.[29] They hid their inabilities behind the predictable foot-dragging of Mexican officialdom and pretended to be guiltless.

Dimensions: The River

Total length of the Colorado River (from the headwaters of its largest tributary, the Green River): 1,700 miles.

Watershed area: 246,500 square miles (157,760,000 acres), representing one-twelfth of the continental United States.

Average rainfall within the watershed: less than 10 inches per year, about one-eighth of which reaches the river.

Highest flow recorded at Yuma gauge: 250,000 cubic feet per second (cfs) on January 22, 1916.

Lowest flow at Yuma, 1901–1905: approx. 2,200 cfs.

Twenty-five-year average flow at international border, 1905–1924: 23,683 cfs (17.136 million acre-feet [maf] per year).

Twenty-five-year average flow at international border, 1974–1998: 5,536 cfs (4.009 maf per year). If the flood years of 1980, 1983–86, and 1993 are excluded, the averages since 1974 become 4,192 cfs and 1.91 maf per year.

Natural sediment load: one-fourth pound per cubic foot, which is 5, 10, and 17 times the load of the Rio Grande, the Nile, and the Mississippi, respectively.

Assumed annual flow at Lees Ferry on which the 1922 Colorado River Compact was based: 18 maf.

Current official estimate of annual flow at Lees Ferry: 15 maf.[1]

Current official estimate of additional tributary flows downstream of Lees Ferry: 1 maf per year.[2]

Estimated annual evaporation and vegetative transpiration from river and reservoir surfaces: more than 2.0 maf.

Presumed average annual flow available for use: less than 14.0 maf.

Total of legal entitlements to annual flow under compact, treaty, and court decisions: 17.5 maf.

Estimated 1996 population in the United States and Mexico wholly or partly dependent on Colorado River water: 23 million.

Projected population in the year 2020 that will be wholly or partly dependent on Colorado River water: 38 million.

DIAGRAM 1 (top):
Intake No. 1, constructed by Chaffey in the U.S. on Hall Hanlon's land, should have been excavated five to seven feet deeper in order to capture low flows and avoid siltation. The CDC excavated intake No. 2 in conformance with its agreement with Mexico. It was little used. Intake No. 3, opened late in September 1904, was the infamous "Mexican cut" through with the two-year flood surged.

DIAGRAM 2 (bottom):
(1) Site of Rockwood's first two attempts to close intake with dams of pilings, brush, and sand bags, March and April 1905. (2) Site of Rockwood's attempted jetty in hopes of deflecting the river to the Arizona side of the island, mid-summer 1905. (3) Site of the first "Rockwood" gate, through which Rockwood hoped to divert the river at low flow, permitting construction of a dam across the rest of the eroded intake. (4) Site of Edinger's attempted jetty, which was destroyed November 29–30, 1905. (5) Areas washed away in the November flood included the north end of the mid-channel island, which was to have anchored the jetties, and the "island" between the first (abandoned) Rockwood gate and the intake. A second and larger Rockwood Gate was next attempted within the much broadened area of the intake, known as the "crevasse." (6) The approximate line of Cory's levees and dams, which by November 4, 1906 had turned water from the merged breach of the eroded by-pass and intake. (7) The CDC levee through which the river broke on Dec. 7, 1906, requiring another furious round of trestle building and rock dumping. Repair of the final breach was completed February 10, 1907.

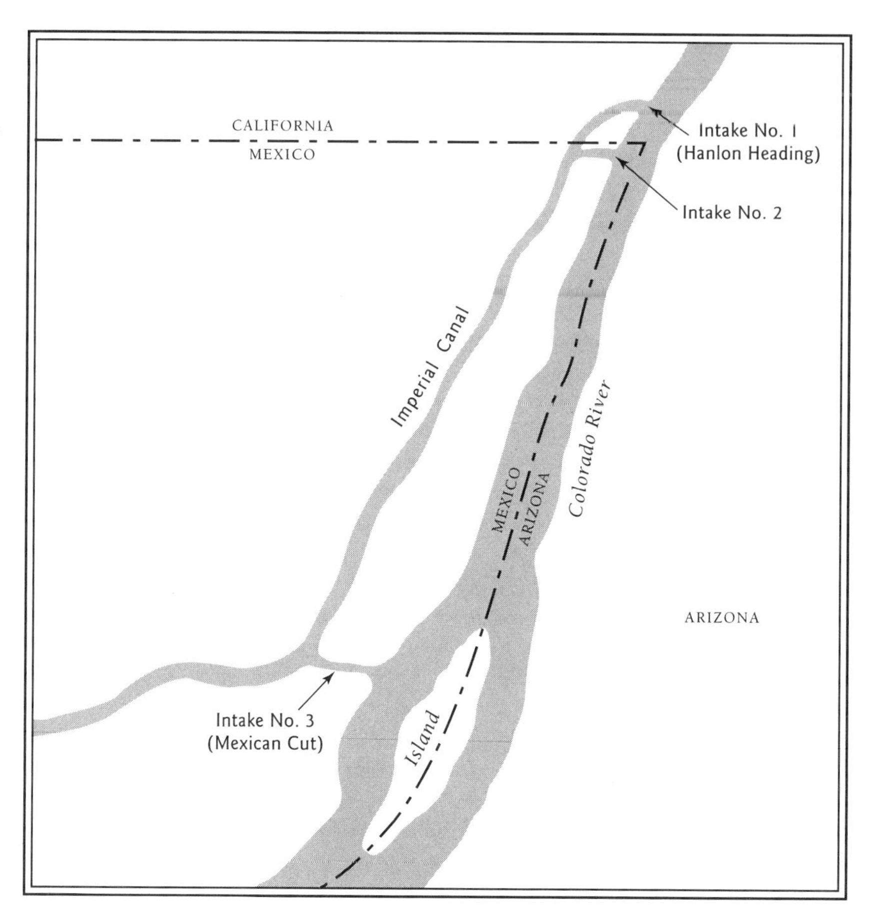

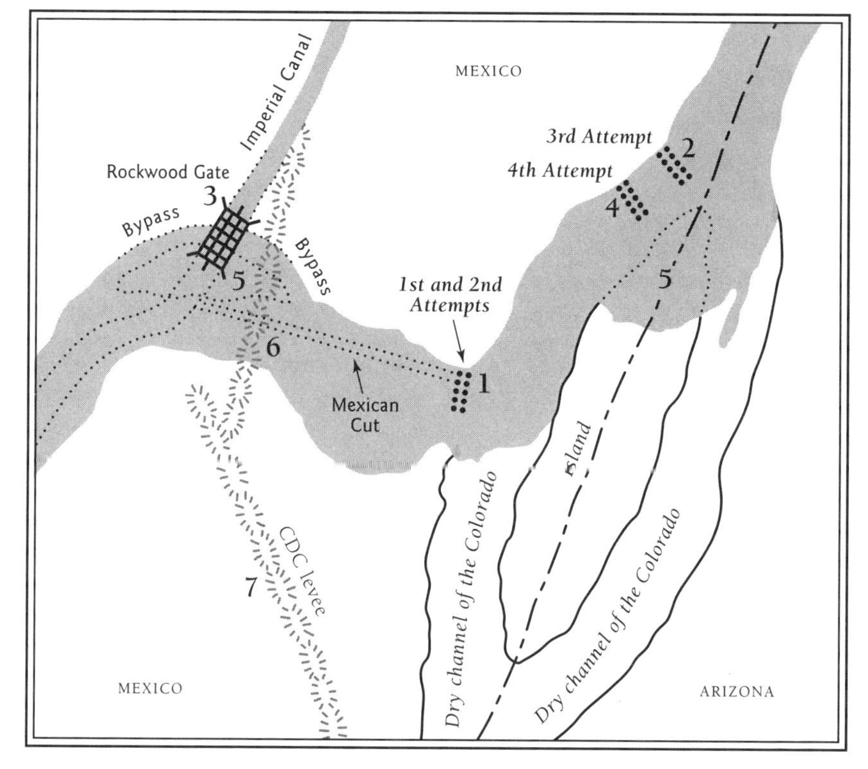

8 | A SEA OF UNINTENTION

Good luck, it is said, is nine-tenths preparation. If the reverse is also true, that bad luck especially visits the unprepared and improvident, then by leaving the Mexican intake unprotected Charles Rockwood and Anthony Heber had laid a handsome sacrifice on the altar of ill fortune. Soon the fates would accept their offering, and unluckiness would infect thousands more whose misjudgments had made them vulnerable to the influence of the California Development Company.

Rockwood later protested that the available twenty-seven-year record of the Colorado's flow (dating back to construction of the Yuma railroad bridge in 1877) suggested no possibility of multiple winter floods. One has to wonder what casuistry he used to conclude that a flood like that of February 1891, which persisted at a high stage for a week or more, posed no danger. That flood, you will remember, destroyed the greater part of Yuma and overtopped the west bank of the river long enough for Godfrey Sykes to build a boat and float from Yuma to the Salton Sink.

Rockwood also believed that the extant record of the river's behavior guaranteed a period of moderate flows in late winter and early spring, when he might build a headgate to withstand the inevitable summer floods. His hopes depended on a delicate chain of events. With the valley's water supply restored, he expected sales of water stock to resume and lawsuits to be forestalled, enabling Heber to secure financing for the construction of a control structure at the intake. By late winter, everything should have been in place: the financing, favorable weather, and formal approval by the Mexican government of the construction plans he had sent them in November.

But Rockwood's plans were far too optimistic. The flooding that he came to see as the capriciousness of nature was probably a by-product of what we today would view as a shift in the El Niño weather cycle. The years 1905 and 1906 would prove to be among the wettest ever recorded in the Southwest. Indeed, scientists using tree-ring data to reconstruct watershed yield for the upper Gila and other rivers have concluded that 1905 may have been the wettest year for that part of the Southwest in several centuries.[1] Combined with the inherent channel instability of a silt-laden river and an aggrading delta, the weather of those extraordinary years doomed all Rockwood's plans.

If he had reviewed the delta's topographic history, he might have learned, as Godfrey Sykes eventually explained to the world, that the location of the main channel of the Colorado was by no means fixed. Through time it had shifted across the delta, now in the east, eroding the toe of Sonora Mesa, now in the west, capturing the Río Hardy and debouching by that channel into the sea.[2]

Cowboy, boat builder, and wanderer extraordinaire, Godfrey Sykes had metamorphosed into a new kind of explorer, a delta rat. Year after year, over a span of decades, he probed the waters of the delta, sometimes alone, sometimes with his wife and small children. Gradually his interests outstripped his homegrown learning. He studied hydrology, read everything available, pored over old maps, debriefed riverboat captains, amassed reams of notes. Ultimately, in 1937, he produced the definitive monograph on the Colorado River delta as it existed when the river ran free. In it, Sykes compared his own detailed knowledge of the lower river with the accounts of European and U.S. explorers who had penetrated its amphibious landscape. He concluded that at the time of visits by Alarcón and Díaz in 1540 and Kino in 1701 and 1702, the Colorado flowed along the eastern margin of its delta, roughly as it did in the second half of the nineteenth century. At other times, its path lay well to the west. This was the case when Father Ferdinand Consag, dutifully extending the missionary reach of Spain, canoed northward from Punta San Felipe in 1746. It was also the case when Lieutenant R. W. H. Hardy of the Royal Navy sailed his schooner *Bruja,* in search of pearls, to the head of the Gulf of California in 1826. Hardy, as any competent naval officer would do, drew a chart of the waters he explored, firmly placing the main river in the west, a position which the journals of the American fur trapper James Ohio Pattie seem to confirm. (Pattie crossed the delta the following year, barely surviving the experience.) Hardy's chart, however, confounded later seamen. The next trained officer on the scene was

Lieutenant G. H. Derby of the U.S. Navy, who concluded from his visit in 1850 that "it is evident that the river must have altered entirely since Lieutenant Hardy's visit." Derby found that the stream he came to call "Hardy's Colorado" lacked water enough to "float a whale-boat at low tide." The Río Hardy received its name from Derby's phrase. Hardy's Colorado was no one else's.[3]

By Rockwood's time, the Colorado had located itself on the east side of the delta for at least a half century and perhaps as long as three-quarters. The mechanics of detrital deposition decreed that the river should soon move again. The unguarded Mexican cut afforded an easy path for the river to shift westward, and an exceptionally wet winter in the uplands of Arizona furnished more than enough fluvial muscle.

The first flood came in the early days of February 1905.[4] It poured through the cut, down the canal system, and into the basin, but no one was alarmed. Winter was the most critical growing season for the valley, yet it was also typically a season of low flows, when siltation of the irrigation canals created problems. The flushing action of a moderate flood was received with joy. When the flood subsided and revealed a newly deposited sand bar in the mouth of the intake, Rockwood ordered a dredge to remove the bar and deepen the cut.

A second flood shortly followed the first, with much the same effect. It silted the intake to a degree that prompted Rockwood to order more dredging, a clear indication that he anticipated no danger: "We still believed there would be no difficulty whatever in closing the intake before the approach of the summer flood, which was the only one we feared."[5] When the summer's expected high water came, the California Development Company would resume diverting water at Hanlon's or, alternatively, at a third cut it had made just below the Mexican boundary, out of reach of U.S. regulators. The valley's irrigation demands would then be low, and the river's heavy flows would likely be sufficient to force an adequate supply of water down the silted upper reaches of the main canal.

But not all was well with CDC. Anthony Heber was having little success in restoring the company's liquidity. A large portion of the valley's residents remained hostile to the company; its creditors clamored for repayment of their loans; and the considerable sums needed for building levees, headings, and other protective structures were nowhere in sight. Rockwood and several other directors became exceedingly anxious. Early in 1905 Heber approached the Southern Pacific Railroad Company seeking a loan of $200,000 with which to refinance debt and fund necessary improvements to the canal system, including, presumably, construction of a headgate at the Mexican cut. Notwithstanding his argument that it lay in Southern Pacific's interest to nurture growth in the Imperial Valley—and thus foster increased freight and passenger revenues—he was rebuffed. A few weeks later, Rockwood and several others rekindled discussions with Southern Pacific, but they mentioned nothing of their contacts to Heber.

A world of significance lies behind this covert rebellion. Rockwood and his allies had lost faith in their company's president. Heber's bluster might temporarily fend off their enemies, but he was incapable of saving the company from financial and operational ruin. Heber was a hustler, or, put kindly, a promoter, but he was not the kind of businessman to whom other businessmen entrusted capital. Rockwood probably knew already that Southern Pacific would order Heber's ouster as a condition of any loan it might make. One wonders if Rockwood did not find this a pleasing prospect.

Rockwood and his fellow directors seem to have taken a dim view of their company's value. Only months after Heber offered the company to the Water Users' Association (and by extension to the United States) for $3 million, the Rockwood contingent agreed in principle to give control of CDC to the Southern Pacific Railroad Company for $200,000 in immediate cash. Whether the transaction constituted a genuine loan or a somewhat concealed sale later became a matter of debate that embroiled even the president of the United States.

While discussions with Southern Pacific continued, the river made a statement of its own. On March 3, yet another flood roared down the Colorado and through the intake, prompting Rockwood at last to acknowledge that "we were up against a very unusual season." As the expected period of high water was now approaching—and with it the promise that the safer, upstream intakes might be used—Rockwood determined to close the unguarded cut. He ordered a barrier of pilings to be driven into the soft sediments of the gap and then had the pilings faced with bundles of brush, mainly arrowweed, and backed by thousands of sandbags, which he thought might hold the mass in place.

In this effort, as in all that followed, CDC's engineers and laborers were working blind. They had no scouts upstream of Yuma to warn them of surges coming down the Colorado or the Gila, and hence they had no way to time their work intelligently so as to attempt their most delicate operations under favorable conditions. As fate and El Niño would have it, a fourth flood slammed into Rockwood's still unfinished dam on March 25 and carried most of it away. Rockwood regrouped and attempted a second dam. A fifth flood sent its timbers floating toward Calexico in early April.

Later in April, Rockwood and fellow CDC directors H. W. Blaisdell and W. T. Heffernan finalized their agreement with Southern Pacific. The president of the line, Edward H. Harriman, personally approved the transaction over the objection of subordinates. According to his biographer, "Mr. Harriman, as a man of imagination and far-seeing vision, was naturally in sympathy with the bold attempt to irrigate and reclaim the arid lands of the Colorado Desert."[6] Charity, however, was not one of Harriman's leading characteristics.

Harriman was then one of the most powerful men in the United States. In a nation dependent on rail transport, he controlled the Union Pacific Railroad, the Southern Pacific, and numerous smaller lines—in total, almost sixty thousand miles of track. Some of the assets, as well as the audacity, of the California Development Company may have aroused his gambler's instincts. Certainly he was aware of the 100,000 acres the CDC controlled in Mexico. These lands held no attraction for the Water Users' Association and even presented an obstacle to the Reclamation Service's purchase of CDC, but their development might have profited Southern Pacific. Mexico had stipulated in its agreement with CDC that half the water diverted from Mexican intakes be used on Mexican land, and CDC's holdings were located where much of that water would logically be used. Another reason for Harriman's interest in the CDC may have been the vulnerability of the Southern Pacific's tracks to flooding where they crossed the Salton basin. Unless the flow of water into the sink abated, the Southern Pacific would have to move its tracks to higher ground, a considerable expense. Harriman may have judged that the cheapest solution to the problem was to fund CDC's repair of the break.

Nevertheless, the far-seeing Mr. Harriman did not see far enough. Like Chaffey, he may have been attracted by the prospect of acquiring an immense fiefdom on the cheap. And like Chaffey, he may have felt he could protect himself from misfortune by driving a hard bargain.

Harriman's people stipulated that Southern Pacific be able to appoint three members (a voting majority) to CDC's board of directors and that one of the three must become the new president and general manager, replacing Heber. The loan was conditional upon the enactment of these measures at the CDC's upcoming annual meeting in New Jersey in June. Once the new board and management were in place, a majority of the company's voting stock would be placed in the care of a trustee named by Southern Pacific. The escrowed stock would serve as collateral for the loan of $200,000, which Southern Pacific would then release. Heber knew nothing of these terms.

It was a handsome, cutthroat deal. Harriman would control the company while it remained solvent and own it if it failed. The only way the company might emancipate itself would be to repay the loan and interest in full, and any decision to do so was entirely Harriman's as long as he controlled the board. In any event, prospects were dim that CDC would soon become profitable, and as the spring of 1905 gave way to summer and summer yielded to fall, they faded entirely away.

AT THE END of April 1905, Southern Pacific's loan remained pending, and the river still surged through the cut. Captain Mellen of the river steamer *Cochran* had watched the river closely and suggested to the CDC that he load as capacious a barge as might be found with rock and earth, tow it into the cut, and scuttle it there to form a plug to which other materials and stabilizing pilings might be added.[7] For the moment, the level of the river had moderated, and Godfrey Sykes, at least, thought the stratagem would work. But opportunity passed it by. Scuttling a barge cost money, and in April and May, the CDC had none. Rockwood resigned himself to waiting until after the summer floods and the CDC board meeting before fighting the river again.

Meanwhile, back in the sink, the sheet of water filling the bottom of the bowl was fast overcoming the levees the New Liverpool Salt Company had thrown up to protect its salt fields. New Liverpool's lawyers sought an injunction against the California Development Company ordering it to cease flooding the basin. After a good deal of wrangling, a federal court granted the order, and when CDC did not obey, as of course it could not, lawyers for the salt works filed the first of many suits for damages against the company, further undermining its credit.[8]

No matter. The real fireworks took place in June, when Rockwood and his allies sandbagged Heber at the CDC's board meeting. No record exists of Heber's protest when he learned his partners had sold him out. No doubt it filled the New Jersey board room with a blast of Imperial Valley heat. Heber soon drifted off into speculator's exile, ultimately dying in a hotel fire in Goldfield, Nevada.[9]

With CDC restructured according to Harriman's dictates, the company's new directors executed the loan agreement with Southern Pacific on June 20. Up to then, the CDC, by Rockwood's reckoning, had built 800 miles of canals, sold water rights on 210,000 acres, and attracted 15,000 settlers to a desert where five years earlier no whites had claimed permanent residence. From this point on, the California Development Company was little but a name. Through its new president, Epes Randolph, a senior executive of Southern Pacific, E. H. Harriman called the shots.

While the corporate machinations proceeded, the river began to rise. In June it crested above 90,000 cubic feet per second (cfs), almost twice as high as its peak the year before. Between 14,000 and 25,000 cfs were pouring through the Mexican intake. The roiling waters gnawed away the light sediments of the riverbank, and soon the naked cut widened to 160 feet, more than triple its original size. A steady parade of engineers began to visit the breach, each offering suggestions as to how it might be closed. Rockwood said, with customary awkwardness, "The break in the Colorado River was a source of great alarm, not only with the people in the Valley, but was becoming so to ourselves." C. E. Grunsky of the Reclamation Service had a look at the intake on June 23 and, with the kind of circumlocution beloved of bureaucrats, characterized the situation as "not serious, but sufficiently alarming to require some attention."

CDC's new president, Epes Randolph, was more direct. Himself an engineer as well as the experienced president of two of Southern Pacific's subsidiary railroads, he came to the cut to assess matters personally. He was shocked by what he found. He telegraphed Harriman that controlling the river would be extremely difficult and that $200,000 was not nearly enough to accomplish it. It was difficult to foresee the engineering tasks that might be required, but the cost "might easily run into three quarters of a million dollars."[10]

Again Harriman's involvement with CDC paralleled Chaffey's. Either man, soon realizing that the situation was far worse than he had imagined, might have walked away. Harriman, for instance, might have ordered the puppet board of CDC to vote repayment of the $200,000 loan before it was spent. But he did not. When Randolph assured him that with sufficient funds, the river might be forced into its former channel, Harriman wired back, "Go ahead and do it."

Rockwood had emerged from the reorganization of CDC as the company's assistant general manager, and Randolph left him in charge of the effort by the river's edge. Upon receiving Randolph's go-ahead, he launched a new plan. Opposite the cut and a little upstream, an island lay in mid-channel. It was a brush-covered sandbar slightly more than a half mile long and a quarter mile wide. Rockwood determined to build a jetty that he would angle down from the west bank of the river to the upstream tip of the island, thus deflecting the Colorado's main flow to the island's eastern side.

Again his crews drove a line of pilings against the current. Again they lashed bundles of brush against the pilings with barbed wire. Slowly and dangerously they forced their wall farther into the river and held back the current. Of necessity, as the jetty lengthened, the level of the river rose, and it flowed with increasing velocity through the remaining gap. Each piling was harder than the last to drive; each one was less stable and less safe as a foundation for driving the next. When only 125 feet remained between the jetty and island, Rockwood abandoned the effort. His men could drive no more piles; the current was too strong. He fell back to another plan.

He elected to build a proper control gate a short distance from the ever-widening cut. When the gate was finished, he would dredge a bypass channel to bring the main discharge of the flood through the new gate, enabling him to dam a now tamer flow through the cut. When this dam was suitably armored, he would be able to open or close the "Rockwood" gate as conditions demanded and use either it or a second new gate being constructed at Hanlon's to divert water to the valley.

By mid-August, preparations for the Rockwood gate were well in hand, and construction materials, shipped from Los Angeles, were stockpiled at the site. Rockwood, however, was called away. Only he and Heber had ever understood the maze of contracts in which the company was ensnared, and with Heber gone, Rockwood was summoned to Los Angeles to untangle the mess. He left E. S. Edinger, Southern Pacific's superintendent of bridges, in charge at the intake, but soon after Rockwood's departure, Edinger abandoned the work.

Edinger, Randolph, and other Southern Pacific officials realized that Rockwood's gate could not be finished in time to prevent the flooding of the railroad's main track through the Salton Sink. Edinger, with Randolph's approval, elected to pursue the quicker expedient of building another jetty, and he ordered Rockwood's stockpile of construction materials moved to the site of the new project. Rockwood, who believed himself to possess some authority in the business of the CDC, received news of this decision days after it was taken. He reacted with outrage, but his protests were ineffectual. Randolph and the others may have avoided opposing Rockwood directly, but when their judgment contradicted his and his egocentric behavior became obstructive, they swiftly went around him.

The decision angered Rockwood and fed his petulance, which was by no means undernourished. The shrinking fraternity of those who still listened to him had heard many times that if only Chaffey had followed his plans, if only Chaffey had built the original intake at the depth he, the original engineer, had specified and had straightened the channel of the Alamo in the manner he prescribed, there would have been no serious siltation and hence no need for the Mexican intake. Now, if only all resources had been devoted to completion of his bypass and gate, instead of squandered on a jetty that was doomed from the start, the river would have been tamed and the flood controlled in the autumn of 1905. Rockwood would have saved the valley.

By Rockwood's account, every one of his countermanded plans would have produced success. Not surprisingly, however, his memoirs are mute concerning the unbroken record of failure of those of his plans that advanced to completion under his direction. These included the Mexican intake, the two pole-and-brush dams with which he attempted to close it, and the first jetty to the mid-channel island.

Contrary to Rockwood's lamentation, it is inconceivable that his gate would have withstood the flood that destroyed Edinger's jetty. Even had the gate's construction progressed with all possible speed, the torrent that ripped down the Gila and Colorado on November 29 and 30, 1905, would have torn it away, just as it tore away everything else in its path. The river rose ten feet in ten hours. In that short time its volume, initially about 12,000 cubic feet per second, doubled, then doubled again, then doubled again, and kept growing. It swelled to a maximum discharge at the railroad bridge of 109,000 cubic feet per second.[11] Worse, the surface of the river was a virtual sheet of driftwood that jammed against any resistance, accumulating in rafts that acted like massive, floating battering rams. Even before reaching its peak, the flood swept away Edinger's jetty. There was no question of saving any of the precious stockpile of materials that had gone into it. All possible effort had to be directed toward rescuing the 240 workmen stranded on the mid-channel island, which rapidly dissolved before the force of the flood. The men, together with thirteen wagons, thirty-eight horses, and two carloads of provisions boarded a barge only hours before the last of the island vanished beneath the flood.

When the river subsided, the northern quarter of the island, including the spot where the jetty was to have been anchored, no longer existed. The intake had widened to six hundred feet, in part because the flood had eaten away the block of terrain between the intake and the channel Rockwood had excavated for his stillborn gate. More emphatically than ever, the Colorado River had asserted its geological and hydrological desire to move west. Its old channel to saltwater in the gulf was now all but abandoned. Its new main channel, carrying 80 percent of the river's flow, led through the intake to the Salton Sink. Much of the river's remaining flow probably also reached the sink by way of overflow channels that circuitously traversed the delta through the Volcano Lake and New River areas.

In the Imperial Valley, alarm spread quickly. The remaining stoics became anxious. The previously anxious began to panic. They feared that the flood might never end, that the river might fill the Salton Sink until it drowned every farm and town in the basin. There was good reason to believe it might. The net change in elevation from the intake to the bottom of the sink was about four hundred feet, and the slope along which the river now flowed was steeper than the slope southward, down the length of the delta to the sea. Every settler knew that rivers follow the path of least resistance, and they saw little hope of resisting the Colorado. They also knew that if they stayed put, the floodwaters would cover their bones only after each of them first died of thirst: their sole source of water was the distribution system of CDC canals, which the flood was eroding, breaching, destroying. Many made plans to move, and a large number of those who owned land offered to sell it for next to nothing. But few were surprised that they found no buyers.

BY DECEMBER 17, Randolph and Rockwood had regrouped, and work resumed on a new and larger Rockwood gate. Specifications called for it to be two hundred feet wide, much larger than previously designed, and it was to be located squarely in the main canal, with a bypass dredged around it while construction proceeded. The new location signaled a retreat to the west. No longer would the engineers battle the river at its edge; instead they withdrew deeper into what they hoped would remain terra firma. A small city had sprung up in the vicinity of the works, complete with a store, bakery, vast dining and dormitory tents, a doctor's office, and an electric generating plant. It was a busy, noisy place where shifts worked day and night and the racket of pile drivers, steam engines, and centrifugal suction pumps never ceased. All through the low-flow months of winter, workers pressed forward with excavation and construction for the Rockwood gate and, upriver, for a permanent intake structure at Hanlon's. Meanwhile, virtually the entire volume of the river poured into the Salton Sea, which rose several inches a day with a flotsam of woody detritus and small-grained pumice floating at its edges. The water crept up the sides of the drying and bagging barn at the New Liverpool Salt Works. Soon only the upper story of the building rose forlorn and sentinel-like from the liquid waste. Eventually even that would disappear. The water licked the clacking wheels of Southern Pacific's locomotives as they huffed across the desert, but only for a while. Soon, sixty miles of track were moved to higher ground, and then the water inched its way up the telegraph poles, stark and uniform, that paralleled the railroad bed. Their sagging wires formed a sort of relict causeway, permitting coded dots and dashes to cross the new sea when nothing else could. The long line of poles was the last thing that marked the former path of the transcontinental rails, and then it, too, slipped from sight.

ON MARCH 9, 1906, George Wharton James, a defrocked Methodist minister who had found a new career as a popularizer of desert tourism, cast off his skiff from an overnight camp at the Mexican intake. His little flotilla of three boats carried five whites in addition to himself and two Indians, probably Cocopa. His aim was to have an adventure and give himself something to write about. He was not disappointed. His four-day descent of the floodwaters provides the closing action in his long-winded rhapsody *The Wonders of the Colorado Desert,* which appeared later that year.[12]

Downstream from the intake and the new Rockwood gate, James and his party floated on a broad, smooth river that had carved a capacious canyon through which its water moved massively, with barely a whisper. But then the channel shallowed, and the waters spread. The river broadened to a thousand feet, then two thousand, after which its dimensions dissolved. It twined in myriad streams through mile upon mile of tangled mesquite. The boaters soon gave up hope of fending off the thorny trees. They crashed into them, head-on, and swung their hatchets as best they could to force a path. The sky darkened and rain began to pour. Cold, raked by thorns, and utterly disoriented, the boaters had no choice but to keep going, following the flow as best they could detect it.

It is axiomatic in literature that heros must suffer before they bring their hard-won wisdom back to the people, and James, now suffering, was getting plenty to write about. It is hard to know what to make of what he says, for his mind was as flat-bottomed as his rowboats. It had no keel, no leverage for direction. James found everything equally marvelous: pristine emptiness and rapid settlement, the wild desert and the tamed desert, irrigation and aridity, stupendous engineering works and stupendous failures. His virtue was that he noticed so much, and even amid the agony of the mesquite he recorded seeing "hundreds of thousands of aquatic birds": a heronry with a thousand nests, the near-constant attendance of gulls, soaring flights of white pelicans, and the nests of eagles, hawks, owls, and "a bird the Indian called a squawk."[13] Clearly the wildlife of the delta was adapting to the new conditions. If the lower delta were to be abandoned by the river, its birds would move where the water was.

James and his company could find no land firm or dry enough to lie upon and passed a miserable night in their boats. The next day they reached Sharp's Heading, where the river collected itself into concentrated streams and started its earnest descent into the sink. Soon they encountered their first white water, and each successive set of rapids foamed wilder than the last. When they spotted the water towers of Holtville, they pulled to the bank. Most of the party had had enough and disembarked. Setting out again, James's two remaining companions manned one boat and James another, which he immediately lodged on a piling of the Holtville railroad bridge and was a long time unpinning. On they scudded, down the frothy river between sheer brown banks that threatened collapse. Here the danger was real:

> The boats were about fifty feet apart. We were in the radius of a great curve. The mad river was here boring under the bank, which was fully forty feet high. No one who has not seen the cutting, or literally auger-like boring power of this river in such places can believe the extent of its work. It cut in deeply and removed the entire foundation of the bank for ten, fifteen, even twenty feet. Then, without a premonitory warning, the whole bank, for fifteen or twenty feet back, dropped with a terrific splash into the river. And it fell off as if cut with some gigantic machine, almost as straight as the cutter slices a bar of soap. Both boats were almost swamped by the great waves that ensued, but fortunately neither of us was immediately under the bank, or this account would have had a more somber ending.[14]

Eventually the river widened into flats, sometimes a mile across, and again James saw geese, ducks, and pelicans by the thousands. He and his remaining companions camped that night near the sand-drifted shack of an abandoned homestead, a scene of desolation: "We saw many pathetic evidences of a woman's presence in the rude and simple efforts to care for a woman's comfort." The next day the river cast them into a headwind on the shallow waters of Salton Sea. Hard rowing won them little progress, and they camped again at a "volcanic butte."[15] In the morning they fashioned their most valuable gear into heavy packs and hiked around the edge of the sea to the rail siding at Imperial Junction.

JAMES WAS fortunate to get off the river when he did. By late March, the discharge recorded by the Yuma gauge stood at 110,000 cubic feet per second, and according to the reliable Godfrey Sykes, "the whole region between Pilot Knob and the head of tidewater was now practically under water."[16]

The velocity of the water, the sheer volume, the irresistible power that swept all things before it must have been terrible to behold. In order to comprehend such a prodigy of nature, one struggles to place it within the dimensions of ordinary life. At the risk of reinforcing a cliché, I will try to describe a flow of 110,000 cfs in relation to that common American standard of measurement, the football field. Let us assume for the moment the existence of a rectangular volume—an office or apartment building, for instance—with the basal area of a football field (300 × 160 feet). Imagine this square-cornered solid as a soaring skyscraper, but untapered in its upper floors and therefore thicker and fatter than any actual building. If a flow of 110,000 cubic feet per second were directed into such a volume, the equivalent of eleven floors would fill with water every minute. In nine minutes, the column of water would pass the height of the Empire State Building; in ten it would top the Sears Tower. No one sprints up stairs that fast, at least not for long. Anyone fleeing the drowning building by racing for the roof would quickly have the rising gurgle at his heels, and then at his throat.

But of course the flow, if it were rolling down the watershed of the Colorado or the Gila, would not stop in nine or ten minutes. In an hour, a column of water fed by the 1906 flood would have climbed more than a mile and a half high in our imaginary building, and it would have kept rising, ever flowing, a great, unstaunched hemorrhage of the planet.

The trouble with the image of the three-dimensional football field, however, is that the water in it rises, and *rising* is a behavior that unconstrained water avoids. The great flow of 110,000 cubic feet per second that surged under the railroad bridge at Yuma was headed downward—as down as it could get, as fast as it could get there. It was the wet embodiment of an enormous load of gravity, which traveled laterally only with the greatest reluctance. Every atom in it sought the core of the earth. If the river had been animate, it would have cried with joy when all its urgent downwardness encountered the Mexican intake and felt the thrill, suddenly, of steeper descent and less resistance. Forty miles farther, it might have shuddered with orgasmic pleasure when the slope at Sharps dipped a few additional degrees and pitched it straight for Salton Sea.

ON APRIL 14, 1906, *Scientific American* explained the predicament of the Imperial Valley to a curious world:

> If the Colorado River continues to flow through the channel which it has been occupying during the last six months, the geography of the Southwest must be radically changed, for at the present time but little water from the river reaches the Gulf of California, which until recently has formed its main estuary. Except when the river is in flood, the bulk of the water flows into what is known as the Salton Sink in southern California—a distance of fully 160 miles from the gulf. The new channel of the Colorado takes a northwesterly course, while the channel it formerly occupied is nearly south.[17]

The magazine further explained that the Salton Sea now stretched for fifty miles over its ancient seabed and that in places it attained a depth of twenty-five feet, while rising at a rate of three inches a day. Readers were assured that at this rate the irrigation district would not be fully inundated for a period of years.

Scientific American received a subsequent dispatch from none other than George Wharton James, fresh from his inspection of the flood and the efforts to control it. James gave a detailed account of the dam building that had thus far failed and of the equally urgent efforts to prevent destruction of the diversion at Sharp's Heading, on which the valley's water system—and its life—depended.

James's tone was confident, even complacent. He wrote with admiration of the 160-man crew working on the Rockwood gate day and night, of the powerful six- and eight-inch pumps that manipulated water level at the site, and of the stream of heavy materials—"740 sheet piles 6 x 12 inches, 22 feet long"—that gave mass and definition to the new structure.

But James's forecast was far too cheerful. Although the high water of late March had passed, summer floods were imminent, and even if the gate were soon finished, grave doubt existed whether it might be possible to dam the rest of the breach against the high water that was sure to come. Accordingly, the work that James had praised so lavishly was suspended.

To make matters worse, on April 18, 1906, a powerful earthquake rocked San Francisco, and the city's ruptured gas mains erupted in flames. The fires burned for three days. Edward Harriman rushed to the stricken city to supervise the repair of his railroads. Epes Randolph rushed there too. While the ruins of San Francisco still smoldered, he told Harriman that the $200,000 loan to the California Development Company was entirely exhausted. Would Southern Pacific advance additional funds? Harriman said yes.

It was an extraordinary decision. Harriman's railroad empire had suffered serious damage at its most important hub, and a city full of customers had suffered even worse. The flow of revenue must necessarily falter. Uncertainty prevailed. Another man might have temporized, saying, "I'll give you my answer in a week when we have better assessed our situation." Not Harriman. He personally authorized the further expenditure of $250,000, although not in the form of a loan to CDC. Southern Pacific had no further use for that hollow shell. From now on, Southern Pacific acted in its own name.

Randolph had still more news for Harriman when they met in what was left of San Francisco. On April 19, one day after the earthquake, Rockwood had resigned, or more likely, Randolph had fired him.[18] The only wonder is that Randolph had not dismissed him sooner. With this small victory to lighten his load of dreary news, Randolph no doubt discussed with Harriman whom to put in Rockwood's place. No doubt they decided quickly. Henry Thomas Cory, a former professor of engineering at the University of Cincinnati, was then assistant to the general manager of Southern Pacific. He was thirty-six years old, cool under pressure, and modestly self-confident. With Harriman and Randolph backing him, Cory would call every shot at the river's edge.

BUT FOR THE MOMENT, Cory had to wait. He could not resume fighting the river until the summer floods had passed. The main question on everyone's mind was, How much of the valley will be left when the waters finally ebb?

In June 1906, flows of 85,000 cfs ripped through the intake, tearing it wider until half a mile of river bank gaped open. The river's old channel to the gulf was all but dry and now sprouted willows and arrowweed. The Imperial Valley lay under a sheet of water eight to ten miles wide, which converged in several streams, the New and Alamo rivers foremost, that churned downward to the Salton Sea. No longer was it possible to duplicate James's journey of two months earlier, for the rivers now crashed not through rapids but over major falls, forty and even eighty feet high and a thousand wide. These were desert Niagaras, and the abrasive waters pouring over them cut back their headwalls at the alarming rate of up to four thousand feet a day. Along the way, they tore out flumes and canals, causing thousands of acres of tilled land to revert to desert.

The retreating headwalls inspired a final apocalyptic vision: if they receded past Sharps, it might never be possible to restore water to the valley. If they receded to the old bed of the Colorado, it might never be possible to direct the main river to the gulf again. Some even feared that the headcut might keep moving—up past Pilot Knob, past Yuma, tearing out the irrigation works at Laguna Dam and destroying the foundation of all agriculture south of the Mohave country.

The people most in the path of the flood scarcely had time to reflect on such nightmares. The dangers of the present occupied them entirely. Engineer C. N. Perry, a holdover from the CDC who served Southern Pacific well, ordered a six-foot levee to be raised around the town of Calexico, whose people rallied to the task with every horse and mule and shovel at their disposal. When they ran out of shovels, they used saucepans. When the waters sucked away part of the levee, they dragged mattresses out of their lean-tos and tents, out of their arrowweed ramadas and their newly built houses, and shoved them into the breach or tethered them against the sides of the channel to break the water's rush. They cut down the few trees they'd coaxed and watered those few years, and the first to go were the ones that had grown big enough to offer cherished shade. They threw the trees in the breaches or tied them to the sodden mattresses to shield the collapsing, gnawed-at banks of what used to be solid ground. They dug up the sandy earth around their homes and shoveled it into gunny sacks, which they filled by the score and then the hundred, working under the remorseless June sun and then by the light of the stars. Most of them being ardent Christians, they sang hymns against the roar of the water. They'd heard many a sermon about the Israelites and the promised land, about the goodness of the Lord who had given the wandering Hebrews and now gave them a land to water and make bloom. But amid the rampant waters that ripped away most of Mexicali and threatened to swallow Calexico, they must have wondered which story in the Bible their fates duplicated. Were they latter-day children of Moses or victims of another Noah's flood?

The people of Calexico moved everything they could inside the circle of the levee, even to the point of dismantling the Southern Pacific depot and reassembling it on dry ground. But some things they could not move. The head office of the California Development Company they abandoned, transferring its precious safe, containing vital deeds and contracts, to the meager shelter of a tent. In the weak light of the first night of crisis they watched their water tower break from its foundation and float ghostly by them, disappearing in the darkness to the north. Then a hot wind rose from the blackness and tormented them further by blowing out the flames of the lanterns by which they struggled to defend their earthworks. By the light of the next day they saw that a great mass of debris had snagged on a remnant of hard-pan clay that resisted the torrent. The debris, a collection of cottonwood trunks, brush, and, in all likelihood, pilings and arrowweed mattresses torn from Rockwood's failed dams, deflected the current in such a way that the recession of the New River headcut headed directly for the town. Calexico would not just be flooded; the very ground on which it stood would be removed. The call went out for volunteers. The only hope was to dynamite the debris, blast it loose, and thereby redirect the headcut away from Calexico's heart. A man named Mobley Meadows, alone, stepped forward. With dynamite strapped to his back, he dove into the roiling waters, fought his way to the snag and planted the charge, then floated to a point of safety downstream. The charge blew. The raft of detritus broke, and the current shifted.

Not long afterward, a team of engineers rigged a cable over part of the flood, enabling men to inch along above the water and bomb the headcut with dynamite so as to steer the fiercest currents away from buildings and ditch headings. All the next day, the charges popped off continually as the town emptied its arsenal of construction explosives.[19] Sunset brought no ease, for no one dared sleep. The plash of levee failure repeatedly called the exhausted townspeople to action.

Sometime in the course of the night, the water began to recede. By dawn the worst was over, but the view from Calexico was hardly cheering. Most of Mexicali had simply vanished—no doubt sacrificed to the very currents that had been deflected from Calexico. In the distance, small groups of survivors might be seen huddled on the few clumps of ground that showed above the sheet of water covering the desert. In the days to come, as rescues proceeded, many a crude boat would be fashioned from boards at hand, and many would be lost, hung up on snags or pinned by the current in the tops of submerged trees. By twos and threes and fours, the stranded were brought in, fed, and succored. Many among them, harrowed by what they had been through, paid a visit to Dr. F. W. Peterson, Calexico's sole medical man. Years later, Peterson composed a third-person memoir of that trying time, recalling, "Many a brave soul did he see finally give up in despair and leave the Valley, never to return. Many had put their all in here and went out penniless."[20]

WHEN THE EMERGENCY had passed, the assessment of damages was severe: four-fifths of Mexicali had been washed away, with a tally of property losses totalling

$75,000; Calexico had suffered $25,000 in damage—a relatively small sum in today's inflated dollars but by no means inconsiderable at the time. Some thirty thousand acres of farmland had been severely eroded, and the valley's prized lakes—Blue Lake and Cameron, the broad, water-holding swales that had saved the lives of forty-niners, nourished the cattle of early ranchers, and provided water for Chaffey's surveying crew—no longer existed. With them disappeared the townsites of Silsbee and Blue Lake. Fifty miles of rail line had been lost (again), and of course the salt industry was no more. Not even the peak of the roof of New Liverpool's tallest building showed above the waves of the new five-hundred-square-mile lake in the bottom of the bowl.

Henry Cory later calculated that the floods of the previous nine months (starting in November 1905 with the deluge that swept away Edinger's jetty) had dug out and washed down to Salton Sea a volume of earth almost four times as great as that which had been excavated for the Panama Canal. The floods scoured out channels with an aggregate length of forty-three miles. Cory estimated the average width of these channels at a thousand feet and their average depth fifty feet. "To this total of 400,000,000 to 450,000,000 cubic yards must be added almost ten per cent for side cañons, surface erosions etc. Very rarely, if ever before, has it been possible to see a geological agency effect in a few months a change which usually requires centuries."[21]

But in July 1906 Cory had little time to reflect on the flood's place in history. He was obliged, instead, to fight it. How to do so was a matter of debate: "Of the 40 or 50 eminent engineers who visited the Colorado delta in 1905 and 1906, hardly any two agreed upon a definite plan of defensive work, while almost everyone found something objectionable in the measures suggested by others."[22] Amid the Babel of advice and criticism, Cory devised his strategy alone.

That he was up against an unprecedented challenge was clear. The Southern Pacific had already moved its tracks twice. Now it prepared to move them a third time. One might ask why a powerful organization, possessed of a topflight staff and privy to all relevant information, would be so foolish as to move its tracks by small, useless increments. Obviously, it should have moved to the highest ground practicable at the earliest opportunity. The answer is that until the summer of 1906 Southern Pacific had failed, like Rockwood, to understand what it was up against. Now matters were clearer. There would be no half measures. Even with all possible resources devoted to the effort, victory was not assured. Informed opinion in the executive offices of Southern Pacific held that "there seemed to be only a fighting chance of controlling the river."[23] At least one thing was clear: if Southern Pacific could not put the Colorado back in its banks, no one could. It was brawnier than the peacetime army and as financially and logistically resourceful as any other entity, corporate or governmental, in the United States.

Cory prepared. He ordered construction of a branch railroad from the main

Southern Pacific line down the west bank of the river to the Mexican intake, which observers were calling the "crevasse." He provided the line with terminal facilities at both ends and sidings intermittently along its length. He borrowed three hundred side-dump "battleships" from the Union Pacific (another Harriman line). These massive cars possessed a capacity of fifty to sixty tons each and had proved their usefulness in the construction of the Lucin cutoff, a rail causeway across part of the Great Salt Lake. He arranged for the use of up to ten work trains and eight locomotives—the equivalent of an entire regional railroad. Also at his disposal was the CDC's "navy": three shallow-draft steamers, plus a motley array of barges.

Fill—a short word with a capacious meaning—was of the utmost importance. Cory prepared to draw upon every rock quarry within four hundred miles of Yuma—an area taking in most of Arizona and much of southern California. He also ordered a new quarry to be opened at the foot of Pilot Knob (which no doubt destroyed innumerable petroglyphs and other antiquities of the Quechan Indians and their predecessors). The working face of this quarry was soon forty feet high and six hundred long. For gravel, Cory enlarged a pit on the main line forty miles west of the spur to the crevasse. There, Mammoth Wash flowed out from the Chocolate Mountains, and the bed of the wash was deep with the stones Cory needed. For clay, which would be broken up and dumped to fill interstices among the gravel, Cory turned to a deposit close by the Mexican boundary.

He ordered a forest of lumber from suppliers in Los Angeles: 1,100 ninety-foot piles and 19,000 linear feet of heavy trestle timbers. To these supplies must be added forty miles of steel cable five-eighths inch in diameter, plus an array of machinery that only a transcontinental railroad could assemble and transport: pile drivers and pumps for river work, steam shovels to operate the quarries, railcar loads of spare parts, miscellaneous tools and stores, mountains of coal to keep the trains moving, and, heaven knows, victuals enough to feed a force of over a thousand workers and a legion of horses and mules.

Finding workers was no small task. A effort to recruit five hundred laborers from population centers farther south in Mexico failed. As a result, Cory "was finally compelled to mobilize all the Indian tribes in that part of the Southwest—Pimas, Papagoes, Maricopas, and Yumas from Arizona and Cocopahs and Dieguenos from Mexico."[24] The railroad men settled the Indians and their families in a camp separate from the rest of Cory's field headquarters. Although the memory of bloodshed between certain tribes was only a generation old, no serious trouble resulted. The Indian laborers soon set to work stockpiling twenty-two hundred cords of brush, which they cut from sloughs and the river bottom. The rest of the laborers consisted of Mexicans from the border area and miscellaneous adventurers and hard-luck drifters from the States. In addition, Southern Pacific recruited and dispatched to the scene enough railroad engineers, mechanics, equipment operators, and other skilled workmen to keep the trains, steam shovels, pile drivers, and other

machinery in continuous operation. Lest disorder among these disparate groups mar the military efficiency of the flood control effort, "arrangements were made with the Mexican authorities to put the whole region under martial law and to send a force of *rurales* with a military commandant to police the camps."[25]

By August, the river had fallen to about 24,000 cfs, exposing sandbars that helped close the crevasse to no more than seven hundred feet. Cory's strategy was to use the Rockwood gate, as had been planned, while closing the rest of the breach with a rock dam dumped on a brush foundation. The brush was necessary to prevent the rocks' simply sinking into the saturated silts that composed the floor of the flood channel. Even so, critics said the brush mattresses would never hold: the tons of rock would pierce and destroy them, and in spite of every effort, the rock would sink uselessly in the near-quicksand of the bottom.

Nevertheless, Cory proceeded. From August 6 to August 26, working day and night, his crews assembled and cabled together 270,000 square feet of brush mattresses. The thick, wire-bound carpet of arrowweed and willow stretched three thousand feet long by ninety wide, with a dozen runs of heavy cable binding the bundles together. A steamer and a barge, steadied by additional cables, pulled the mass into place and allowed it to sink, while keeping it tethered to a line of pilings upriver. In the deepest part of the channel, where the mattress would have to bear the greatest weight, three thicknesses were laid down. As quickly as possible, piles were driven through the mattresses, nailing them to the river bottom. Next, the piles were joined by cross members to form a ten-foot-wide trestle, and ties and tracks were laid on the trestle. By September 14, Cory's trains were rolling on a precarious bridge above the river's breach, and the dumping of rock had begun.

The mattresses held. For nearly four weeks the dumping never ceased, the river continued to fall, and all went well. On October 10 the Rockwood gate was passing 13,000 cfs, and fewer than 1,500 cfs were flowing over the new dam. But the gate showed signs of strain. On October 11, under the pressure of water and accumulated driftwood, two-thirds of it pulled loose and floated off toward Mexicali. Instantly the bypass channel became the main channel of the river, and the engineers were once more powerless against its flow.

The Rockwood gate, like nearly everything else Rockwood built, had failed, and as in every other case, Rockwood soon proffered an excuse: Cory and his men had been too slow in rip-rapping the wings of the gate. It was their fault that the two-hundred-foot-wide gate, costing $122,000 and four months of intensive labor, had yielded to the pressure of only a moderate flow. His design and engineering, Rockwood said, had been correct.

In all likelihood, no one listened. There was too much to be done. The new dam had proved that a brush mattress foundation would neither sink nor break under the load of rock. Attention now shifted to completing the new headgate at Hanlon's and to dredging the silted main canal—a supply of water had to be assured for the valley before shutting off the flow through the breach. Soon all was ready, and the first of three thousand railcar loads of rock, gravel, and clay began rumbling toward the break. The cars kept coming, an average of six every hour, hour after hour, night and day, every day for three weeks. Dredges and scraper teams, meanwhile, moved 400,000 cubic yards of earth as they fashioned levees and dams to buttress the system. By November 4, a little more than two years after it opened, the infamous Mexican cut had been closed.

The resultant wave of congratulatory good feeling lasted a month, during which the people of the valley reacquired the habit of sleeping soundly through the night. Experts who previously had disputed the merits of various dam-building schemes began to speculate on the future of the Salton Sea. How long would it take to evaporate? Might it, should it be retained? W. M. Reed of Chicago assured *Scientific American* that "if the Salton Sea is retained, climatic conditions in the West and Southwest will experience great if not wonderful changes. . . . With sufficient rainfall the section now known as the arid Southwest would develop into the most wonderfully fertile region in the entire country."[26] But before Mr. Reed's letter could be published, the river spoke again.

ON DEC. 7, 1906, a sudden flood surged down the Gila and lifted the flow of the Colorado at Yuma from 9,000 to 45,000 cfs. The old main channel leading to the Gulf of California, unused for the better part of two years and now clogged with sandbars and willow and arrowweed thickets, could not easily pass such a volume. At the newly dammed Mexican intake the water slowed, spread, and deepened. It infiltrated the soils of the channel bank and seeped into the material of the levees. This was a matter of little concern where the levees had been built to Cory's specifications. But the CDC's old levees, built under Rockwood's direction, were another matter. These levees lacked ballast and were not designed to resist infiltration. They were little but heaps of the same fine silt that the river so easily dissolved and redeposited whenever it grew restless. And it was restless now.

The rising waters saturated a section of CDC levee a quarter mile south of Cory's dam. The levee began to leak and then to melt. It finally gave way. Within three days the entire river was pouring through the new break, which it had opened to a width of a thousand feet. All the struggle and expense of the previous four months seemed to have been in vain.

Cory, Randolph, and their colleagues at Southern Pacific assessed the situation. It seemed clear to them that the tendency of the Colorado to flow to the west was increasing, that floods of 50,000 cfs or more could occur at virtually any time of year, and that the defensive dikes built by the CDC were worthless. Southern Pacific had already spent about $1.5 million in its efforts to control the river, and the tasks that lay ahead would be astronomically expensive. Not only would closing the present breach consume $300,000 to $350,000, but permanent control of the river would require, at a minimum, the reconstruction of some twenty miles of levees at an additional cost of $1.2 million an amount roughly equal to the gross value of all the goods shipped to the Imperial Valley in a year. These shipments generated less than $35,000 annually in fees to the railroad, making the survival of the valley a matter of dubious financial benefit to the company. The railroad had already committed itself to removing its tracks across the Salton basin to a new route safe from all possibility of inundation. Edward H. Harriman, having received the reports of his engineers and managers, concluded that his railroad had done enough.

On December 13, 1906, Harriman outlined the situation in a telegram to President Theodore Roosevelt, with whom he was not on the best of terms. Harriman was then under investigation by the newly invigorated Interstate Commerce Commission, and Roosevelt, showing little concern for prejudicing a possible prosecution, had publicly excoriated Harriman only weeks before, calling him "an undesirable citizen."

Harriman's telegram painted the direst possible picture: renewed flooding of the Colorado threatened to destroy the Imperial Valley, and the progressive upstream retrenchment of the river would in due course destroy agriculture in the Yuma Valley and render worthless the government's considerable investment in that valley's reclamation. In addition, the government would further lose the potential development value of the unconveyed public domain in the region. Harriman did not ask outright for government funds with which to continue battling the river. He simply concluded: "In view of the above it does not seem fair that we should be called to do more than join in to help the settlers."[27]

There followed a rapid exchange of messages between the two men. Harriman offered the engineers, organization, and equipment of the Southern Pacific but insisted that the work be carried out under the direction and at the cost of the Reclamation Service. On December 20, Roosevelt replied that the "Reclamation Service cannot enter upon work without authority of Congress and suitable convention with Mexico. Congress adjourns today for holidays. Impossible to secure action at present. It is incumbent upon you to close break again."

Roosevelt further asserted that responsibility for the break lay squarely with the California Development Company: "I wish to repeat that there is not the slightest excuse for the California Development Company waiting an hour for the action of the Government. It is its duty to meet the present danger immediately and then this Government will take up with it, as it has already taken up with Mexico, the question of providing in permanent shape [*sic*] against the recurrence of the danger." Harriman wired back his response the same day. He said that Southern Pacific did not own CDC (a debatable proposition), and Southern Pacific should not be held liable for CDC's errors (equally debatable). Nevertheless, he pledged that Southern Pacific would resume its battle with the river: "I am giving authority to the Southern Pacific officers in the West to proceed at once with efforts to repair the break, trusting that the Government, as soon as you can procure the necessary Congressional action, will assist us with the burden."[28] A relieved Roosevelt replied that he would seek funding as soon as Congress reconvened.

THE CREVASSE was now 1,100 feet wide and 40 feet deep. The entire river was pouring through it at a rate of 45,000 cubic feet per second. Cory's strategy was muscular and simple. He judged that there was no time to waste on the construction of brush mattresses. He would build a pair of trestles across the gap and try to dump rock into the breach faster than it might sink into the sediments or be washed away. But building the trestle proved harder than expected. The river stayed high and gave no relief; it washed away three successive attempts. Finally, on the fourth try in as many weeks, the first of Cory's two trestles stretched across the gap, and the dumping began.

The side-dumping "battleships" of earlier efforts were less important this time than the simple flatcars, a thousand of them, that provided the backbone of Cory's rock-hauling fleet. The river had to be raised eleven feet before it would begin to flow down its old channel, and the logistical effort required to force such a change was staggering. For two weeks all rail traffic in and out of Los Angeles ceased, as every available engine and every car capable of hauling boulders or ballast was thrown into the fight against the river. The quarry at Pilot Knob shipped rock day and night. So did the quarry at Tacna, up the Gila sixty miles by rail to the east. And so did the quarry at Colton, in the Los Angeles basin west of San Gorgonio Pass. Rock came from as far away as Patagonia, Arizona, southeast of Tucson, almost five hundred miles away. And again the greedy sediments of the delta swallowed a forest of northern timber: 1,200 ninety-foot pilings and 16,000 linear feet of 8-by-17-inch pine timbers, chased by a boxcar or two of bolts, drill bits, spikes, and reinforcing flanges.

The railcars streamed in, some of them burdened with boulders so large they had to be blasted apart with dynamite while still on the flatcars. Only then could the broken pieces be levered over the side. The sound of dynamite "pop-shots" cracking open the rock, the whistles of the locomotives and the huffing engines with their blasts of steam, shunting up and down the lines, the rasp of steel wheels on rail, and then the shouts, the background chug of pumps and generators, the ceaseless clang of tools, the splashes and the rush of the river—all this rang on for fifteen days without pause.

Beginning on January 27, 1907, 80,000 cubic yards of rock, at a rate of almost 225 yards an hour, went into the river. As soon as one train left the trestle, another took its place, with workers scrambling to heave its contents overboard. The fatigue of the frantic labor and endless rotation of shifts took its toll. Two Mexican workers lost their footing from one of the cars and fell into the flood. If their bodies were ever recovered, the accounts of the day do not mention it.

After the rock came the gravel—in 5,285 railcar loads. And after the gravel came the clay—another 500 loads. The voice of the river changed as the dam took form against it. From a guttural objection it rose to a high, seething pitch, like the angry drone of a moon-sized hornets' nest. Slowly the seething declined to a hiss, and the hiss to a sigh, and on February 10, the sigh gave out. The Colorado River again flowed within the old banks it had tried so hard to leave.

Four attempts to contain the river had failed utterly: Rockwood's two brush and piling dams and both Rockwood's and Edinger's attempts to build a jetty to the sand island opposite the cut. Cory's first two attempts had succeeded tactically but still failed when the river burst through anew, first at the Rockwood gate and later at a weak stretch of CDC levee. When finally his third dam plugged the last crevasse, he did not stop. His crews kept dredging and digging until they moved 900,000 cubic yards of earth and built a double row of dikes southward from Hanlon's Heading for more than twenty miles. (Such a volume of earth, incidentally, would fill the aforementioned football-field-sized skyscraper to a depth of over five hundred feet—just a little shorter than the top of the Washington Monument in Washington, D.C.). Cory also installed a telephone line down the levee road with call boxes every mile. Watchmen in "gasoline track velocipedes"—contraptions approaching what people would soon call "cars"—patrolled the levees constantly. In case a leak, let alone a break, were spotted, a train loaded with rock waited on a nearby siding, ready to be rushed to the scene.

Such precautions were appropriate. El Niño—or whatever genie was brewing the region's weather—had not finished its mischief. In June 1907, rain and rapidly melting upstream snowpack combined to produce yet another powerful flood. At the crest, 116,000 cubic feet of water rushed beneath the railroad bridge at Yuma every second. It was the highest flow the river gauge had ever recorded.[29] Yet Cory's dikes and dams withstood the strain.

WHILE THE immediate battle was over, the war for dominion over the Colorado had scarcely begun. In decades to come, it would be fought as much in Washington, D.C., as in the delta and desert. The first legislative skirmish involved Harriman's bill for saving not just the valley but, to believe Roosevelt's message to Congress, the railroad bridge at Yuma, Laguna Dam, and all agriculture south of Fort Mohave. The cost of Southern Pacific's effort subsequent to Roosevelt's December 20 telegram to Harriman exceeded $1 million. Added to the $1.5 million the railroad had already spent by that time, the total charge climbed, with interest, toward $3 million. Congress was unmoved.

William Howard Taft resumed the crusade for recompense when he became president in 1909. In 1911, a congressional committee finally recommended reimbursement in the amount of $773,647, but the matter never came to a vote in the House. Elwood Mead, a respected early commissioner of the Reclamation Service, characterized the dalliance of Congress as an example of "supine indifference and irresponsibility." Ultimately, in 1930, two decades after Harriman had gone to his grave, the federal government paid Southern Pacific a judgment of $1,012,665, representing at best a partial recompense.[30]

In the aftermath of the flood, the courts were busy. The New Liverpool Salt Company kept pressing for payment of its $458,000 judgment against the CDC, while Southern Pacific sued that shell-shocked and hollow entity for $1.5 million. Valley growers, meanwhile, sought damages that ran to fantastical numbers. It was all abstract, all rhetorical. No matter how much the assembled carnivores argued their rights to the best cuts of meat, they had little to divide among them but a rotten and eviscerated carcass. Still they squalled for the bones.

The people of the valley pressed on in somewhat new directions. Before 1907 was out, they split from their old county seat of San Diego to form Imperial County. Straight away they voted the county dry, an act of legislative irony for a recently flooded desert and a substantial gift to the saloonkeepers of Mexicali. Mobley Meadows, the hero of the defense of Calexico, became their first sheriff and promptly undertook to build a proper jail. In 1911 the people of the valley took the further step of forming an organization to replace the old Water Users' Association. They called it the Imperial Irrigation District. In due course, the district purchased the CDC's irrigation system from Southern Pacific, which had assumed title to all of the assets—if they can be called that—of the now defunct California Development Company.

These events were among the most predictable results of the years of flood, but not necessarily the most important. The events of 1905–1907, which Godfrey Sykes wryly dubbed "the Great Diversion," had consequences so profound they might almost be said to have restarted the region's history. It was as though the receding floodwaters left behind a deposit of silt covering every possible thing, visible and invisible, tangible and abstract. After the flood, nothing looked the same, not landscapes, institutions, politics, or ideas.

PART THREE

CONSEQUENCES

MARTINEZ
INDIAN SCHOOL & AGENCY

9 | THE UNDERWATER RESERVATION

The great flood rocked the valley's sense of safety, and in its aftermath the cheerleaders of settlement strove to assure a doubtful world that the disaster would never be repeated. They rhapsodized about the reliability of the levees and dams that Cory had built. They extolled the vigilance of the men of the Southern Pacific, who patrolled the levees day and night. They marveled at the telephone call boxes, spaced at mile intervals along the main levee, from which a train might be summoned to rush a load of rock to any incipient breach.[1]

Everything was up to date along the levee, and the good news did not stop there. Boosters soon found reason to applaud the flood and to proclaim that its massive work of erosion had been good for the valley. The canyons cut by the flood, up to sixty feet deep along the New River and thirty feet deep along the Alamo, might now serve—"in a better way than anything that could be devised by engineers"—as drainageways to carry spent irrigation water from the fields. That was not all:

> The sewerage problem of the towns is also solved, and the actual damage caused by the runaway river is not felt to be a great price to pay for the immediate and prospective

advantage of these great dry channels. They are available for all time and deep enough on each side of the valley to keep the land perpetually sweet.[2]

The channels were the New and Alamo rivers, which today carry a steady flow of irrigation drainage and municipal sewage to the Salton Sea. In 1907 and the years immediately following, however, no one considered that the bottom of the sink might become the site of a permanent sump. Valley leaders and government scientists generally agreed that the Salton Sea was "destined to disappear from the cutting off of its source of supply." And good riddance to it, they muttered. The lake was a reminder of the nightmare they had narrowly survived. The sooner it evaporated into the sky, the better they and all their neighbors might sleep. And so they promised themselves that the lake would soon be gone and that the river would not threaten them again: "There is no ground for fear of injury to any farm lands from this quarter. . . . It is certain as anything human can be that the farmers of Imperial Valley will not be disturbed again by the rise of the river. . . . There is in the Valley to-day, an absolute sense of security, and it is affecting prices and stimulating investments in all realty."[3]

Fortunes, along with crops, began to grow anew. The quarter million acres under cultivation by 1913 swelled to 360,000 acres five years later.[4] The transformation of the region proceeded as fast or faster than anywhere else on the continent. From Hispaniola to Plymouth Rock and onward to the California coast, Europeans and their New World descendants had reworked the environmental character of large portions of North America in only a generation or two. But in Imperial Valley, they quickened their pace. A vast agricultural plain materialized from the desert in little more than a decade. But as the former wastelands bloomed, the swamps and marshes of the Colorado River delta commensurately dried up, and John Van Dyke's beloved "bottom of the bowl" remained submerged.

FEW PEOPLE thought much about either the delta or the Salton Sea in the first years after the flood. So certain was the U.S. government that the new lake would soon vanish that it made a gift of some of the land beneath it. In 1909 the Department of Interior reserved, in trust, ten thousand acres lying under the sea for the benefit of the Torres-Martinez Band of Desert Cahuilla Indians. The new "lands" added greatly to the size of the tribe's holdings, but of course they did so only on paper, for they could hardly be considered terra firma. Then again, neither could most of the land the tribe already controlled: between eight thousand and nine thousand acres of the existing reservation had been inundated by the Great Diversion.

The Torres-Martinez Cahuilla of 1909, like their five hundred descendants today, would probably not feel insulted to be called a "patient" people. But even Job might take exception to the predicament they have endured. Nine decades after the greater part of their land was flooded, and after the government bestowed its curious "gift" of ten thousand additional acres of flooded land, nothing has changed. Their land is still under water and they have been able to acquire no land to replace it. Members of the tribe like to joke, not without bitterness, about borrowing a boat and cruising out on the Salton Sea "to look at my land."

At first blush, the granting of a reservation beneath a sea might seem a cynical act. In this case, however, the evil of government resulted not so much from malice as from a nearly perfect absence of foresight and vigilance.

In the nineteenth century the Cahuilla people inhabited several dozen villages scattered through San Gorgonio Pass and the foothills and valleys of the Santa Rosa Mountains. Today they hold claim to ten small reservations, some of which nestle among the golf resorts of greater Palm Springs and the farmlands of the Coachella Valley. The word *coachella* derives from a variant of *cahuilla*.

The lands of the Torres-Martinez band lie farthest south of the Cahuilla reservations. Ancestors of the band probably inhabited the north end of Salton Sink since time immemorial. Presumably it was they who built the rock traps in the bluffs west of Mecca in order to capture fish from the receding waters of Lake Cahuilla. And it was they who entertained William Phipps Blake in 1853 with legends of earlier inundations.

In 1876 the U.S. government conferred on the band a reservation consisting of one section (640 acres) of land. In 1891 the government withdrew from the public domain almost 20,000 additional acres for the benefit of the tribe, but nearly half of this area was inundated by the floods of 1905–1907. Under the Indian policies of the day, these lands were to have become the farms that would support the tribe's transition to a more Euro-American style of life. For the farms to function, however, the tribe needed two things. First, they needed the newly formed Salton Sea to evaporate. Second, they needed a good supply of freshwater with which to leach salts from their prospective farmland and then to irrigate their crops. The superintendent of the Martinez Indian School notified authorities in the Department of Interior's Bureau of Indian Affairs that "a fine belt of artesian water" lay under certain lands that the sea had inundated.[5] He recommended that the lands be reserved for the tribe in order that, when the sea finally evaporated, its members should have the water they needed.

The superintendent's recommendation became the basis for the additional reservation of twenty-two sections in 1909, which brought the tribe's total holdings to more than thirty thousand acres. The Cahuilla now presided over the nation's largest submerged Indian reservation.

And so the tribe, then and now one of the poorest in the region, waited for the sea to evaporate. At its greatest height, in 1906, the surface of the sea lay 195 feet below sea level—over 30 feet higher than today. The arithmetic of prediction was simple: given that the lake stood 80 feet deep and would evaporate at a rate of slightly less than 6 feet per year, it should vanish in about fourteen years. (Percolation into the ground was rightly deemed negligible, owing to an impervious stratum of clay underlying the basin.) Allowing for the occasional winter storm and the in-flow of irrigation tailwaters at pre-flood rates, those who studied the matter predicted that the Salton Sea would disappear by 1923.[6]

But of course it did not. What the seers of the sink failed to take into account was the continued expansion of agriculture in Imperial Valley, as well as the reclamation of northward-draining lands in the Mexicali Valley, where 118,500 acres were being farmed by 1918.[7] Although the level of the sea declined steadily from 1906 to about 1924, it was clear by 1920, if not sooner, that the sea had become a permanent feature of the landscape.

The agricultural empire that had formed in the Colorado Desert contributed water to the Salton Sea in three ways: as tailwater, or surface runoff from irrigation, as drainwater, which percolates through the irrigated soils and seeps into the drains, and as spillwater, which runs through the canal system and into the drains without being diverted onto a field. These flows, plus municipal discharges and runoff from storms, are the inputs to the sea, while evaporation is the only appreciable output. As the level of the sea rises, its surface area expands, increasing the volume of water that evaporates from it. At some point, if inflow remains steady, evaporation will offset inflow entirely and the level of the sea will stabilize. If inflow decreases, the reverse occurs. The sea surface falls, the sea's area shrinks, and total evaporation diminishes until it comes in balance with inflow again. The level of the sea has ranged between 220 and 235 feet below sea level for most of the period from 1924 to the present.[8]

THE EVIDENT permanence of the sea became an issue of considerable interest to the Imperial Irrigation District not long after it was formed in 1911. Following the collapse of the California Development Company, the affairs of the valley became more chaotic than ever. Water deliveries failed as mutual water companies warred with each other and with Southern Pacific, which controlled the remnants of the CDC. The federal Reclamation Service, never popular with valley settlers, watched for a chance to step in and take over, a prospect that few in the valley welcomed.

Hence the referendum in June 1911 by which voters in Imperial County approved the organization of an irrigation district. Once chartered, the IID set about acquiring the CDC's water delivery and distribution system. The task was long delayed, however, by the maze of litigation arising from the Great Diversion. Finally, in 1914, the IID voted a bond issue of $3.5 million, of which $3 million went toward purchase of the CDC's former assets, including its Mexican property, and $500,000 remained available for improving the canal and levee system.[9]

Over the next decade, the IID bought up the valley's remaining mutual water companies and consolidated its position as the most powerful political and corporate entity in the region. In about 1920 it turned its attention toward the northern limit of its territory, to the lake that refused to go away. The IID began buying railroad and other private lands that lay below the sea, and it soon also approached the Department of Interior, which was responsible for the nation's public domain, to ask that it be granted an easement over the public lands that the lake had inundated. Federal attorneys, however, were reluctant to convey permanent rights in the submerged lands to the IID and proposed an alternative: the federal government, by executive order of the president, would set aside the submerged lands as a permanent drainage reservoir, a sump. IID agreed. The first withdrawal was made in 1924 by President Coolidge; it was supplemented by a second withdrawal in 1928 by President Harding. The explicit purpose of both actions was to provide an "evaporative pan for surplus and waste water from Imperial Valley irrigation development." Effectively the two executive orders included all the federal land in the basin up to an elevation of −220 feet.[10]

Unfortunately, no one at the Department of Interior troubled to reconcile these two withdrawals with the earlier withdrawal of a reservation for the Torres-Martinez Cahuilla. Certainly someone should have noticed. In the management of the nation's public domain, the government bears a certain periodic obligation to produce maps, and cartographers within the Department of Interior had duly recorded the 1909 reservation, as well as the formerly dry 1891 reservation lands that now lay beneath the sea. One might wonder who was minding the business of the department and its constituent agency, the BIA.

ONE DAY IN 1990, Albuquerque attorney Tom Luebben received a phone call from a stranger in southern California. The caller introduced herself as Mary Belardo and said she was chairman of the Torres-Martinez Desert Cahuilla Indians. She had just received a letter from the federal government, which she did not understand. The letter notified her that the government had awarded the tribe a grant of $140,000 for undertaking hydrological studies. Neither Belardo nor anyone else in the Torres-Martinez community was accustomed to receiving official correspondence from Washington, D.C., let alone notification of six-figure grants. She had heard from officials at another tribe that Luebben was pretty good at representing Indians. Would he be interested in helping the Torres-Martinez Cahuilla find out what the letter was about?

Luebben made some calls and learned the basic elements of an improbable story. Eight years earlier, in 1982, an attorney representing the U.S. Department of Justice had filed suit in the Southern District of California on behalf of the Torres-Martinez Cahuilla against the Imperial Irrigation District and the Coachella Valley Water District (CVWD).[11] The suit alleged that the two irrigation districts had trespassed on Cahuilla property for many years by storing their drainage water on the reservation. The suit further held that the irrigation districts owed payment to the tribe for loss of use and damage to its land. There was a final element of Wonderland logic in the government's argument: the plaintiffs asked that the irrigation districts be ordered to cease their deposit of trespass water on Cahuilla land, which order, if granted, would require the cessation of irrigation in the region. Surely no one thought this was likely to happen.

The government initiated the suit as a place-holding action. Otherwise, the tribe's right to seek legal remedy for its injuries would have expired under a statute of limitation. Once the filing was made, however, nothing more happened for seven years. No briefs or motions were filed; no discovery was undertaken; no expert testimony was prepared. Worst of all, no one in the Justice Department troubled to inform the tribe that the suit had been joined, nor did the Cahuilla hear from the Bureau of Indian Affairs—if anyone there even knew. For the rest of the decade, the docket of the case fit easily on a single page. It consisted solely of repeated notations that, by mutual consent of the parties, the case had been continued.

While the government dawdled, other adversaries of the IID, possessing less cause but more energy, busied themselves in litigation. Throughout the 1980s they obliged the IID, as never before in its history, to defend itself—unsuccessfully in many cases—against claims for flood damages. The argument behind the claims carried an echo of *Chinatown;* it held, in essence, that the IID intentionally wasted water. Further, the wasted water produced unnecessarily high water levels in the Salton Sea, which inundated valuable real estate, waterlogged it, or poisoned it with salt. Eventually IID would pay over $30 million in damages to a host of claimants, and millions more to its attorneys.[12]

The Cahuilla case languished until the summer of 1991, when a new district judge, Judith Keep, made efforts to clear the backlogged calendar of her court. She summoned the various parties and asked how soon they could go to trial. The experienced attorneys of the water districts, already versed in the issues to be argued, answered that they would need a year to develop their expert testimony, assemble their exhibits, refine their theories, and school their witnesses. Judge Keep then turned to the young government attorney representing the Cahuilla, a newly minted Juris Doctor. Might the government have its case ready in six months, asked Judge Keep? The government attorney, scarcely acquainted with the matters at issue, replied, "Absolutely."

It was a dismal start for the Cahuilla's cause and set the tone for what followed. Judge Keep rejected a motion by the Luebben firm that it be admitted to the case as counsel for the tribe, thus tying the fortunes of the Cahuilla to attorneys from the Department of Justice whom the tribe had not selected and in whom it had no faith. Judge Keep set the trial for January 1992, and on October 29, 1991—a mere seventy-five days before the trial was to start—a government lawyer set foot on the Torres-Martinez reservation for the first time.

The trial was eventually postponed until July 1992, in which month, in Judge Keep's courtroom, government attorneys commenced their arguments to recover, on behalf of the Torres-Martinez Cahuilla, $30,313,059 for lost income, $32,617,730 to rehabilitate damaged land, and $6,632,450 for future lost income during the period of rehabilitation—a total of $69,563,239 in damages. And, they added, the flooding would have to stop and the Salton Sea be dried up.[13]

These were the arguments of only one tentacle of government. The Bureau of Reclamation, which delivers water to the IID and CVWD through the All-American Canal, would have presented a different opinion. The Department of Agriculture, which nurtures crop production from the half million acres served by the districts, and the U.S. Fish and Wildlife Service, which maintains a wildlife refuge at the south end of the sea, would have added their own perspectives.

The trial featured some unusual twists. The government put forward not a single Cahuilla witness to testify to the impact on the tribe of the submergence of its land. Nor did the government call a witness from the tribe for any other purpose. In closing arguments, the government seemed to switch sides by admitting that it would be unfair for the court to enjoin the defendants from releasing water, for such an action would impact too many people and too much property.

The attorneys for the water districts found the government's arguments on this point so cogent that they later used them as a basis for their appeal of the case. Lest this seem logical, one should remember that the arguments of plaintiffs and defendants are not ordinarily interchangeable. Even Judge Keep noted in her decision that considerable irony was at work—she pointed out that the case might have been better and more fairly tried if the government had been the defendant.[14]

In spite of her misgivings, Judge Keep ruled in favor of the plaintiffs, but she awarded the tribe only $3,910,062 in damages. Not surprisingly, she refused to order the districts to cease discharging water into the Salton Sea.

Most people will acknowledge that a payment of $3.9 million is better than nothing at all, but in view of the fact that prior to the trial the defendants had offered to settle out of court for $5.3 million, the Cahuilla had obvious cause for disappointment. Nevertheless, the government declared a victory.

As is usual in complex legal matters—especially where water is concerned in the Southwest—things did not stop there. The attorneys for the water districts prepared their appeal of Judge Keep's decision. The Luebben firm filed suit against the government for malpractice, an act that ratcheted the general spirit of hostility and dysfunction yet higher. Luebben and his colleagues also sought the illusive injunctive relief that would cause the Salton Sea to disappear and much of the region's agriculture to cease. The government filed a motion to sanction the Luebben firm for improper and unethical practice. And quietly, independent of the posturing, acrimony, and formality of court proceedings, real negotiations began to take place. They involved the Luebben firm, on behalf of the tribe, the Department of Interior (not Justice), and the irrigation districts. The tribe instructed the Luebben firm to say they wanted land in replacement for the lands that had been lost through submergence, and they wanted money for damages.

By September 1992, the three parties had reached tentative agreement that the tribe should be allowed to select 11,800 acres of government land administered by the Bureau of Land Management to replace the land they had lost, and they should be paid roughly $11 million dollars for damages. These monies would include the $3.9 million court settlement, plus government payments of $3 million dollars for past damages and slightly more than $4 million to fund the rehabilitation of agricultural land that was waterlogged and salinized but not submerged. In return for the land and cash, the tribe would relinquish its claims against the government and the irrigation districts.

The Justice Department, unfortunately, still smarted from its clashes with the Luebben firm and refused to approve the deal. Soon the impending presidential elections of 1992 put all serious government business on hold, obliging the Cahuilla to resume doing what they had done since the days of the Great Diversion: wait.

As it waited, the tribe made no secret of what it would do with the replacement land it would select. Public lands near the retirement and recreation communities of Palm Springs and Indian Wells looked attractive, and the casino business looked good, said Mary Belardo. Indeed, the tribe would have entered the gaming business long ago, if it could have: "You know, if the sea wasn't so smelly and trashed, with dead fish, dead birds all over the place, we could have done a riverboat casino and drove around the Salton Sea."[15]

But opposition to Torres-Martinez gambling has presented a formidable obstacle to effecting the kind of settlement which the principals outlined in 1992. Chief among the opponents are other casino operators, including the nearby Cabazon Cahuilla band. Allied with gambling interests from Nevada, they helped prevent a settlement bill from passing Congress in 1996.

And so the Torres-Martinez Desert Cahuilla wait. Mary Belardo is no longer tribal chairman, but she still works as a receptionist at the Indian Health Service Clinic next to the old Torres-Martinez Indian school. Other tribal members struggle with jobs in Indio or nearby towns in the Coachella Valley. Under a new tribal chairman, the Torres-Martinez Cahuilla have hired counsel to replace Tom Luebben, who gnashes his teeth when he thinks of the devastating Mississippi floods of 1993 or the 1997 floods in both the Midwest and northern California. Says Luebben, "Everywhere else you see politicians tripping over each other to get flood relief for white people, but here is a group of Indians who have been waiting without help for nearly a century."[16] And the Cahuilla still joke among themselves, as they have for generations, about how nice it would be to borrow a powerboat, if only they knew somebody who had one, and to cruise out on the Salton Sea to look at their land.

Side Trip: Port Isabel

In Godfrey Sykes's day, as now, mudflats at the farthest limit of the Colorado River delta stretched horizon to horizon. The weak-willed earth vacillated: when tide was in, saltwater ruled; when tide was out, mud was all the world.

With two exceptions: something like a channel cut through the mud not far east of the mouth of the Colorado, and an acre or two of drier mud peeked up like a crocodile's eye from the surrounding flat. This was Port Isabel.

Port Isabel never thrived, but for a brief time it subsisted. Riverboatmen gave it a name in the 1860s. It offered a place where the dreaded tidal bore roared less fiercely than in the actual river, allowing ocean vessels to transfer passengers and cargo to shallow-draft steamboats that plied the river up to Yuma and the mining camps beyond.

A shipping company diked off a swatch of mud and built a dry dock complete with a carpenter's and a blacksmith's shop, a kitchen, and a mess hall. A old steamer, lodged in the goo, became a kind of barracks. Other defunct hulls, as they accumulated, served as storage sheds or yielded parts or fuel.

After the railroad came to Yuma in 1878, the world had no further use for steamers on the lower river, or for Port Isabel. The tides soon wore the dikes away and collapsed the sheds, leaving the detritus of the place to the slow work of water, sun, and salt.

Gulls visited, but only crabs dwelled at Port Isabel when Godfrey Sykes tied the first of his delta-probing sailboats to a piling there in 1891. He and his temporary partner were dozing in the pilot house of one of the wrecks when they became aware of a shift in the stillness.

A file of eight Indians approached, on foot, slogging through the mud. They were, Sykes noted, "extremely unclothed." The Indians proceeded to one of the larger steamboat hulks and began to pry up sheets of zinc plating from its upper deck. They rolled the plating into bundles, as though to carry it away. Sykes, ever curious and unaware of the reputation of the Sand, or Areneño, Papago for preying on outnumbered wanderers, inquired of their purpose.

The leader of the band spoke little English. He produced a parcel wrapped in sacking, which appeared to be the band's sole possession. He unwrapped it, solemnly and with reverence. Within the sacking lay an ancient smoothbore musket with a flintlock firing mechanism, such as one of Anza's soldiers might have used. The original stock, long gone, had been replaced by a crude one of mesquite. With a gestural narration spun from mangled English, Spanish, and onomatopoeia, the Indian pantomimed how he and his fellows would build a hot fire; how they would select shafts of willow or arrowweed that matched the bore of the musket; how they would use the shafts to probe deep holes in firm, damp sand; how they would melt the zinc—in what vessel was unclear—and pour the molten metal into the holes, there to let it cool; and finally, how they would extract the cooled cylinders and cut them up like sausages. How then they would have bullets.

Sykes did not attempt to ask how they might ever hope to hit anything with so crude a missile.

The Indians solemnly restored the musket to its sackcloth swaddling. They lifted their bundles of zinc, and, as naked as Adam in the first week of creation, they filed away eastward across the limitless flats.[1]

10 | THE DELTA, HUNG OUT TO DRY

Even before the Great Diversion ended, the brimming new sea attracted the notice of waterfowl. Ducks, geese, coots, and cormorants flocked by thousands to the tepid lake, although the game birds soon departed, finding little vegetation on which to subsist. But minnows and other small fish must have thrived, for three species of cormorants stayed. Baird's (known today as the Pelagic), Brandt's, and the double-crested cormorant all nested at Salton Sea in 1908. (Of the three, only the double-crested is still present.) Such unusual birds as pigeon guillemots, tufted puffins, and black oystercatchers, none of which has been sighted at the sea in recent memory, were also present in 1908, as well as phalaropes, curlews, sandpipers, and great blue herons. Gulls, of course, soon found the place, but most impressive was a colony of two thousand or more white pelicans that nested on the hot, sun-varnished volcanic cones that comprised the lake's few islands.[1]

These birds only duplicated the movements of their ancestors, which by the hundreds of thousands had colonized the habitats created with every recurrence of Lake Cahuilla. If the immediate cause for creation of the Salton Sea was human blunder, the birds did not care, and if the sea's habitats have persisted since then thanks solely to irrigation runoff, the birds care still less. The Salton Sea remains a vital breeding and resting place on the Pacific flyway, a fact recognized in 1930 with the establishment of a national wildlife refuge at the south end of the sea. Since then, the Salton Sea National Wildlife Refuge has reported the presence of 384 species—the most of any national wildlife refuge in the West—and the total number of individual birds using the Salton Sea probably exceeds two million a year. It is an oasis on the western flyway as important to them as the putrid waterholes of the Colorado Desert were to thirsty gold seekers in 1849.

Unfortunately, the comparison between the Salton Sea and those contaminated wells is altogether too apt. In order to survive, Durivage risked debilitating illness to gulp down the waterholes' "tincture of bluelick, iodides of sulphur, Epsom salts, and a strong decoction of decomposed mule flesh." The avifauna of the Pacific flyway struggle with equally superlative pollution. The Salton Sea brims with high concentrations of salt and organic compounds, and in certain areas it bears levels of selenium, pesticide metabolites, and other possible toxics that warrant grave concern. In ways only partly understood, the cumulative foulness of the sea has triggered spectacular die-offs among the birds that depend on it. In 1992 and 1994, epidemics of mysterious origin claimed over 170,000 eared grebes. In 1996, avian botulism killed 20,000 birds, nearly half of them pelicans. In 1997, an outbreak of Newcastle's disease swept through a colony of nesting cormorants. All of these afflictions remain present in the Salton ecosystem and contribute to a steady toll of death even without epidemic breakouts. By the end of 1997, low levels of botulism, avian cholera, and the mysterious grebe ailment helped push the year's count of feathered bodies to 10,000. In the next year the death count had already matched that figure by the end of March.

The chronic problems of the Salton Sea derive from its closed-in, aging character, and they give every indication of growing more formidable year by year. The freshness and purity of an evaporative sump do not improve with time.

A later chapter details the sea's environmental problems. For the present, let us count the lake a blessing because it provides a place of relative rest and safety along the arduous Pacific flyway. From the moment of its creation, the importance of the sea has increased as the vitality of other such places has declined.

Our story now turns to one of those beleaguered "other" places, the one most closely linked to the Salton Sea. Of all the oases along the Pacific flyway, none was more an avian paradise, none was larger or more productive, and none harbored a greater diversity of finned, furred, and feathered creatures than the shifting, amphibious landscape of the Colorado delta. Although the obliteration of other wildlife oases may have been more complete, none in its demise has presented a greater loss. One can measure the magnitude of that loss through any number of means. Aerial photography can show the loss of wetlands. Bird censuses can attach numbers to that diminution. The long-term decline of shrimp harvests from the upper Gulf of California may measure overall ecosystem health. But to appreciate what the delta used to be, and thereby to judge the shadow it has become, one cannot do better than see it through the eyes of one of the greatest nature writers of any time or place.

ON OCTOBER 25, 1922, Aldo Leopold and his brother Carl crossed the international border south of Yuma in a Model T pickup with a canoe tied diagonally across the box. They planned to spend three weeks in the delta, hunting, camping, and paddling its labyrinth of channels. In the border town of San Luís, the experienced delta hands they consulted duly warned them of the tidal bores, waves taller than a man, that would reduce their craft to kindling and of the hungry sharks that followed the tide far up the river. Though such tales wore the trappings of folklore, the dangers they foretold were real. In the years before upstream dams slackened the river's flow, the mechanical contest between incoming tide and downstream current produced the continent's highest tides. One of the earliest descriptions of the bore comes from fur trapper James Ohio Pattie, who in January 1828, while encamped in the delta, wakened from sleep to the sound of rushing water. In moments his camp was under three feet of swirling water, and Pattie and his fellow trappers escaped only by leaping to their canoes and holding fast to whatever vegetation they could reach. They were fortunate compared with others. In September 1921, just a year before the Leopold brothers put on the river, an exceptionally heavy bore capsized a small river steamer, and 130 of its passengers drowned.[2]

With cautionary tales fresh in mind, the Leopolds hunted deer for a week from a land-based camp on the Sonora side of the delta. But in time, they tired of probing thickets of arrowweed and spear-tipped *chachinilla* bushes on foot. They loaded their canoe with supplies and shoved off into the waters of a slough known as the Rillito.[3]

The Rillito lies well down the eastern side of the delta, and its brackish waters attest its proximity to the gulf. Except for their starting point, however, the route of the Leopolds remains unknown. As they made their way, it remained unknown even to them, which in truth was their purpose. They were hunting game, but more than geese and ducks, more than deer, they hunted wilderness, and in the delta they believed they found it in its truest form. Securing water fresh enough to drink proved the only real difficulty in a journey that became, in Leopold's later telling, the penetration of a hunter's Eden. The delta was timeless, trackless, teeming. To the Leopolds it was a place beyond nations; it was Mexico no more than it was the United States. The only governor was Nature, in the purity of its power, and its embodiment, *el tigre,* the jaguar, lurked in every shadowy thicket. Here was true jungle: tall cottonwoods draped in vines, tangles of willows all around them. Farther ahead might lie grassy mudflats or raw sandy scours; still farther, brush country as tight and thorny as a briar patch. Twining everywhere and interlacing everything was water: leads and sloughs that divided, rejoined, and doubled back; channels that were choked with vegetation; channels that broke open to the wide brown back of the river.

Aldo and Carl followed the sloughs and their own inclinations wherever they led. They possessed neither map nor plan. Like the Colorado itself, they dallied, wandered, got lost, and were glad of it. "For the last word in procrastination," Leopold wrote, "go travel with a river reluctant to lose his freedom in the sea." Theirs was a journey into mythology as well as geography. They sought, like many generations of American wanderers and explorers before them, to touch the untouched continent. They carried with them a hunger for the smell and taste of the untrodden wild, and the delta, without requiring of them more effort than Eden first required of Adam and Eve, gave them all they sought. Rounding bend after bend they saw

> egrets standing in the pools ahead, each white statue matched by its white reflection. Fleets of cormorants drove their black prows in quest of skittering mullets; avocets, willets, and yellowlegs dozed one-legged on the bars; mallards, widgeons, and teal sprang skyward in alarm. As the birds took the air, they accumulated in a small cloud ahead, there to settle, or to break back to our rear.[4]

In those waters of ease and plenty, where the dreaded bore never did appear, even the obligation to find food hardly stirred them from their ease: "At every camp, we hung up, in a few minutes' shooting, enough quail for tomorrow's use." Or they fired a volley at a cloud of circling geese, and instantly, "all the geese we could eat lay kicking on the bar."

In this paradise, "all game was of incredible fatness. Every deer laid down so much tallow that the dimple along his backbone would have held a small pail of water, had he allowed us to pour it." Nature was profligate in the delta; the fecundity of this "milk and honey wilderness" was everywhere in evidence. In grove after grove, the branches of honey and screwbean mesquites hung with nutritious pods. Panic and salt grasses carpeted the mudflats, their seed heads bowed over with milletlike grains that might be "scooped up by the cupful." Leguminous, pod-bearing shrubs grew in such tangles that "if you walked through these, your pockets filled up with shelled beans."

Only twice as he drifted through this land of bounty did Leopold detect the least sign of human presence: a path that could have been a wagon track and, later, a single, rusting can—"it was pounced on as a valuable utensil." Nowhere did he encounter fresh sign of human passage. One may wonder how hungry people, native or newly arrived, can have overlooked such a place, but Leopold's account of his trip carries one along so swiftly on a river of beautiful words that even to drag a questioning finger in the water seems ungrateful.

Leopold's delta sojourn became the basis for one of his finest essays, a nostalgic rhapsody on the wild heritage of North America.[5] Written in 1944, "The Green Lagoons" culminates with one of his most memorable and bittersweet laments:

> All this was far away and long ago. I am told the green lagoons now raise cantaloupes. If so, they should not lack flavor.
>
> Man always kills the thing he loves, and so we the pioneers have killed the wilderness. Some say we had to. Be that as it may, I am glad I shall never be young without wild country to be young in. Of what avail are forty freedoms without a blank spot on the map?

It is hard to argue with poetry like that, but to be truthful, the delta was not the unexplored fastness that Leopold thought he saw. By 1922, it had already been deeply affected by history, and that history had been visited more roughly on the Cocopa than anyone else. One reason Leopold saw neither ax-cut nor footprint when he and Carl probed the green lagoons was that the Cocopa had already begun to withdraw to the margins of their former homeland.[6] The Great Diversion, which robbed the delta of both water and nourishment for the better part of two years, was catastrophic to a people who depended on annual floods the way inhabitants of other lands depend on snow or rain. Driven, on one hand, by changes in the land, and drawn, on the other, by the attractions and temptations of town life, the Wi Ahwir Cocopa moved north out of the Hardy and Pescadero drainages and resettled in and around Mexicali. Completion of Laguna Dam above Yuma in 1909 further limited the vigor of the Colorado and added to the array of ecological and social forces that led two other Cocopa bands, the Hwanyak, who dwelled in the southeastern delta, and the Mat Skrui, in the northeast, to trade their fishing and hunting camps for shanties in Somerton, Arizona, and its twin town across the border in Sonora, San Luís del Río Colorado. Of major Cocopa groups, only the Kwakwarsh stayed behind, and their descendants still cling to their home in the Río Hardy wetlands at the delta's western edge.[7]

Leopold came to the delta at an opportune time. The previous summer's flood had been too big to have been much limited by the hydraulic structures then in place. In the month of June, peak flows at Yuma reached 117,000 cubic feet per second, and some six million acre-feet of water passed downstream to the delta and the sea, flooding, if not forming, the mudflats that Leopold observed.[8] As floodwaters receded, panic and salt grasses sprouted in abundance. Had the Cocopa still abided in the portion of the delta where Leopold wandered, he would have seen them harvesting the annual grasses that had provided them sustenance since time immemorial.

Other changes, all but invisible, were also under way. As Aldo and Carl paddled the placid lagoons, they likely failed to notice the tiny but abundant sprouts of a cedarlike tree on the same mudflats that so impressed them with their carpet of grasses. Tamarisk, or salt cedar, a shrubby tree introduced from Eurasia for erosion control and ornament, had found its way into the delta, and the flood of 1922 no doubt hastened its spread. In the years ahead, favored by the increasingly regulated flow of the river, salt cedars spread rapidly, supplanting native flora and effectively transforming large portions of the delta into simple yet tenacious monocultures. Today the tamarisk can rightly be considered the bane of warm-weather river systems throughout the Southwest, where its thickets choke as many as a million acres of former floodplains, including vast alluvial stretches along the lower Colorado.[9]

IT IS IMPOSSIBLE to guess how much of the delta Aldo and Carl Leopold actually saw. Perhaps they drifted downstream and spent their days hard against the Sonora side in the area known today as the Santa Clara Ciénega, which remains (thanks to accident, not intention) an important natural area. Perhaps they never crossed the main stem of the river. They could not have floated down the main channel of the Colorado without realizing that the wilderness that enraptured them was succumbing to settlement.

Following the Great Diversion, the Colorado adopted a former distributary called the Río Abejas as its main channel, by which it flowed to the western side of the delta and deposited so much silt in the Volcano Lake area, near the delta's northwest limit, that the greater part of Volcano Lake filled in. In 1908, a levee was built to keep the waters of the Volcano Lake district from draining northward, and in order to keep pace with sedimentation, workers periodically raised the height of the levee and armored it with stone. The physical buildup of the area meant that the Colorado was flowing ever more precariously down the crest of a low ridge, and the danger to Imperial Valley continued to grow. The levee offered, at best, a thin line of protection. If a big flood came, the river might as easily spill to the north as to the south.

In order to prevent this, the Imperial Irrigation District and Mexican engineers dredged a channel four miles long from a bend in the Abejas southward to another stream, the Pescadero, which took form near the center of the delta. They then built a dam, which became known as the Pescadero Diversion, across the Abejas so as to force the main flow of the Colorado into the Pescadero and thence, theoretically, to saltwater. Workers completed the diversion and the new channel by April 1922. Construction of the Pescadero Levee, which flanked the new channel in order to prevent the westward escape of waters, probably continued for months after the summer's flood subsided.

Had Aldo and Carl Leopold floated down the main channel of the Colorado in 1922, they would have noticed, no doubt with surprise, the hard left turn of the main channel at the Pescadero Diversion, to say nothing of the formidable bulk of the diversion structure itself. They would also have noticed the unnaturally straight alignment of the channel into which the current carried them. And they might have remarked with surprise, a half dozen or more miles later, the devastated landscape of the lower Pescadero, which the summer flood of 1922 had buried under forty-five million cubic yards of silt.[10]

The river had dashed the engineers' hopes again. Their gamble had been that the river might scour out the tiny Pescadero into a channel capacious enough to carry its floodwaters virtually to the gulf. Instead, the Colorado filled in such channels as existed, and the Pescadero all but disappeared. As a result, more surveying, dredging, levee building, and general construction soon followed, even including the laying of railroad track atop the Pescadero Levee in order to bring in supplies of yet more fill and rock.

Aldo Leopold's wilderness of green lagoons was actually a busy place in 1922, and this fact was not lost on another hunter and writer, Lewis Freeman, who descended the delta a month before Aldo and Carl. Evidently, the lands and waters they visited did not overlap. Freeman drifted down the main stem; the Leopolds, miles to the east, seem to have paddled sloughs at the foot of the Sonora uplands. Freeman's 1922 trip was his second through the delta. Years earlier the erstwhile big-game hunter had floated the Hardy, shooting everything he saw. He sought to collect heads and skins enough to furnish "a trophy room," and toward that end, the scow in which he drifted became a floating abattoir. Dead pelicans furnished him a mattress. The corpses of two deer, a large bobcat, and a giant green sea turtle, which Freeman strangled with a rope, occupied the stern. Various beaver, waterfowl, and other unlucky animals filled the bow. Much of Freeman's cargo was ungutted, all of it unsalted, and the sun overhead was hot. His trophies bloated and putrefied. Reluctantly, he put them over the side.

When Freeman returned in 1922, the delta had changed. In the same autumn that Aldo Leopold reported finding a hunter's paradise, Freeman saw that it was a paradise lost:

> Between floods from the river and hunters from the increasingly populous Imperial Valley, the decimation if not the extermination of the game was swiftly and surely accomplished. . . . One will look in vain for the swarming wild life of other days. . . . It is still one of the queerest, weirdest regions that fancy can picture, but no longer a paradise for the hunter, no longer a Golden Land of Enchantment for the youth who would furnish a trophy-room.[11]

First the Cocopa departed, then the game animals followed, and ultimately the vegetation and even the topography of the land were transformed.

The conversion of Mexican land to agriculture began with the first diversion of water from the Colorado and proceeded simultaneously with the transformation of the Imperial Valley, although more slowly. Reclamation initially focused on desert tracts close to the border. It should come as no surprise that this land soon fell under control of U.S. interests

The government of Porfirio Díaz had conveyed virtually the entire area to General Guillermo Andrade in 1888. Andrade's empire was fabulously large, comprising nearly all of the Mexicali Valley and most of the upper delta. The general sold a portion of his empire to the California Development Company, but eventually the company that he formed to develop his holdings, the Compañía Mexicana Industrial y Colonizadora de Terrenos del Río Colorado, repurchased the bulk of these lands from successors of the bankrupt CDC and reconsolidated the vast Andrade fiefdom. Andrade long served as Mexican consul general in Los Angeles, and in 1904 he sold out to Harrison Gray Otis, the publisher of the *Los Angeles Times,* and his son-in-law, Harry Chandler. They became the "dueños absolutos" of the Mexicali Valley, controlling some 862,000 acres.[12]

Their plan was simple: to grow cotton with free water and cheap labor and sell it in the United States for top dollar. This they did for the better part of three decades, operating possibly the largest cotton farm in the world, even through the Mexican Revolution, which was a relatively calm affair in northern Baja California.

But not entirely calm. In 1923, after the Leopold brothers had safely returned to U.S. soil, a Villista by the name of Marcelino Magaña led a group of campesinos in the brief seizure of 13,800 acres belonging to Chandler's Colorado River Land Company.[13] Magaña and his cohorts justified their action in the rhetoric of land reform, one of the central tenets of the revolution, whose ideals were now codified in the constitution of 1917. But Magaña's insurrection soon failed for want of broader support, and land reform remained a purely theoretical concept in the Mexicali Valley for a decade more. It took the convergence of the Great Depression, completion of Hoover Dam, and strong initiative on the part of Mexico's most forceful and popular president, Lázaro Cárdenas, to trigger meaningful change, and with it, the conquest of the delta.

When the depression hit the U.S. economy, some of the first to lose their jobs were migrant Mexican farmworkers. Thousands of them, obliged to return to Mexico, settled in Mexicali, where they demanded land and agitated for reform. Cárdenas came to power in 1934 and began to launch programs of land redistribution in various regions throughout Mexico. Northern Baja took its turn in 1936, when Mexico's secretary of agriculture signed a contract with Chandler's Colorado River Land Company stipulating that the company must survey, divide into small tracts, and sell under favorable terms nearly three-quarters of its holdings. Theoretically, a portion of these lands would be newly reclaimed from uncleared portions of the delta, the settlement of which was now feasible because Hoover Dam, completed in 1934, provided security from floods.

The promise of change, however, far surpassed its delivery. Chandler's company dissimulated and stalled. By January 1937, it had sold only one thousand acres in eight transactions. The campesinos again rose up. Following Magaña's earlier example, they "invaded" the company's lands and seized control.

This time, the outcome was different. Cárdenas invited the leaders of the uprising to Mexico City and met with them himself. He did not then take long to act. In a year's time, Cárdenas would leap onto the world stage by expropriating and nationalizing Mexico's foreign-owned oil resources. In the Mexicali Valley in 1937 he executed a smaller version of the same maneuver by essentially nationalizing the Colorado River Land Company. Within one hundred days of his meeting with the campesino leaders, Cárdenas ordered the formation of 67 collective farms, or *ejidos,* totaling nearly a quarter of a million acres. These agricultural communes were to provide homes and livelihood for 16,000 campesinos. An additional 150,000 acres were conveyed as private holdings to independent farmers and settlers, while division and sale of the rest of the company's lands were ordered to continue under the terms of the 1936 contract.[14]

The Cárdenas reforms, with their promise of land for landless people, triggered a wave of migration to the upper delta. New villages and crossroads sprang up bearing names like Michoacán de Ocampo, Toluca, Querétaro, and Veracruz, names reflecting the origins of the people who gambled the little they had on starting a new life in Mexico's far northwest. Much of their hope was founded on the promise of the *ejido,* the centerpiece of the Cárdenas land reforms.

Almost from the start the *ejidos* bore the taste of salty tears, a taste well known to the Mexican people. The *ejido* concept was deeply rooted in colonial tradition, but its modern purpose was similar to that of collective farms in the communist world. *Ejidos,* it was hoped, would permit Mexican agriculture to reach levels of productivity and efficiency comparable to those of the industrialized world without the anti-democratic concentration of ownership and capital that characterized Mexico's pre-revolutionary past. The *ejido* movement came to embody one of Mexico's dearest dreams, a dream in many respects similar to the (North) American dream of landed independence and economic freedom. Its slogan was "haciendas sin hacendados," an evocation of great farms unspoiled by despotic landowners. The slogan proved all too true, but in a darkly unintentioned manner. Every ill known to centralized, state-run agriculture was visited on the *ejidos:* undercapitalization, incompetent bureaucratic administration, artificially low, state-controlled prices for staple products, and more. In the end, the *ejido* movement did indeed produce *haciendas sin hacendados:* the farms were large, if not productive, and there were no arrogant landowners to flaunt their wealth before the aggrieved campesinos. On the other hand, thanks to an overcontrolling and often corrupt system of agricultural administration, the peasant workers were scarcely less poor and exploited than they had been in the day when they served the *hacendados.*[15]

These problems persisted into the early 1990s, when Mexican president Carlos Salinas, in the name of reform, opened the *ejido* system to the so-called free market.

The Mexican constitution was amended to allow members of *ejidos* to separate their share of common lands from the holdings of their group. Once separated, the lands might be rented or leased, used as collateral for loans, or sold outright. The amendment cleared the way for expanded real-estate trading and speculation throughout Mexico. How it will serve the interests of impoverished campesinos on marginal delta lands remains to be seen, but if history serves as a guide, the more things change, the more they are likely to stay the same.

COLLECTIVIZATION failed worldwide, but the *ejidos* of the Colorado delta faced especially tough odds, given their limited access to water and its poor quality. In dry or average years, when the only water Mexico gets from the Colorado is the 1.5 million acre-feet the United States is obliged by treaty to deliver to it (in theory, about a tenth of the river's total flow), Mexican farmers must make do with less than half as much water as the Imperial Valley uses to irrigate an area of equal size. Their share is further diminished by municipal withdrawals for Mexicali and Tijuana, which receives its supply from a pipeline across the coastal mountains. Mexican farmers also know well that the water that trickles down to Mexico tends to be relatively high in salts, having already run a gauntlet of fields and drains before it is diverted at Morelos Dam into the *canal alimentador central*.[16]

The farther south you go in the delta, the more you note the decline of the crops: the cuttings of hay that, north of the line, would be too meager to warrant baling; the small-headed lettuce; the dull-looking broccoli; the corn, short and pale. On a November visit to the delta, driving beside fields of stunted stalks, Joan Myers, the Iowa-born photographer with whom I was traveling, observed only half in jest, "That's not corn. Where I come from, respectable people would not allow such plants to grow."

Salinization is a major factor. As the elevation of the land declines toward sea level, the soil drains less well, so that salts fail to wash out. In the Imperial Valley, farmers install expensive subsurface drains and flush the salts from their fields with abundant amounts of water, which they pump away if gravity will not carry it. But the struggling *ejidos* of the lower delta lack the capital for drain tiles and pumps and cannot hope to claim the needed water. And so harvests suffer, and telltale, crusty blooms of white sprout at the edges of the fields and in the low places where water pools and evaporates.

JOAN MYERS is at the wheel of our tiny car, dodging potholes. We are headed south toward country that was once a "blank spot" on Aldo Leopold's map.

We pass Nuevo León, Delta, Oaxaca. Tecate beer signs announce each village. We see jumbles of junked cars and old tires, we smell the acrid smoke of woodfires and burning trash, and finally we come to the houses. Chickens strut by the roadside. Tethered goats bleat. People stand before their houses, cooking over fire pits or barbecues of stacked-up bricks. Curtains flap from the windows, blowing outside. In some cases, the roof also flaps, consisting partly of plastic sheeting. We pass a faded government billboard, half obscured by eucalyptus, that proclaims, "¡Con la limpieza, la cholera tropieza!" With cleanliness, cholera stumbles.

We cross the Río Pescadero, a lazy green-water slough, ten or fifteen feet wide. Tamarisk, arrowweed, and cattails crowd its edges. We see a coot and an egret. Three palms bend above the water, the wind dryly rustling their fronds. A dozen feet from the water's edge, the stark fields begin again.

According to the map, we are coming to Ledón, a crossroads where we will meet a paved road running east to the Colorado. We plan to cross the river, zigzag south a half dozen miles, then bend west on a rare paved road. The route describes a rectangle, as the road ultimately turns north and angles back across the Colorado to Ledón. But in the delta, maps and plans often prove to be approximate.

In Ledón, we check our bearings and learn that no one has heard of Ledón; the community is called Colonia Carranza. Eventually we speak with an old-timer who offers an explanation: years ago, there used to be a *gasolinera* by the name of Ledón at this crossroads; the central government's mapmakers must have taken the name from him.

The man who remembers the filling station says this part of the delta was settled in the 1960s. The land clearing and agricultural conversion set in motion by the Cárdenas reforms were limited in their southward spread by periodic flooding—not deluges on the scale of the Colorado's unharnessed days, but flows that exceeded the retention capacity of upstream dams and reservoirs. As more dams came on line, the pulses of surplus water grew rarer, until in the sixties they ceased. A significant pulse did not again reach the delta until 1980, and then only briefly, for the system was dry again in 1981 and 1982. This prolonged period of desiccation was due primarily to the completion in 1963 of Glen Canyon Dam, upstream of Grand Canyon, and by the subsequent filling over seventeen years of Lake Powell, its immense, canyon-drowning reservoir. During those years, the U.S. held back every possible thimbleful of Colorado water. Mexico got its treaty allocation, and that was all.[17] Nothing was left to nourish the river, the delta, or the fisheries of the upper gulf.

With the Colorado mastered, settlement pressed southward, past Colonia Carranza, to elevations as low as a few feet above sea level. With no opposing flow of freshwater, the saltwater of the gulf moved north. Pushed by tide and storm, the brine moved up the rivers, into the sloughs, over the marshes. Not until 1983, with Lake Powell filled and deep snowpacks in the Rockies producing heavy spring runoff, were the few remnant wetlands of the delta refreshed with a renewed flow of river water. By then, the riparian gallery forest of the lower Hardy, formerly a jungle of native cottonwood and willow, had died of thirst. Hundreds of square miles of low thickets, swatches of panic grass, and fresh and brackish marshes—the kinds of green lagoons that drove Leopold to rhapsody—had given way to dusty, naked mudflats, salt-cedar thickets, and near-monocultures of salt-loving pickleweed and iodine bush.[18]

The renewed flow of the Colorado that began in 1983 lasted a few years and ceased again. It resumed briefly in 1993, the year Joan Myers and I first passed through. Heavy rains in January and February had caused the normally docile Gila River to flood, washing out crops in Arizona, closing roads, causing thousands to flee their homes, and even threatening collapse of Painted Rock Dam, the last barrier on the river.[19]

What was bad for Arizona was worse for the lower delta. The town of Colonias Nuevas, which our map identified as "General Francisco Murguía" (the smallest places bear the longest names), lay behind the protection of a levee, but even so, it flooded. As we drove to its center, past the usual *cerveza* placards and traffic signs tilted at drunken angles, we noticed the mark of floodwaters well up the walls of the houses. Another government billboard, the paint fresher this time, exhorted, "¡Evitamos la cholera!" In the recent crisis, with all the area flooded and the land so flat the water would not drain, the prevention of cholera was doubtless not a simple matter.

Our official if somewhat fictional road map showed that here the highway crossed the main channel of the river, on a route parallel to that of the region's solitary railroad.[20] We drove across a levee, weaved slowly through muddy intersections, and followed a line of cars and trucks that thickened toward the river. A drizzle began to fall. Ahead we saw a bridge, but only one. We wondered where the railroad bridge might be. Our road approached and then climbed a berm, at the top of which we were surprised to see a single line of railroad tracks. A large truck immediately ahead of us blocked our view. The drizzle now increased to a steady rain, and our crawling line of cars and trucks soon stopped. Moments later, a procession of cars and trucks began to pass us from the opposite direction. Evidently these vehicles had come across the bridge. Then it was our turn: following the truck ahead, Joan inched our car forward. Our "road" merged with the railroad tracks, between and beside which heavy planks had been laid to provide an even surface. But it was less even than originally planned, for the timbers were pitted deeply with rot, and the pounding of traffic had broken large splinters free. The truck ahead belched black exhaust and inched forward. We could only follow. In moments we were on the bridge.

Our car slid side to side on the rain-slick timbers and the glistening, slicker rails. Our narrow tires slipped now into this rotted crevasse, now into that one. Some of the timbers were loose and slid with us, slamming against the rails. The "roadway" was scarcely wider than its ten-foot railroad ties. Although guardrails lined pedestrian walkways on either side of the bridge, our attention focused on the considerable gap—*chasm* is perhaps the right word—between the edges of the roadway and the handrails of the walkways. Looking down we could see the brown river sliding beneath us. I glanced at Joan. She stared rigidly ahead through our rain-spattered windshield and gripped the steering wheel with a ferocity that caused the veins to bulge from her thin forearms. Over and over, in a soft but extremely firm voice, she recited a mantra of unprintable words.

It was a crossing less heralded and less exotic than that of Father Kino, who, nearly three hundred years earlier, bobbed across the Colorado not far from here in a basket propelled by Cocopa swimmers. Nevertheless, our crossing, thanks to the mortal simplicity of the bridge, retained a certain flavor of the region, as well as, to our lights, a measure of mystery. We crept across the bridge in our file of vehicles, the turbid waters winking below. Soon after we reached terra firma on the other side, the oncoming file responded to an unseen cue and began to move across the narrow span. We marveled that so much life and traffic, moving across the river, depended on so slender a link. Later we learned that the railroad bridge carries rubber-tire traffic only when the river drowns a low-water crossing a quarter mile downstream. But on that rainy day, with the Colorado still swollen from Gila floods, we drove on, mystified, impressed, and delighted, after it was done, to have earth and not a river beneath us.

Heading south, then west, we passed through land that only three decades earlier had been a kind of Mexican Amazon: a wet, low world from which land-hungry people hacked fields and strove to build an approximation of home. Their fields were by far the poorest we had seen, the most ill-drained and salt-rimed. Their standing crops of corn, pasture grass, and sorghum looked starved and bleached. Their homes were made of cardboard, plywood, and plastic sheeting that sagged and flapped in the rain. The most substantial structure in one shanty cluster was the stout signpost to which a fading placard had been attached. A young girl and a smaller boy stood bareheaded, holding hands in the drizzle, and stared at us as we slowed to read the brave name of their hamlet: Héroes de la Patria.

A few miles farther the road turned north, marking the last leg of our circuit back to Colonia Carranza. We had not seen another vehicle for some time, and the fields appeared abandoned—seedling trees and low shrubs grew among a cover of weeds. Suddenly the fields yielded to tamarisk thickets that closed upon the road. Potholes riddled the pavement. Then the road gave out. Yard-high humps of sand forced us to stop. We got out, heard the sound of rushing water, and threaded our way through the disordered hummocks of sand. At last we stood where the last slab of pavement broke off and yielded to brown water. At our feet ran the Río Colorado.

We retraced our path back through the hovels of Héroes de la Patria, back past fields of failing crops and the vacant eyes of children. We recrossed the unnerving railroad bridge, sliding white-knuckled on rotted timbers and glistening rails, and made our way to Colonia Carranza. A passerby told us that the river channel, dry for years, had overflowed in the floods of the previous winter and washed out the crossing between Colonia Carranza and Héroes. He said that even before the floodwaters receded, government officials had promised to rebuild the road and dredge the channel, but nothing had been done. The promises had been made with an election imminent, and the election had since been held. No one, we were told, expected remedy to be swift.

SO MUCH IN the delta wants attention, investment, and repair that one doubts if even the most essential human services will receive them. Under such circumstances, the restoration of the area's once fabulous ecological systems must rank low on the list of public priorities. Yet because the acreage susceptible of immediate restoration is small and the resources needed to restore it are limited, a hope for constructive action may not be farfetched.

At least two great natural areas, present or potential, exist in the delta. One, Santa Clara Slough, is an accidental by-product of U.S. efforts to dispose of irrigation tailwaters too saline for release into the main channel of the Colorado. These waters are produced in the Wellton-Mohawk district of Arizona, along the lower Gila. Farmers there use Colorado River water to irrigate their crops, and because of peculiarities of soil and subsurface geology, the waters that drain from their fields are extremely salty. In the 1950s and 1960s, when the district's farms were developed, these drainwaters flowed to the Colorado and were re-diverted for irrigation in Mexico, where they damaged crops and sometimes killed them outright. The Mexican government took exception to this practice, and its protests led to a treaty revision establishing minimum water quality for the 1.5 million acre-feet the U.S. must deliver annually at the border.[21]

Congress provided the means for meeting those standards by passing the Colorado River Basin Salinity Control Act of 1974, which authorized the construction in Yuma of what would be the world's largest reverse-osmosis desalination plant.[22] While the plant was being built, the U.S. excavated a fifty-mile canal and lined it with concrete to carry Wellton-Mohawk waters from Yuma to the Rillito Salado, close to where Aldo Leopold set off on his adventure of 1922. The Rillito then carried Wellton-Mohawk drainage to a five-hundred-acre wetland called Santa Clara Slough, whence, in theory, the salt-laden water would discharge into the sea. Later, with construction complete, the desalting plant would treat the Wellton-Mohawk effluent, delivering purified water to the Colorado to help meet U.S. obligations to Mexico and discharging a caustic waste stream into the concrete channel that flowed to the Rillito.

Fortunately, this plan has encountered problems. Santa Clara Slough is not really a slough, for it lacks a true outlet. In Mexico it is known, with greater accuracy, as a marsh: La Ciénega de Santa Clara. In 1977, when flows began moving down the concrete canal, the *ciénega* retained much of the water conveyed to it. In a few years a vast system of wetlands developed, covering close to fifty thousand acres. Since then, the extent of the wetlands has varied with fluctuations in inflow, but even in periods of relative drought, the complex has shown impressive resilience. It has become a haven for birdlife. Geese, ducks, Virginia rails, soras, avocets, ibises, herons, and terns thrive there, and the *ciénega* is home to perhaps the world's largest population of the endangered Yuma clapper rail. Desert pupfish also abound in its marshes.[23]

As the Yuma Desalting Plant neared completion in 1992, conservationists on both sides of the border feared that the wetlands were doomed. At full operation the plant would have pumped into the Colorado the freshwater it extracted from the Wellton-Mohawk flows, and it would have dumped the leftover hypersaline brine into the canal leading to the *ciénega*. In this way the plant would have simultaneously starved the wetlands for water and poisoned them with waste. But after eighteen years of development and construction costing more than a quarter billion dollars, the Yuma plant went not into service but into mothballs.

El Niño was responsible. Soon after the plant was declared complete, the Gila floods of 1993 destroyed portions of the canals that delivered Wellton-Mohawk water to it. The cost of their repair would have boosted the exorbitant cost of the plant to a level of pure absurdity. Moreover, operation of the plant was estimated at $30 million a year, an avoidable expense. Full reservoirs throughout the Colorado system enabled the Bureau of Reclamation to meet U.S. obligations to Mexico without operating the plant at all. And so the plant today stands idle.[24]

The Santa Clara Ciénega, meanwhile, continues to be nourished by Wellton-Mohawk waters, as well as by local agricultural drainwaters that flow to the Rillito. In 1993, the Mexican government decreed the establishment of a biosphere reserve for the "Alto Golfo de California y Delta del Río Colorado," of which the Ciénega de Santa Clara is the centerpiece. The designation was followed in 1996 by promulgation of a management plan and in 1997 by actual staffing of the reserve. Only nine federal rangers patrol the two million acres in their charge, but the commitment of the Mexican government assures continued pressure to keep Wellton-Mohawk water flowing to the wetlands.[25]

ANOTHER MAJOR natural area lies at the delta's western edge.[26] You can visit the area, or at least its periphery, if you drive down the main highway linking Mexicali to San Felipe and lower Baja. On your right will be the treeless, rocky sierra of the Cocopa Mountains and behind them, the taller, chromatic ranges that form the spine of the peninsula. Tall and sheer, their bright ridges recede through shaded distances: red mountains, yellow mountains, purple and shadow-blue mountains, a kaleidoscope of geology. On your left are the fields of the delta, thriftier here than in the dismal flats below Colonia Carranza. You pass the turnoff for Ejido Cucapa Mestizo, one of the earliest *ejidos* established in the delta. Its original purpose was to rescue the fortunes of the delta's remaining Cocopa, but the Cocopa proved little interested in mechanized agriculture. By 1940 they had yielded control of the *ejido* to mestizo immigrants from other regions.[27]

A few miles south of the turnoff, the fields give way to bramble, and then close by the highway a great edifice appears, a sprawling extravagance out of proportion to all else in the delta. It appears to have been designed as a chicken-growing operation for broiler or egg production. Like the *ejidos,* the Yuma desalt plant, and much else in this corner of the world, the giant installation must have grown from ambitious dreams but now stands utterly defunct.

A few miles farther and you note that the jungle to your left is thickening even as the mountains verge toward the gulf, limiting the highway to a narrow track between vertical rock and what you presume to be bottomless mud. A sign appears for "Campo Mosqueda." If you follow it and take the dirt track along the top of a spoil levee, as Joan Myers and I did, you will meander among lagoons where egrets, avocets, and herons preen. We were looking for a quiet place, away from the drone of the highway. We continued along the levee into ever-denser vegetation until stopped by a cautionary sign: "Cuidado," it read. "Zona de Víboras. Dirección de Bomberos y Protección Civil, Ayuntamiento de Mexicali." It was emphatic and official in its tone, and it certainly gave pause. On the one hand, we were pleased to receive the warning of the good firemen of Mexicali and to have found our way to a wildlife area so fecund that its profusion of snakes posed danger to life and limb. But on the other hand, we did not wish to vie with rattlers, moccasins, or other *víboras* for the small amount of real estate we needed to open our cooler and have lunch.

The *zona de víboras* in which we ate the unappetizing last of our supplies belongs to a system of wetlands that stretches along the Río Hardy. Fed by agricultural runoff and wastewater from a geothermal complex in the old Volcano Lake district, the wetlands as recently as 1983 spread over more than 150,000 acres. In that year, however, floodwaters from the Colorado breached the natural sediment dike that had impounded water in the wetlands. The loss of the dike allowed waters in the area to drain more efficiently to the sea, and the wetlands shrank to less than a fiftieth of their former size. They have since partly recovered, and in 1993 were estimated at about 60,000 acres.

Víboras notwithstanding, the Río Hardy wetlands provide important winter habitat for waterfowl migrating on the Pacific flyway—a fact of considerable significance to Ducks Unlimited, the U.S.-based conservation group. A consultant for the organization, J. M. Payne, has drawn up a plan for stabilizing the wetlands by replacing the breached dike with a gated structure that might hold back waters for the benefit of the wetlands or release them, as necessary, for flood control. The plan would also require attentive management of the irrigation and wastewater plumbing of the delta to ensure that tailwaters and surplus flows that descend the Colorado in wet years be allowed to spread over such wetlands as need them. Payne estimates the cost of his proposed dam at only $250,000, a modest sum for keeping alive so productive and diverse an ecosystem as the Río Hardy's.[28]

The Río Hardy might also benefit if the city of Mexicali builds a reliable sewage treatment plant. The effluent from the plant, amounting to 100,000 acre-feet a year, might be pumped to the Río Hardy rather than discharged into the New River. So copious a flow of relatively clean water might bring back life to thousands of acres of delta land. Claiming those waters for ecological use, however, poses great challenges. If the water is clean, it will be wanted for agriculture, and if it is not, damage from pollution may offset its benefits. In the competition between agriculture and ecology, Mexican officials will need to weigh the economic benefits that a vibrant delta might produce by way of eco-tourism, already a mainstay of the economy of lower Baja California.

The danger of pollution, however, seems to be high: the geothermal installation at the head of the Hardy presents a chilling sight. In the way that Holtville calls itself the carrot capital of the world, the Volcano Lake geothermal complex might claim to be the planet's capital of corrosion.[29] Scarcely a living plant can be seen in any direction, except, when I was there, a single blue oleander sprouting like a miracle in front of a pair of rusting exhaust stacks. All else consisted of gray industrial buildings and endless pipelines. I scanned the horizon and in one long pivot of 360 degrees counted 89 plumes of steam and smoke. One grows used to the rotten-egg smell of the waste gases and natural fumaroles, but not to the contents of the open drains. One ditch, which stretched across the landscape to the limit of sight, held a viscous fluid the milky color of a blind eye. From sources such as this, the modern Hardy is born.

RESEARCHERS IN the United States, using digitized geographic information systems and complex hydrological models, have estimated that restoration of the thousand-square-mile delta to something like its native condition would require 500,000 to 750,000 acre-feet of river water a year—up to half of Mexico's entire allocation. Although evidence exists to suggest that large throughputs of freshwater can translate in only a few years' time to vastly increased harvests of shrimp and other fish in the upper gulf, water resources on such a grand scale are not soon destined to be devoted to environmental purposes.[30] Nevertheless, where more modest flows are possible, U.S. water managers would do well to consider the needs of the delta on those rare occasions when, compelled by law, they seek to mitigate the adverse environmental effects of their operations. It is unlikely that any wetland system in the region of the lower Colorado would respond more quickly or more productively to modest, well-targeted inputs of freshwater than would the relict natural areas of the delta.[31] Certainly the Salton Sea could not generate a comparable ecological return on gallons or dollars invested. Nor would the meager shoreline habitats of the reservoirs that swell like aneurysms along the river's main stem above the border.

Those of us who live north of the international line would also do well to remember that the demise of the once-magnificent Colorado delta is not wholly or even mainly attributable to the actions and decisions of Mexicans. In an average year the people of the United States divert and withhold from the river over nine-tenths of its natural flow. Imagine for a moment, perhaps fantastically, that the United States and Mexico might once again revise their 1944 treaty and cooperate in the restoration and protection of delta wetlands. Imagine that an average annual supply of only 130,000 acre-feet or so, apportioned between base flow and larger "flood" pulses roughly every four years, were somehow squeezed from the jubilee of river exploitation and reserved for nourishment of the delta. One quickly falls to dreaming: perhaps, by restoring some fraction of the natural pattern of freshwater moving to the sea, a portion of the delta's natural bounty of wildlife, such as Leopold observed, might also be restored. Small amounts of water might not help the endangered *vaquita,* the world's smallest porpoise, whose population in the upper gulf has declined to a few hundred. Nor would they do much to change the brackish shallows of the delta where the *totoaba,* a nearly extinct, steel-blue fish that grows bigger than a linebacker, used to breed in significant numbers. But almost certainly, even with small dedications of water, some of the species that ply the Pacific flyway might flourish nearly as they did a century ago. Perhaps the most vulnerable among them might withstand the disasters and epidemics of the Salton Sea with little, or at least less, permanent loss. Perhaps even the few remaining delta Cocopa, who cling to their traditional ways in the sun-battered village of El Mayor, at the foot of a mountain that bears the same name, would again fill their boats with fish and their hearts with joy, even though they don't expect to.

The prophesy of the Río Hardy Cocopa is hardly to be envied, except in its generosity and breadth of spirit. According to Onésimo Gonzales, a grizzled elder, the people of his tribe believe that they and their culture will soon pass from this earth: "We know from our prophesies that our people will disappear. We also know that we must pass our knowledge to your culture so that it may survive."[32]

Onésimo Gonzales may be more realistic than those who dream of restoring the delta. The people on both sides of the border who argue that every possible liter of water must be reserved for cities or for cotton, corn, and broomstraw far outnumber those who tout the dividends of environmental renewal. Those dividends, being unsusceptible of immediate quantification, are easily doubted. No matter what path our two nations take into the future, the truths we encounter there will always be messier, stranger, and more uncertain than anyone predicted, and the actual result of what people do in and to the delta will be messier and stranger still. Time will pass, policies and populations will ebb and flow like tides, and the silent lands where the Colorado seeks the sea will drowse on, sphinxlike, greedy for water and ever ready, whether now or in ten thousand years, to be the palette of a new creation.

11 | UPHILL TOWARD MONEY

Imagine yourself on a mountainside far from the dry, cracked mudflats of the Colorado delta. Imagine yourself, for instance, suddenly transported to the lush, brief summer of the Wind River Mountains of Wyoming, at the farthest headwaters of the Colorado system, seventeen hundred river-miles from the salty Gulf of California. The meadow before you, choked with sedge and pedicularis, drains to the Colorado's largest tributary, the Green River.

If not in Wyoming, you might be perched on the side of any of a thousand high valleys on the west slope of the Colorado Rockies. This is snow country: the Colorado west slope may comprise only 15 percent of the watershed of the Colorado River, but it produces 86 percent of the river's flow.[1] Even late in summer, remnants of snowpack linger on the tallest peaks and feed the streams below. From your hillside perch, you gaze down toward one such brook. The water falls, as smooth as hair, over glacial cobbles, spray flying like sparks where it strikes a fallen tree. In obedience to gravity, the water tumbles, garlanded with froth, always down, down, down, headed for the sea.

You may think, as you reflect upon this wilderness tableau, that what you see is the quintessence of nature at work: water moving through the land, yang through yin, time through eternity. A sense of peace, almost like an audible sound, comes to you. Your cares and worries begin to fall away.

But now comes a memory, as sudden as a slap. And you snap to.

You look at the creek again. It takes mere seconds to recall, good westerner that you are, that this creek is no symbol of sublime eternity. Given the pervasive human manipulation of the system to which it belongs, it can hardly be considered a limb of Nature. In the gimlet-eyed view of western American law and governance, this creek is a delivery system. What flows in it is no mere water; it is *property.*

Every drop of fluid in it wears an invisible brand. Even as moisture deliquesces from the snowpack, it belongs to someone, or more likely, to several someones in succession. The contents of the creek are liquid chattel and may be bought or sold, bartered or borrowed against. In particular, they may be converted into our age's most incontestably universal solvent: money. In the American West, not only is all land subject to ownership, all water is, too.

Assuredly, this durable tenet of our time will one day seem as strange to our successors as belief in the divine right of kings now seems to us. But lest you doubt the doctrine's authority, consider these words of an Imperial Valley farmer, spoken in 1996 in condemnation of the sale of water to the Metropolitan Water District of Southern California. "I just feel it's really bad business," said farmer and labor contractor Steve Scaroni of Heber, "to sell your most important raw product—water—for cost."[2]

Never mind, for the moment, that in 1996 the water Scaroni put on his agricultural fields cost him $12 per acre-foot in payment to the Imperial Irrigation District and that the sale of water to which he objected carried a price of $128 per acre-foot.[3] The remarkable element in Scaroni's contention is the notion that water is the "most important raw product" of the Imperial Valley. It takes no small amount of temerity to argue that the nation's driest desert basin *produces* water, let alone that the water produced surpasses lettuce or beef as the basin's most important product. But while Scaroni may have put the matter more bluntly than most, he was by no means alone in his argument. In seeming contravention of laws of nature as well as logic, the former desert sink now known as Imperial Valley produces cattle, silage, and vegetables of every conceivable kind, to the tune of a billion dollars in sales yearly—and it does this because of the wealth of water it controls. The valley may not *produce* that water, but it has the power and the legal right to cause approximately a fifth of the natural flow of the Colorado River to depart the river's channel and appear within the basin's arid precincts. Such is the reach and power of Property.

THE POLITICAL history of the Imperial Valley, at least since the end of the Great Diversion in 1907, has mainly involved the assertion and defense of the valley's claims to the river. Its economic history has had two themes: converting that water into money by turning it first into food and, not less important, capturing the potential for that income in the value of real estate. The future promises a change from these main themes. As one might infer from Scaroni's comment about the valley's most important "raw product," the conversion of water into money will less and less require the intermediary process of agriculture. The simplest and ultimately, perhaps, the most profitable thing to do will be to sell it, direct, to the thirsty multitudes of coastal southern California, whose number is sixteen million and growing.

The valley's claims to the river, of course, originate in Rockwood's and Chaffey's early overachievements in river diversion, which produced the floods that formed the Salton Sea and nearly destroyed the valley. Water law in the western states and territories was then young, but its tenets were solidly established. One of them held that whoever first diverted river water and applied it to a "beneficial" use, such as agriculture, thereby earned the right to continue that diversion and use in perpetuity. Water rights thus established enjoyed protection under the maxim, "First in time, first in right." Undeniably, the massive diversion of water to Imperial Valley was very early in time.

In 1911, when Imperial Valley's water rights were salvaged from the wreckage of the Colorado Development Company and transferred to the newly formed Imperial Irrigation District, the valley still had its just-got-here, pitch-a-tent-and-start-plowing spirit. The paint was fresh on the schoolhouse, and farmers were scraping, ditching, leaching, and bringing thousands of acres of blistered desert under cultivation each year. The Imperial Valley rightly claims to be one of the nation's last agricultural frontiers, and its settlers thought of themselves, proudly, as pioneers. Soon after the IID was formed, one of their finest, Judge Finis C. Farr, undertook to memorialize his neighbors by compiling a history of the county.

Despite a weakness for overblown prose, Farr produced an important record of his time and place. The final two hundred pages of *The History of Imperial County, California* consist of biographical sketches of the leading citizens of the valley. If these portraits fail as realism, they nevertheless tell us how their subjects hoped the world would see them.[4]

These newly vested claimants to a fifth of the Colorado River were pleased to be described as *self-made, up-to-date, enterprising, influential,* and *pioneering*—such terms being the most common of the laudatory adjectives that crowd Farr's profiles. The valley's favored few were also *progressive*—which is to say, their lands were *highly cultivated,* and they freely lent their energies to the collective efforts necessary to make the desert bloom and to build a society where none, by their standards, had been before. The best of Farr's select were further said to be *representative*—a curious term for praising men presumed to be exceptional. What did it mean? Clearly, the men of the first generation of settlement had to be strong and

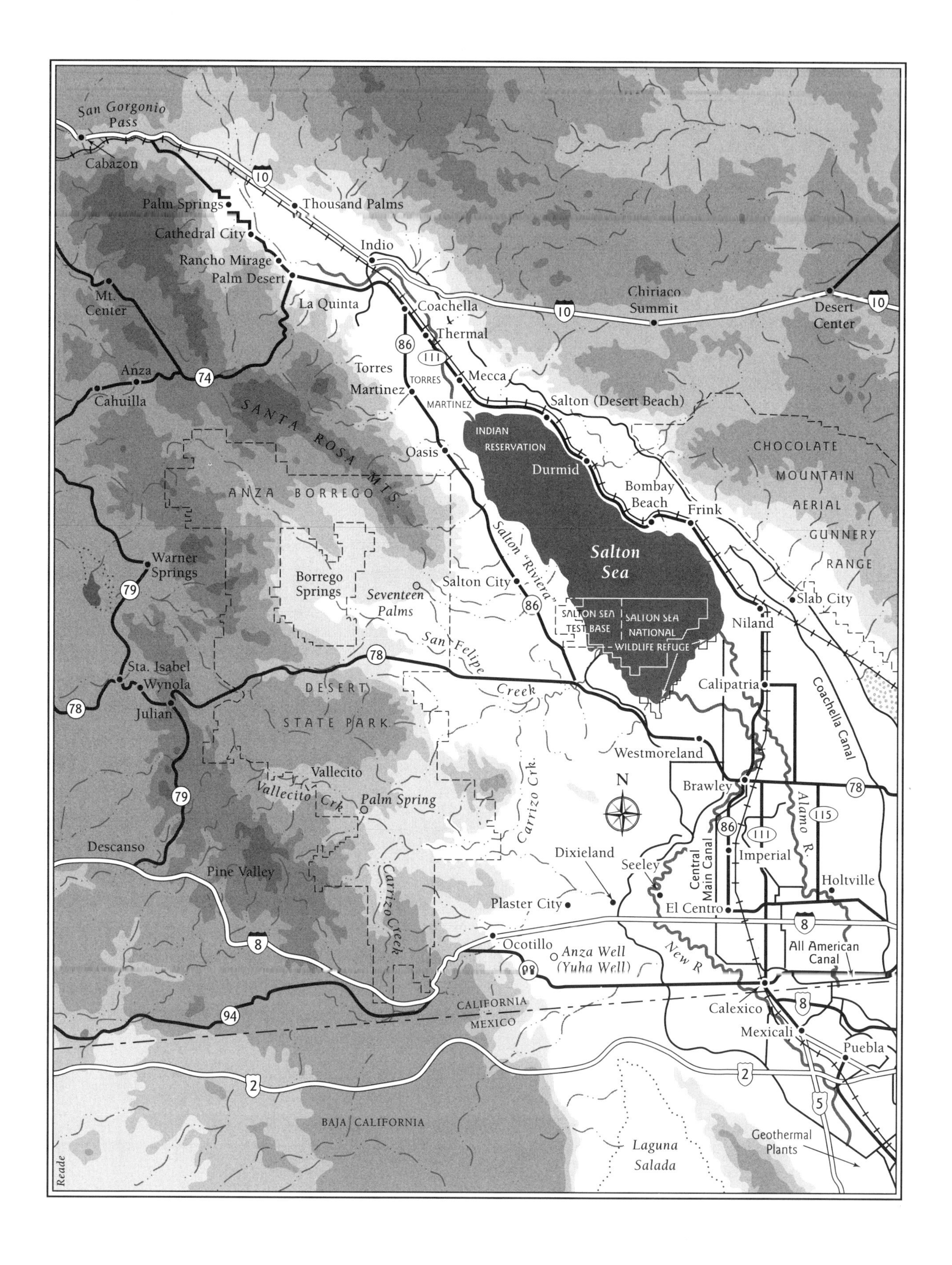

San Gorgonio Pass
Cabazon
10
Palm Springs
Thousand Palms
Cathedral City
Indio
Rancho Mirage
Palm Desert
Mt. Center
La Quinta
Coachella
Chiriaco Summit
Desert Center
Thermal
86
111
Torres Martinez
TORRES
MARTINEZ
Mecca
Anza
74
Cahuilla
Salton (Desert Beach)
SANTA ROSA MTS.
INDIAN RESERVATION
Oasis
Durmid
CHOCOLATE
MOUNTAIN
AERIAL
GUNNERY
RANGE
ANZA BORREGO
Bombay Beach
Frink
Salton "Riviera"
Salton Sea
Warner Springs
79
Borrego Springs
Seventeen Palms
Salton City
Slab City
SALTON SEA TEST BASE
SALTON SEA NATIONAL WILDLIFE REFUGE
Niland
San Felipe Creek
78
Sta. Isabel
Wynola
Julian
DESERT
STATE PARK
Calipatria
Coachella Canal
Westmoreland
Vallecito
Vallecito Crk
Palm Spring
Carrizo Crk.
N
Brawley
Alamo R.
115
Descanso
Dixieland
Seeley
Central Main Canal
Imperial
Holtville
Pine Valley
Carrizo Creek
Plaster City
El Centro
8
Ocotillo
Anza Well (Yuha Well)
New R
All American Canal
94
CALIFORNIA
MEXICO
Calexico
Mexicali
Puebla
2
5
BAJA CALIFORNIA
Laguna Salada
Geothermal Plants
Reade

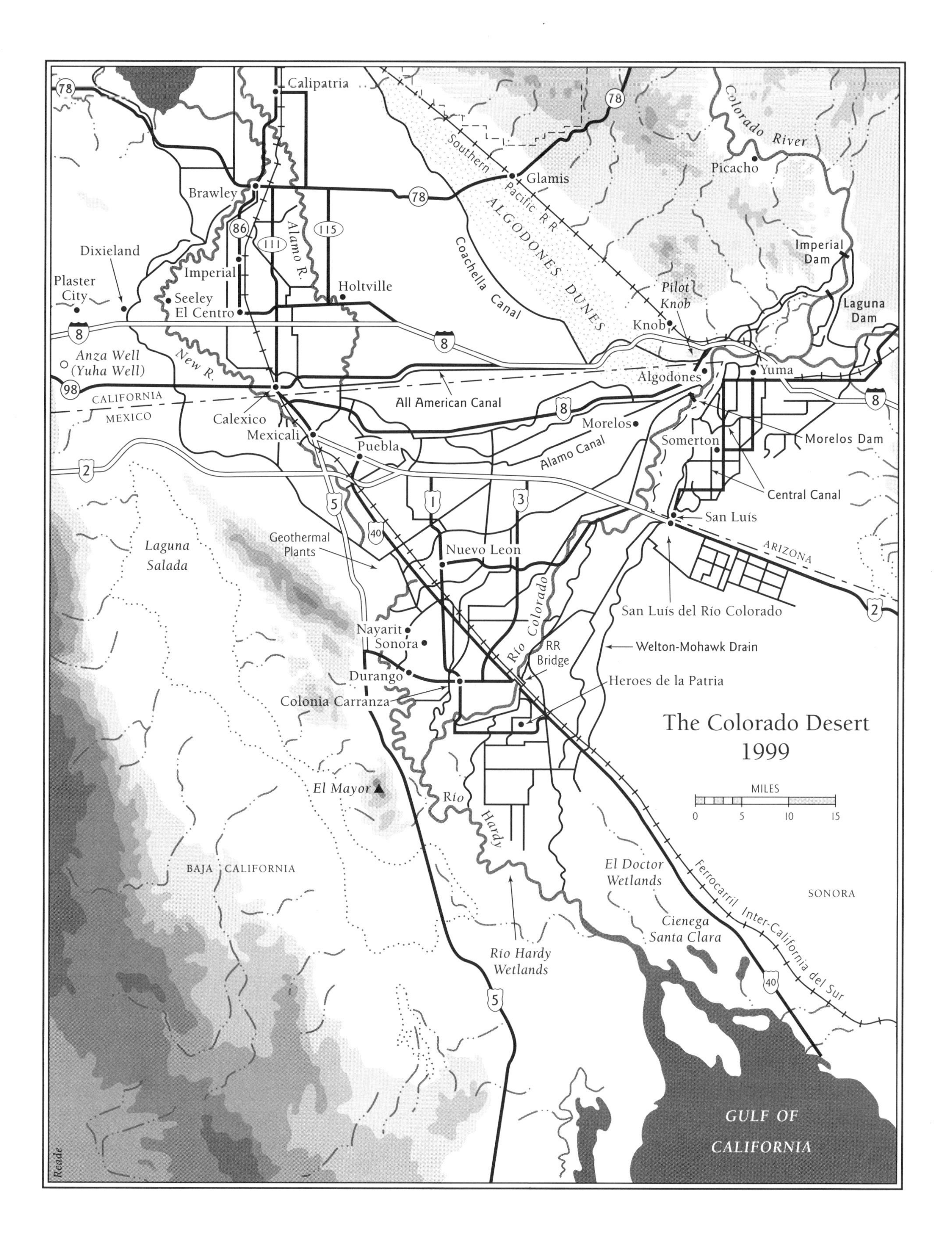

The Colorado Desert
1999
Calipatria
78
Glamis
Southern Pacific R R
ALGODONES DUNES
Colorado River
Picacho
Brawley
86
111
Alamo R.
115
Coachella Canal
Imperial Dam
Laguna Dam
Dixieland
Plaster City
Imperial
Seeley
El Centro
Holtville
Pilot Knob
Knob
8
Anza Well
(Yuha Well)
New R.
98
Algodones
Yuma
CALIFORNIA
MEXICO
Calexico
Mexicali
All American Canal
Morelos
Morelos Dam
Somerton
Puebla
Alamo Canal
2
5
1
3
Central Canal
San Luís
ARIZONA
40
Geothermal Plants
Laguna Salada
Nuevo Leon
San Luís del Río Colorado
Nayarit
Sonora
Río Colorado
Welton-Mohawk Drain
RR Bridge
Durango
Heroes de la Patria
Colonia Carranza
MILES
0
5
10
15
El Mayor
Río Hardy
BAJA CALIFORNIA
El Doctor Wetlands
Ferrocarril Inter-California del Sur
SONORA
Cienega Santa Clara
Río Hardy Wetlands
GULF OF CALIFORNIA
Reade

resilient, but not to a fault. It was better to stand shoulder to shoulder with one's neighbors than to go it alone. Irrigation demands cooperation, and so one might be rugged, but not too much an individualist.

Farr's heroes very well knew who they were: they saw themselves reflected in each other. First, they were men—only one woman earned admission to Farr's elite, which otherwise included 242 men. And their heritage was overwhelmingly northern European. Twice as many came from Sweden (4) as from nearby Mexico (2), although from the outset of settlement, brown hands had dug most of the ditches, hoed the fields, and picked most of the crops. More of Farr's elect hailed from Missouri (23) than from any other state, although the rest of mid-America was also well represented, including Illinois (19), Ohio (17), Iowa (13), and Kansas (10). Twenty-two leading citizens came to the valley from other parts of California; whether native or adopted, they were restless sons of that most restless state. A dozen came from Canada, and slightly more from Europe; nearly all the rest were U.S. born. A fair number proudly cited linkage to the Mayflower or to fathers, uncles, or other relations who had fought and bled in the Civil War. One Imperial Valley pioneer fought with Custer but missed the Little Big Horn. Another escaped national classification, for he was born at sea.

In keeping with the social darwinism of the day, the verdict on these self-made successes was that, ipso facto, they were the best of their kind. Inclusion in Farr's unaristocratic blue book of valley society was their diploma of survival and fitness. Most of these men were young and ambitious, or had been when they came to the valley. No doubt many had failed elsewhere, or at least failed to get a break. But in Imperial Valley, they made it. Never mind the many who didn't, those who came and tried and went away defeated. Of them, Farr's tome of congratulation scarcely breathes a word.

Together with their less illustrious neighbors, the valley's leading men formed the Imperial Irrigation District in 1911, and the district, or IID as it is better known, set about to address the valley's interest in water. An earlier chapter mentioned IID's purchase in 1916 of the remnant assets of the California Development Company. Necessarily, the district devoted much energy to the valley's irrigation infrastructure, the administration and further development of which dominated its internal affairs. Its external relations, however, are what concern us here.

The valley clung to life by a narrow and vulnerable thread of water, a condition of exposure that bred three consuming fears. One concerned water quality. The occasional bloated cow floating down a main canal or, now that the Mexican Revolution was in full swing, the discovery of a waterlogged and unrecognizable human corpse provided regular reminders that every gallon of water destined for field or home flowed first through Mexico. A second fear concerned supply. It was not hard to imagine circumstances that might one day deprive the valley of needed water. As mentioned earlier, a syndicate headed by Harry Chandler, publisher of the *Los*

Angeles Times, controlled a vast territory on the Mexican side of the border. The upstream position of those lands and their steady conversion to agriculture, according to one valley spokesman, "menace us like a great sponge."[5]

The obvious solution to both problems, for which the IID began agitating from the day it was formed, was to construct an "All-American" canal to carry water from the Colorado to the valley without touching Mexican soil. Such an intervention would place Chandler's holdings and all of Mexico downstream of Imperial Valley and would eliminate the worst threats to water quality.

Construction of such a canal would be a massive undertaking. Eighty miles long, it would have to cross the porous and shifting Algodones dunes, a miniature Sahara that bounded the valley on the east. Even so, the canal would not remedy all the valley's concerns about its water supply, for on occasion during the winter growing season the natural flow of the Colorado dropped below 3,000 cubic feet per second—and Imperial Valley farmers claimed their crops would fail with less than 6,500 second-feet. Worse, even if all interests south of the border were shunted aside, the expanding irrigation operations of the Yuma and Palo Verde valleys would remain upstream of IID's diversion, and at times of low flow, their withdrawals might leave too little water in the river to keep the valley wet.

A third fear, however, overshadowed all else. The red and muddy river that had been known to dwindle to 2,200 cfs had also memorably swollen to more than a hundred times as large—to 250,000. The river attained this stupendous volume on January 22, 1916. On at least a dozen additional occasions in the decade and a half following the Great Diversion, the river crested higher than 100,000 cfs at the Yuma gauge, and most of these floods were more powerful than anything the Great Diversion had seen.[6] No matter if the levees holding back the Colorado surpassed the walls of Hadrian or the Mings, there were many in the valley who would never sleep easily when the river surged. Only one thing could quiet such fears: a great dam that caged the river's brawn as though it were a lion in a zoo. Once caged, the river might be regulated as well as tamed, releasing precise quantities when needed.

The champion of big dam construction was Arthur Powell Davis, head of the Reclamation Service. His uncle, John Wesley Powell, had not only made the first descent of the Colorado by boat but also preached a version of the irrigation gospel at the highest level of government and helped bring the Reclamation Service into being. Although Powell, like William Smythe, had evangelized for a West of small, family-run, irrigated farms, Arthur Powell Davis failed to make the Reclamation Service an effective instrument of that vision. The fault may have lain mainly with the discouraging economic realities of early-twentieth-century agriculture, but Davis nevertheless found himself directing an agency that, after almost two decades, had little to show for its existence. Gradually, he became convinced that survival of the Reclamation Service depended on dramatic, large-scale achievements. After commissioning and considering feasibility studies for a main-stem dam on the Colorado River, he pinned his hopes on building a massive concrete barrier in Boulder Canyon. It would be the largest dam in the world.[7]

At first, Imperial Valley resisted joining its cause to Davis's. Connection with a government project would entail entanglements and regulations. The Reclamation Service might, for instance, attempt to extend its limit on landownership to the farmers of the valley. The Reclamation Act of 1902 specified that no beneficiary of a reclamation project might receive water for more than 160 acres. The social and egalitarian ethos of the act held that rivers were to be tamed and lands reclaimed for the sake of the family farm, not for large holdings.

In the early years, the founders of Imperial Valley availed themselves of reclamation rhetoric when it served their purposes, but few among them believed that a limit of 160 acres could or should be maintained in the valley. Even by 1918, some of Finis Farr's heroes possessed holdings greater than 500 acres. A decade later, ownership of 500 acres would be merely average, and only those with six times that amount would have attained the topmost tier in size and power.[8]

Valley leaders also hesitated to cast their lot with Davis's most powerful constituency, the cities of California's southern coast, whom they feared as competitors for the Colorado's water. Valley growers had watched William Mulholland conquer the Owens Valley and deliver to Los Angeles the entire flow of the Owens River by means of a 233-mile aqueduct. As the Owens Valley dried up and succeeded back to desert, they worried that such a fate could also be theirs.

In time, however, the people of the valley judged that they had more to win than to lose by joining with Davis and the cities. Congressman Phil Swing, a former IID attorney who never ceased to represent his old client, collaborated with California's Senator Hiram Johnson to draft legislation authorizing both the All-American Canal and Boulder Canyon Dam (later Hoover Dam). The Swing-Johnson bill, first introduced to Congress in 1922, also called for a generating plant to be built at the dam, so that the sale of electricity, mainly to coastal cities, might pay the enormous construction costs of the project.[9]

Opposition came from many quarters. Power companies feared federal competition. Arizonans feared California might monopolize the river before they claimed their share. Upstream states feared downstream states for the same reason. Under the doctrine of prior appropriation and beneficial use—"First in time, first in right"—Colorado, Wyoming, Utah, New Mexico, Nevada, and Arizona all worried that California's early initiative might limit their prospects for growth.

Impelled by the threats implicit in the Swing-Johnson bill, water commissioners from the seven states met in Santa Fe, New Mexico, late in 1922. Secretary of Commerce Herbert Hoover mediated their deliberations and elicited agreement on what emerged as the Colorado River Compact. Essentially, the negotiators agreed to suspend the doctrine of prior appropriation and to divide the river's average annual flow as follows: 7.5 million acre-feet for the upper basin, defined as the watershed upstream of Lees Ferry, near where the Colorado enters Arizona from Utah; 7.5 million acre-feet for the lower basin; and an additional 1 million acre-feet for the lower basin should it be needed (this somewhat cloudy provision concerned the yield of the Gila River and other tributaries of the lower Colorado and lay outside the interest of upstream states). Additionally, Mexico would take any subsequent treaty allocation it might receive from the excess above the 16 million acre-feet already allotted. If this were not enough, the two basins would have to make up the deficit—although the mechanism for such a division remains to this day a matter of debate.

The compact was deeply flawed—not least because the calculation of river flow on which it depended was based on measurements collected during an exceptionally wet period and resulted in an overallocation of the river by 20 to 30 percent, depending on whose data one accepts.[10] This realization, however, lay many years in the future.

For the present, only Arizona refused to ratify the compact. While the other states shared Arizona's jealousy of California, they recognized that a flood might come at any time, devastating areas like the Imperial Valley and causing Congress to approve hasty flood control legislation that would favor California even without a compact. Under such a scenario, no force would check California's seizure of the river. For this reason, the legislatures of Colorado, Nevada, New Mexico, Utah, and Wyoming joined California's in approving the Colorado River Compact. Their action laid a groundwork for the ultimate passage of the Swing-Johnson bill, now recast as the Boulder Canyon Project Act, in 1928.

In its final form, the bill authorized construction of Boulder Dam, the All-American Canal, and Imperial Dam, which serves as headgate for the canal. It also limited California's share of the river to 4.4 million acre-feet, with additional allowances to be made in wet years. To no one's surprise, the promise of a tamed river further set in motion the next phase of William Mulholland's grand strategy for getting water to the coast. By 1928 he had completed formation of the Metropolitan Water District of Southern California (MWD), a union of several dozen water utilities serving communities from the Los Angeles basin to San Diego. (To San Diego's eventual dismay, Mulholland structured the MWD's board in a way that forever guaranteed Angelenos control of the organization.) In 1931, the district's voters approved issuance of bonds totaling $220 million—a colossal sum for the time—which would finance construction of yet another aqueduct to carry water across the desert to the coast. In 1935 Congress would instruct the Bureau of Reclamation to build Parker Dam, not far downstream from Boulder Canyon, impounding Lake Havasu, from which water would enter the 242-mile Colorado River Aqueduct. A source for the energy to lift the water over the topographic spine of

California already existed: MWD had contracted for 36 percent of Boulder Dam's electricity. Everything meshed: flood control, cheap power, water for the coast, and a canal for the Imperial Valley with a dam for its headgate. Arizona could gnash its teeth and weep. Thanks largely to the munificence of the federal government, southern California was poised for a future few could imagine.

SLOWLY THE colossus of the dam rose from the bowels of Boulder Canyon. It has become customary to praise that staggering enterprise with an invocation of very large numbers: 3.25 million cubic yards of cement, 3 million board feet of timber, 662 miles of copper pipe to carry coolant through the hardening concrete. The project required the newest and largest concrete batching plant in the world, which dispatched a 16-ton bucket every minute for 22 months. It put thousands of men to work and housed them in a new city that rose entire from the naked desert. It generated 1,500 new megawatts of cheap energy to fuel the growth of a region. And it caused 110 deaths: some from injury, some from heart attack and sunstroke, and some, so the tale is told, from live burial in the viscous river of concrete that mounted and jelled until it formed a wall 726.4 feet high and 1,244 feet wide. The dam outmuscled the river and turned the wild and leonine Colorado into a kind of giant housecat.

The miracle was dedicated in September 1935. It was the biggest thing Americans had ever made, and in the dark years of the Great Depression the building of Boulder Dam became an apotheosis of national pride and economic hope. Woody Guthrie and Franklin Delano Roosevelt sang its praises. Nearly everyone agreed it was the start of something new.

What exactly it was the start of, however, depended on the perspective of the observer. For the companies that actually built the dam, it was the start of a career in giant water projects. In little time they would add Parker Dam, Bonneville Dam, Grand Coulee Dam, and others to their list of profitable ventures.

For Los Angeles and other cities of the coast, it was the start of a new cycle of growth, this one longer, larger, and less logical than any that had preceded it, perhaps in the history of humankind.

For the Bureau of Reclamation, now emerged from the chrysalis of the Reclamation Service, it was the start of a redefined mission and identity. Forget William Smythe's romantic arguments about small farms and a new American Holland. Forget John Wesley Powell's hopes for an egalitarian democracy based on irrigation agriculture. Forget the Jeffersonian idealism underlying both men's visions. The bureau was now purposed toward grandeur: big projects on big rivers funded by huge budgets in the service of powerful interests.

For Imperial Valley, completion of Boulder Dam inaugurated a new era of physical and economic security, thanks also to the All-American Canal, which first delivered water in 1941. The dam, however, also brought a new kind of insecurity. The water now feeding the valley was cheap, safe, and abundant. Notwithstanding the hundreds of millions of dollars invested upstream, valley farmers had to pay only the IID's cost of diversion and delivery. But the water was now *federal* water, and so it required the propitiation of a new and potentially fickle power.

For the American West, completion of the dam marked the emergence of a new hydraulic society organized on a vast scale and dependent on the highest technical achievement and immense stores of capital. One thing had not changed since the earliest debates over settlement of the Northwest Territory: the fortunes of the West still depended on the political will of the entire nation—and by depending, helped to shape it.

And for the river (let us not forget the river), it was the beginning of the end. The metamorphosis of the Colorado was well advanced. Once a river, it was becoming something else: a delivery system for liquid property.

ALL THIS CHANGE demanded a new vision of what was good and right and a new story about how the river, the desert, and Americans should coexist. Not that people had to believe that the new utopia would or even should come about, but they had to have a view of things that made them comfortable, that told them, "All this is for the good. Keep going."

The old vision of the irrigated West had offered a fantasy mosaic of clustered small farms tilled by virtuous families. The new vision had loftier aspirations. It was Promethean in range and breadth. It served not social ends but economic ones: growth was the grail that it pursued. Even more, the new vision revered mastery for its own sake. It strove for completeness in its control of nature and in the utilization of resources. Such mastery promised not just wealth but greatness.

Deep in the bowels of the power plant at Parker Dam, there hangs a portrait of the ideal landscape of the new hydraulic West. Tourists may freely visit this spotless industrial basement where turbines hum and recorded narrations, at the touch of a button, recite the miracles that dynamite, concrete, and slide rules performed. Wherever one looks, order prevails. Paint as thick as your tongue, in institutional yellows and greens, seals this world of steel from the chaos of rust and decay. The lineation of ramps and ladders and the methodical joinery of pipes confirm that nothing exists here that did not first exist in a blueprint. One sees no hammer dings, no grease spots, no sign of anything born of chance.

A portrait of reclamation heaven, titled "Conservation and Full Utilization of Water," hangs under glass. Elements within its landscape are clearly labeled, lest the uninitiated fail to comprehend its marvels. Clouds at the top of the tableau, renamed *atmospheric water resources,* are shown to submit to the hand of the modern rainmaker. The words *forest management* and *range management* give as-

surance that the upper watersheds of this mythical land are properly tended. Elsewhere, no possible modification of the land is omitted: the labels identify a transmountain diversion, a multipurpose reservoir, diversion dams and power plants, transmission lines and rural electrification. There is a fish hatchery and a fish ladder; there are navigation locks, canals, drainage ditches, irrigation agriculture, tile drains, wells for groundwater pumping, even a desalinization plant on an ocean shore. Scientific management is everywhere at work: *weed control* is under way in the foreground, *erosion control* in the background. Yonder is the *snow survey*. Closer are the *measuring flume* and the site of *evaporation studies*. Everything fits and connects. The picture presents a festive, interlocking dance of the many ways, as its legend describes, "man asserts his ingenuity to conserve and utilize his most precious resource—water."

Absent from the picture is any suggestion that subversives or malcontents might lurk round the edges of the festivities, threatening to spoil this most perfect reclamation party. One such rowdy is the natural world, which persists in manifesting injury as its processes are disrupted. The stubborn problems of the desiccated Colorado delta, described in the previous chapter, and of the increasingly saline and toxic Salton Sea, about which more in a later one, are cases in point. Endangerment and extinction of species is another vexing outcome of full usufruct of the Colorado, although one would not know it from the visitor's brochure at Parker Dam. Even in the 1990s, two decades after passage of the Endangered Species Act, the brochure confidently states, "Before Parker, Davis, and Hoover dams, only a few nondescript species of fish with little commercial or recreational value could survive the seasonal changes wrought by the temperamental Colorado." The dams improved matters, according to the brochure, by creating habitat for striped bass, large-mouth bass, and other species more suitable for catching and eating. The brochure takes no notice that the Colorado basin formerly sheltered thirty-seven species of native fish, of which twenty-three were found nowhere else, yielding the highest rate of endemism within any North American watershed. Of the natives, three species are now extinct, and seventeen are listed as threatened or endangered. The decimation of the once plentiful bonytail chub, one of the latter group, prompted one scientist to write, "If it were not for the stark example provided by the passenger pigeon, such rapid disappearance of a species once so abundant would be almost beyond belief."[11]

Another angry voice at the reclamation celebration belongs to the landless poor in whose name the entire hydraulic experiment was first begun. The gradual abandonment of social goals in reclamation became complete long ago, and Imperial Valley led the abandonment at every step. Far from fulfilling William Smythe's breathless promise that irrigation agriculture would build a bulwark of middle-class democracy, the Imperial Valley became instead one of the most feudal landscapes in North America.

Smythe had prophesied that in the Imperial Valley settlement would begin "on comparatively large areas, but it must tend inevitably and swiftly to the very smallest farm unit on the American continent."[12] He could not have been more wrong. From the first years of settlement to the present, the average size of Imperial Valley holdings steadily increased, and the gap in wealth between the rich who owned the land and the poor who worked it commensurately widened. The 1990 census found nearly one in four Imperial County residents living in poverty—one of the highest rates in the nation.[13] If one includes in such a calculation the circumstances of approximately eight thousand farmworkers who live in greater Mexicali but work seasonally in Imperial Valley, the gulf between rich and poor grows even broader.

The small-farm fiction of irrigation agriculture is a kind of creation story, a fable about the way things got started, but it is as irrelevant to the contemporary reality of Imperial Valley as the story of Adam and Eve is to natural selection. In 1980 the Supreme Court overruled the fable by giving Imperial Valley growers permanent dispensation from the 160-acre limit of the 1902 Reclamation Act. The court judged that the limit, which Imperial Valley farmers had always ignored, never properly applied to valley lands.[14] By 1996, the holdings of the valley's largest landowner exceeded that toothless old standard by a factor approaching three hundred. If one is to understand the future of the valley and its "most important raw product," one must look closely at the identity of that modern agricultural potentate.

THE LANDOWNER was a corporation, Western Farms. Beginning in about 1993 its owners, the billionaire Bass brothers of Fort Worth, Texas, pumped $50 million to $100 million into land purchases within Imperial Valley. They assembled ownership of forty-two thousand acres, representing slightly less than 10 percent of Imperial Valley's irrigated farmland. Company officials declared that their only aim was to fatten low-weight cattle from Mexico.[15] Of course, no one believed them.

People instead believed that the Bass family, whose original fortune derived from one liquid—petroleum—now aspired to a new fortune based on another—water. They read as proof the support Western Farms showed for large transfers of Imperial Valley water to the thirsty cities of the coast, in exchange for a return flow of cash. Such an idea was hardly original with the Basses. Water marketing had loomed on the horizon for as long as anyone cared to remember, and in 1989 the IID and the Metropolitan Water District had sealed a deal to transfer 106,000 acre-feet per year.[16]

The entry of the Bass brothers into the water transfer business stirred fears because of their size and clout. The valley was already accustomed to outsiders—by the mid-1990s, 60 percent of it was owned by absentee landlords. None of them, however, was as big as the Basses, who owed nothing to entrenched valley interests.[17] Word soon spread that Western Farms was taking control of the IID, a suspicion that an Imperial County grand jury strengthened in 1996 when it concluded that IID management provided Western Farms with information unavailable to the general public and that the process resulting in the selection of a new general manager for IID, a former consultant to Western Farms, was "seriously flawed."[18]

The jury's condemnation would have been far harsher had it learned of Western Farms' confidential negotiations through the summer of 1995 to sell water and water rights to the San Diego County Water Authority (SDCWA), the water utility for one of the thirstiest cities in the West.[19] Western Farms assured SDCWA that all members of the IID board but one "supported the proposal." Western Farms intended to idle its own lands, and together with other farmers whose participation it would arrange, it promised to make available up to 500,000 acre-feet of water for transfer to San Diego at an annual cost of over $90 million—not a bad yearly return on an original investment of about the same amount. During the course of negotiations, an attorney for SDCWA pointed out the obvious: "Western Farms believes that this area of the state has not been particularly progressive in dealing with water as an asset."[20]

Before it could reap such a bonanza, Western Farms needed to guarantee San Diego the long-term security of its new water supply. Toward this end, Western Farms offered to sell actual water rights, a proposition of dubious legality given that the IID had long defended its claim to hold all Imperial Valley water rights as trustee for the entire valley. It may have been concern on this point that caused San Diego to suspend negotiations with Western Farms and open talks on a possible water transfer directly with IID.[21]

No matter; the Basses were not discouraged. San Diego had no monopoly on thirst. The Basses' next move was definitive of the valley's future. They sold Western Farms, or rather traded it, to the US Filter Corporation for stock valued at about $250 million, plus a seat on the new owner's board. Based nearby in Palm Desert, US Filter was already the world's largest water treatment company. Its 1996 sales of $2.3 billion exceeded those of its nearest rival, Culligan, by seven times, and within a year it would buy Culligan, growing still larger. Even so, US Filter remained a small player in the worldwide business of water treatment, which the company estimated to have sales of $300 billion a year. Its chairman, Richard Heckmann, frequently observed that the water business, in spite of its size, had no General Motors or IBM. Achieving such a status was US Filter's goal. Toward that end, between 1991 and 1996, Heckmann guided US Filter through the acquisition of no fewer than seventy-five companies, including thirty-nine in the year he took over Western Farms.[22]

The Basses wrung a premium from US Filter in their sale of Western Farms. The forty-two thousand acres changed hands at about three times normal market value, but normal value was obviously not what attracted either the Basses or US Filter to the valley. One might discount the US Filter purchase by arguing that the company's stock, with which the Basses were paid, was overvalued, but the New York Stock Exchange did not concur with this analysis: the stock rose after the purchase. One might argue that the company was out-negotiated by the Basses and simply paid too much to obtain a showplace property, yet the fact remains that, far from running back to Fort Worth with their pockets leaking cash, the Basses willingly became US Filter's biggest investors and agreed not to sell their position for several years. What can they have been thinking?

The unavoidable conclusion is that the Basses and US Filter agreed not that the valley's land was undervalued but that its water was, and grossly so. Said Heckmann of his new possession, "The water we have is 81 billion gallons a year in perpetuity."

Figure it this way, in round numbers: Western Farms's 42,000 acres commanded the use of almost 250,000 acre-feet of water a year. Ignoring potential agricultural returns, one can view US Filter's acquisition of Western Farms as a purchase of water rights at close to $1,000 per acre-foot. If US Filter annually sells the water to which those rights entitle it at $100 an acre-foot, it earns a gross return of 10 percent on its original investment, although that certainly is not its goal. No young company in a young industry goes hunting for 10-percent returns.

US Filter had other ideas about the value of a potable gallon of water, and so did the rest of southern California: the water transfer that IID and the San Diego County Water Authority ultimately approved in April 1998 called for sales to start at $249 per acre-foot, and the price would eventually rise as determined by an elaborate formula—no one knew how far or how soon.[23] This much was certain: as the price of water climbed, the physical currents of rivers, canals, and ditches would turn to follow it.

A FEW NUMBERS and facts paint the edges of the big picture of water use in southern California. Within these limits, IID, US Filter, MWD, SDCWA, and others jockey for position and profit. One thing is certain: the picture will never be finished, for the struggle for desert water will never end.

California's share of Colorado River water is 4.4 million acre-feet. The Colorado River Compact established this number, and the Supreme Court confirmed it in 1963 and 1980. But California has never honored it. Ever since the Colorado River Aqueduct began delivering water to the Los Angeles basin in 1941, California has diverted far more than its entitlement. Through the 1990s California's diversions exceeded its allocation by 800,000 acre-feet per year. California got away with such excess by diverting the unused shares of other states' apportionments and by persuading the secretary of the interior, who serves ex officio as water master for the basin, to declare conditions of "surplus," which temporarily lift allocation limits.

California's days of excess, theoretically, are numbered. Other states have gradually developed the capacity to exploit their allocations, envious Arizona first among them. Arizona began diverting water for its Central Arizona Project (CAP) in 1985, and as the century drew to a close, it was diverting all its lower basin allocation of 2.8 million acre-feet, though it still lacked the ability to put so much to use. No problem, said Arizona. It would "bank" its Colorado River water by providing it to farmers to substitute for the groundwater they otherwise would have pumped. In theory, the river water was thereby stored in the state's aquifers where it might safely stay until needed. In practice, Arizona's farmers received a boost in their already considerable government subsidy, as CAP water was cheaply provided to them with no abridgment of their ultimate right to pump groundwater in the future. Effectively, Arizona devoted its Colorado River allocation to extending the life of highly subsidized undertakings of questionable value. Nevada, meanwhile, began to admire Arizona's shell game. Lacking significant aquifers of its own, Nevada asked Arizona to consider storing some of its 300,000 acre-feet allocation in the fictional Arizona water bank.[24]

As other states launched similar initiatives, a rare thing developed among members of the Colorado River Compact: consensus. All of them save California agreed that they would support no further declarations of "surplus" until California came up with a credible plan for living within its means—a "4.4 plan" in the parlance of Southwestern water buffaloes.

Here's the squeeze: the Colorado River water rights of southern California's agricultural block (the Imperial, Coachella, Palo Verde, and Yuma districts) are senior to the rights held by the MWD. This means that if California were suddenly and stringently forced to honor its water allocation, the entire reduction—800,000 acre-feet—would come out of the 1.2 million acre-feet MWD diverts into the Colorado River Aqueduct each year for municipal and industrial use.

Realistically, such a draconian measure is about as likely as nuclear war. Those 800,000 acre-feet of water sustain approximately 6.4 million men, women, and children—a population equivalent to that of the entire state of Virginia. Moreover, those 6.4 million people are part of a water-consuming community of 16 million served by MWD, among whom the burden of a system-wide shortage would be shared. No state in the union, except California itself, has more people. The primary fact of West Coast life is that the coastal megalopolis of southern California makes up in votes and sheer numbers what it lacks in ecological resources. Accordingly, the political realities are these:

The Colorado River Aqueduct must be kept full, which means diverting into it about 1.2 million acre-feet annually.

California must eventually reduce its withdrawals from the Colorado River by an amount that at least approaches 800,000 acre-feet.

The greater part of this reduction must be achieved by squeezing water out of agriculture and delivering the saved water to the urban coast.

THERE ARE basically three ways to squeeze water out of agriculture. The simplest—and most controversial—is not to irrigate at all, but to send the water that would have been used directly to the cities. This is what Western Farms intended in its secret 1995 discussions with San Diego. Giving new meaning to an old term, such a process is called "fallowing." It raises a host of concerns, including the possibility that land retired from cultivation would become a dustbowl. Depending on the scale of fallowing, the effect on a region's economy might be equally devastating. Farmworkers, food processors, truckers, sellers of equipment, seed, and chemicals, and everyone else connected to farming—which in Imperial Valley includes nearly everyone—would have less to do, less to earn, and less to spend. No one in Imperial Valley openly champions fallowing.

But like most highly charged issues, a prohibition against fallowing may mean something quite different from what it seems to mean on the surface. The groundbreaking 1998 water transfer agreement between IID and San Diego, for example, strictly prohibits fallowing but leaves the term undefined. An advisory panel of farmers proposed that IID define fallowing as the practice of "withholding water from and not growing a crop on farmland for a year or more."[25] The operative verb here is *growing*, not *harvesting*. If a farmer sells most of his water to the coast and then plants a crop of rye as green manure, waters it stingily, and plows it back into the soil, he has not fallowed the land in the technical sense of the proposed definition, yet neither has he much tickled the farm economy.

Less blatantly, a farmer might grow a crop of lettuce but forego planting a second crop of, say, sudan grass after he has harvested the lettuce. The water he saves by not double cropping might then be sold to San Diego. With 80,000 to 100,000 valley acres growing sudan grass as a second crop and consuming 300,000 to 500,000 acre-feet to do so, there is every possibility that the valley's obligation to San Diego might be satisfied by an abridgement of double cropping. Don Cox, a longtime IID director and the lone dissenter to the San Diego deal, points out that "people who understand farming know that the definition of fallowing being proposed opens the door for people to get paid for not growing sudan."[26] The threshold question is not whether double cropping must be preserved but whether communities that suffer as a result of decreased farming activity should share in the proceeds of water sales.

The second approach to squeezing water out of agriculture is conservation, which means using irrigation water less wastefully. Conservation can take place at either the "system" level—the process of delivering water from the river to the fields—or in the actual application of water to cultivated land. In the latter instance, drip irrigation or sprinkler systems can nourish the same amount of plant life with much less water than flood irrigation, which is still the norm in the valley. Selection of crops less thirsty than alfalfa and cotton, two of the valley's heaviest drinkers, and the collection and reuse of tailwater can also produce large economies in the consumption of water. The question is, who will pay for these economies? When farmers buy their water for only $13.50 an acre-foot (IID's 1998 price), their incentive for conservation is at best meager. Knowing the value of their "most important raw product," the growers of Imperial Valley remain steadfastly determined to avoid what they see as a "worst case scenario": the obligation to pay for water-conserving infrastructure themselves.

Saving "system" water entails another set of issues and challenges. Much can be accomplished by lining canals to prevent seepage and by installing interceptor ditches to capture seepage and waste flows. The delivery of water can also be made more exact so that a higher percentage of water actually reaches thirsty fields, with less flowing through the system as transmission waste. The transmission waste that remains can be captured in reservoirs and held for release back into the system. The 1989 water transfer agreement between IID and MWD, under which MWD paid for the conservation of 106,000 acre-feet, focused on saving system water in these ways. Contrary to expectation, however, it produced no net saving of water. During the agreement's period of implementation, consumption of water in Imperial Valley actually increased. Farm prices were good, farmers double cropped with greater frequency, and so they asked for—and IID delivered—more water than ever. In the absence of an enforced cap on IID's diversion from the river, the IID-MWD water transfer generated no net progress toward meeting California's compact limit of 4.4 million acre-feet.[27]

Whenever "system" water in the Imperial Valley is discussed, the question arises whether the IID "wastes" water. The district maintains that every drop diverted into its system is used as efficiently as conditions allow and that the *apparent* waste of water results from the high salinity of Colorado River water and the periodic necessity to cleanse fields by leaching their salts away (and flushing them into the Salton Sea). Few of the IID's neighbors, however, have agreed with this position, and some have pressed their arguments about IID waste in courts of law and before regulatory bodies, where they have at times prevailed. In 1984 the State Water Resource Board concurred that, indeed, IID wasted approximately 400,000 acre-feet a year.[28] But winning an argument is one thing. Changing the institutional behavior of the IID, given the avenues for appeal and obfuscation, is like nailing Jello to a wall. While the water board's finding pressured IID to pursue a transfer with MWD, the deal that resulted failed to limit IID's water consumption or to benefit anyone but the principals.

It remains in IID's interest to divert as much water from the Colorado as it can persuade the courts and the Bureau of Reclamation to let it. What it diverts, it says it *needs,* and it justifies its needs with data it develops for the purpose. IID's chief mission is to protect the water rights of the valley, which it holds in trust for all its customers. The more IID can prove the valley needs, the higher the ultimate quantification of its rights is likely to be. The higher its quantification, the more "raw product" it will have to sell to the cities of the coast. It has never been in IID's interest to limit its consumption of water.

The third approach to squeezing water from agriculture involves recycling wastewaters for reuse in irrigation or even for human consumption. US Filter stands ready to do the former on the 42,000-acre fiefdom it bought from the Bass brothers and to recycle the water of the valley's major drains, especially the Alamo River, which lacks the municipal and industrial contaminants that foul the New.[29] MWD and other utilities equally anticipate competition for wastewaters, no matter how polluted they may be. At least in theory, the possibility exists that the murky fluid that people euphemistically refer to as the "water" of the Alamo might one day be purified for human consumption, although the infrastructure for delivering it to population centers does not now exist. Unappetizing as such a prospect may be, the certainty of thirst will have its way. Several California municipalities are already planning to cycle outflow from their sewage treatment plants back into their drinking water systems. Some authorities say the state might meet one-third to one-half of its needs this way, and hired spinmasters have offered such terms as *repurified* to help the public get over the "yuck" factor.[30]

US Filter, meanwhile, claims that it can turn anything wet into something drinkable—and has positioned itself in Imperial Valley to do just that. It bases its expectation of wealth on the belief that its reverse-osmosis purification technology will provide the alchemy for a drought-stricken, overtapped future. Given the chance, US Filter will spin liquid gold from the dreck of the Alamo, from the drains of its own vast fields, from anything containing enough molecules of "dihydrogen monoxide" to render extraction profitable. Delivering the water thus won to the coast, or (more likely) exchanging it for water farther upstream that might be more easily delivered, is a matter of details.

In the United States, sixteen million thirsty people will get what they need. The problem for IID, US Filter, and the other water brokers of the valley is to be *just kind enough*. They need their 16 million hostages to pay a handsome ransom, but if they overplay their hand, the hostages may rebel and, through one means or another, simply take what they need. According to retired grower Don Cox, whose nine years on the IID board of directors places him in a position to know, these tensions divide the farm community of Imperial Valley. There are many, says Cox, who feel that water is like oil, "a natural resource that can be used any way it puts money in your pocket." And there are those who see it as a public resource to which the farmers of the valley possess rights of use, not ownership. The former group tends to ignore the contribution of the rest of the nation toward delivering the water of the Colorado River abundantly, safely, regularly, and for free to the headgates of the All-American Canal at Imperial Dam. By contrast, the latter group assents that the gift of the river may require some thanks. Even if its members deny moral obligations toward that end, they accept the political necessity of not going too far in exploiting water for personal gain—lest they lose or suffer restriction in their rights of use.[31]

AT THE TURN of the last century, a novel by Frank Norris described what was wrong in the fallen Eden of the Golden State. First published in 1901, *The Octopus* depicted the railroad monopoly that kept a stranglehold on central California's lifeline of grain. It told a tale of remorseless corporate power, trampled virtue, brutality, and greed.

Today California looks to movies more than books to tell its stories, and no film better probes the undergirding of California society than Roman Polanski's dark masterpiece, *Chinatown*. If bread is the staff of life, water is its blood. Water is so integral to life that to toy with it, to connive, dissemble, and lie about it, to treat it as a base commodity and stain it with greed, becomes a sin against the spirit, against life itself. Wars fought over oil are mere commercial wars. A war over water becomes holy. Accordingly, we are not surprised, viewing *Chinatown*, when an inquiry into the depraved manipulation of water leads—inexorably, it seems—to a revelation of the profoundest evil.

California's struggle to reduce its Colorado River withdrawals to 4.4 million acre-feet will elicit every conceivable behavior that humans are capable of bringing to enterprises of pitch and moment: leadership and selflessness, venality and greed, honor and reason, perversity and stupidity. Mere humans will be in charge from start to finish, and they will craft a set of flawed agreements that people will justify as being "as good as might have been hoped for." The agreements will have been shaped by the accidents of the region's historical inheritance and by the accumulated illogic of western water law. They will be grounded in limited knowledge and false expectations, and while their authors will at best be only faintly aware of those limitations and falsities, the rest of us will be even less so. The agreements will be further shaped by the leverage of wealth and the outcomes of multiple intrigues. They will submit, in the end, to political expediency. They will, in summary, embody the character of the society they serve as perfectly as any other of society's creations. Future historians may see them as a Rosetta stone of the era that produced them, and although the historians may flatter themselves that they can fathom the mazelike journey of the agreements' evolution, they will do well to remember the closing wisdom of Polanski's film: "Forget it, Jake. It's Chinatown." Where western water is concerned, there is always more than meets the most perceptive eye. A portion of the truth will always remain deeply shadowed, immune to general understanding, lost in a hydrological Chinatown.

California's water agreements will assuredly fallow fields, though perhaps not initially in Imperial Valley, and perhaps even then only in dry years. The Palo Verde area will be the first to see fallowing.

The agreements will surely feature conservation of system water: complete lining of the All-American and Coachella canals might save 100,000 acre-feet of seepage annually. Trouble is, Mexico depends on seepage from the All-American to provide groundwater for more than seven hundred irrigation wells on its side of the border. The international implications of the lining project remain unresolved.[32]

Mexico, as a matter of fact, can expect little from the emerging water markets of the north. It will get its 1.5 million acre-foot allocation, as provided by treaty, plus a share of declared surpluses, and no more. The superintendent of the All-American Canal's river division, an IID employee who opens the gates that allow the Colorado to dribble across the border, puts the matter succinctly: "The thing in the back of your mind is *never over-deliver to Mexico*."[33]

California's water agreements will certainly include transfers that depend on combinations of strategies. The 1998 transfer agreement between IID and the San Diego County Water Authority specifies that at least 130,000 acre-feet of transferred water must be conserved through on-farm conservation. At the discretion of IID, the transfer can be expanded to 200,000 acre-feet or, under another set of schedules and conditions, to 300,000. Yet the signing of an agreement is only the beginning of the transfer process, not the end. The IID must settle its considerable differences with the Coachella Valley Water District over their respective shares of Colorado River Water.[34] San Diego must maintain agreement with MWD on the cost of "wheeling" transferred water through the Colorado River Aqueduct and the rest of MWD's elaborate delivery system.[35] Both Coachella and MWD can be expected to use their considerable leverage to shape the deal to their liking, or to try to kill it. Additionally, substantial numbers of farmers must subscribe to long-term water conservation obligations. The entire deal must also pass environmental review, which is no small matter when its likeliest impact will be a dramatic reduction of fresh inflows to the already stressed Salton Sea. Moreover, the cost of mitigating such impacts must be tolerable—the agreement obliges IID to pay no more than $15 million for mitigation. On top of everything else, the transfer must win the blessing of the state of California and the secretary of interior.[36]

Even with so many difficulties, a prospective IID-SDCWA transfer of 200,000 acre-feet has become the centerpiece of California's "4.4 plan," and still the plan falls hundreds of thousands of acre-feet short reaching the 4.4 goal.[37] Where will the rest of the water come from? Some may come from interstate transfers or from "creative" accounting for water storage in Lake Mead. Ultimately, the squeeze on agriculture will tighten further, and representatives of the cities and of agriculture will tensely cut new deals of increasing magnitude, with billions of dollars and whole seas of Colorado River water at stake. Lies will be told and heroes honored as fortunes are made and destroyed. Winners will gloat, losers weep, and still others will find their interests served, but not too well. The environment will occasionally benefit, according to the priorities of the day and the amount of loose money the big deals leave behind for birds and fish. More often it will suffer.

The most certain outcome will be the lack of change in the most fundamental circumstance of the region: the unending competition for water, the dearest property in the West. The myths and stories of the future will revolve upon it. In the politics of every possible future, the water of the Colorado will flow toward the sea only for so long as the gravitational field of planet's mass can hold it in its power. As the water nears California, it will enter the still stronger field of human want and ready cash, at the command of which the mighty river will change course, as it has for many a year, and flow uphill to money.

PORTFOLIO II

Fish traps, in Cahuilla territory

On the Torres-Martinez reservation

Mexican geothermal plant in the Colorado River delta, where Volcano Lake used to be

New River, flowing north from Mexico toward the Salton Sea

Imperial Valley canal

Irrigation equipment

Imperial Dam and beginnings of the All-American Canal

Imperial Valley farmworkers

Calexico-Mexicali
border crossing

Johnny and Joe R. Hernandez

Snowbirds

SLAB CiTY.
WELCOME

BIBLE
HOLY GHOST AND
WITH FIRE.
ST
MATT
3-11
BIBLE

Leonard Knight

At home at the Slabs

SOLAR PANELS HERE
WARNING

MASTER

Chocolate Mountains Aerial Gunnery Range

The Navy's Salton Sea Test Base, abandoned during the 1970s, was used for military target practice in the 1980s

The Salton "Riviera"

Pool at the Salton Bay Yacht Club

View from the dining room of the defunct Salton Bay Yacht Club

Salton City golf course

Norm Niver at his dock in Riviera Keys

View from corner of Don Avenue and Barbara Drive, Salton Sea Beach

Fishing for talapia

12 | THE THEORY & PRACTICE OF BORDERS

The pedestrian gate leading into Mexicali from the United States is, literally, a revolving door. It consists of a tall metal spindle set in a fence of vertical pipes. Spokes cut from the same one-inch pipe form vanes that radiate from the spindle, and when you fit yourself between them and push lightly, the door deposits you in Mexico. Naturally, it spins only one way.

I follow Johnny Hernandez through the gate. A hinged steel rod repeatedly lifts and falls against the topmost spokes, clanging like the hammer of a drunken blacksmith. The rod's purpose is to jam the door if it rotates in reverse. Like much else along the border, this handy piece of backyard welding cost next to nothing, needs no maintenance, and makes a lot of noise. Also, like nearly every device intended to limit movement across the border, it exists fundamentally in vain.

I am crossing the border today to learn about the kind of society the diversion of one-fifth of the Colorado River has brought into being. The river has brought life not just to new landscapes but to new homes and communities, new sanctuaries and battlegrounds. It is at the border that life grows most complex. The border is a barrier across which two very different kinds of social energy arc and flash, and the lightning that results ignites all kinds of things.

I have to admit, the border makes me nervous. Its swirl of color, smell, and language; its separate and unclear rules; its cynical and sour authorities unsettle me. Then too, I'm tall and white, as conspicuous as a clown without a circus. I ask Johnny what he feels crossing the border, and he smiles, "Why should I feel anything? I look just like they do." He is right, of course, but there is more to it than that. Borders by their nature do injury, favoring some, rejecting others. Few other places play host to the arbitrariness of chance and power so often or so much. In few other places do so many dreams collide.

Johnny Hernandez is a son of the border, and he has agreed to guide me through its web. One needs a guide in such a place, amid its complexities and contradictions, and my suspicion is that a good guide would have to be nearly as complex, if not as contradictory, as the place. John L. Hernandez, I am finding, is a very good guide.

From a distance, it is easy to observe that the border is a land of its own, that it is neither Mexico nor the United States but truly a third world reaching some ill-defined distance north and south and obeying rules and customs of its own. Inhabiting that world, as Johnny does, and navigating its tiered communities and intertwined relationships of blood, money, and fate are no simple matters. "You must be bicultural to survive," he says with a hint of pride. "You must be able to live in English and in Spanish, to go back and forth and operate on either side."

Johnny lives where he grew up, a dozen miles north of the border in El Centro, a gritty town of modest bungalows much like a rust-belt mill town, only here the factories are the fields. Johnny is in his mid-forties, and flecks of gray soften the black thatch of hair that stands up off his scalp—trimmed, I can't help thinking, after the fashion of César Chávez. There is a softness to his eyes, a kind of pained receptivity not often seen in men his age, which may partly explain the restlessness behind those eyes and the earnest way he searches a listener's face as he speaks. He's lived in other places enough to know that life in the middle ground is bittersweet: "You can think you are assimilated, but eventually reality will strike and you realize you are not really accepted anywhere else."

Past the clanging gate, in the no-man's-land between U.S. and Mexican customs, we descend into an arcade beneath the port of entry's main highway. Shops line either side. There is a newsstand, a snack bar, and a *farmacía* dispensing drugs unavailable in the United States. Johnny moves through the crowd like a street-smart kid, scanning faces alertly. People used to call him Peanuts because he was so small—less than five feet and a hundred pounds when he graduated high school. He compensated with quickness and savvy. "I stayed a head below and a step ahead," he says. On his paper route in downtown El Centro, he used to slip into a pool hall on Main Street and wheedle adults into playing him for a dollar a game. Then he'd run the table.

Johnny's drill instructors in the Marine Corps forced him to eat double rations, and because of their bullying—or in spite of it—he shot up to five-six and filled out. Now, in the arcade beneath the port of entry, he still has the weight-forward edginess of the little guy among the trees. He sees a friend from El Centro. Then he recognizes a farmworker he once counseled—Johnny used to be a caseworker for the state's Employment Development Department. Twenty yards farther, he spots a neighbor of his mother-in-law's, who lives in Mexicali. He hails each, shakes hands, exchanges greetings, and moves on. "I am not always good with names," he says, "but I never forget a face." Others around us are also meeting, greeting, making connections. This tight passage between two nations, designed to serve two customs bureaus and the U.S. immigration police, doubles as a kind of village square. Sooner or later, everyone from Mexicali, Calexico, and points far beyond passes through.

On average, 38,000 people walk this passage every day, counting traffic in both directions. More than double that number pass back and forth in vehicles on the avenue overhead. The port of entry is the sole artery joining two communities that really are one: scruffy Calexico, with fewer than 20,000 people, and giant, burgeoning Mexicali, with 700,000 officially counted souls and probably a great many more.[1] From an airplane you look down and see the colossal sprawl of Mexicali, the neighborhoods, shantytowns, roads, and factories spreading out in seventeen shades of brown against the pale green of irrigated desert. The most extraordinary thing about the city is its shape: it forms a ragged half-oval jammed against the razored line of the border. South, east, and west the ovoid shape trails off with edges as organic and irregular as a culture of cells. But on the north lies only the dwarfish splotch of Calexico, which is joined to Mexicali by a pulse of movement stronger than anything either city contains alone.

The existence of this transnational community, with all its familial and economic connections, is one of the two great truths of the border. The other is that whereas most *norteamericanos* imagine the main currents of human history to run east to west across the continent, today the strongest current flows south to north. Irrespective of the walls Americans may build along the border, this powerful human tide will determine the future of these lands. The facts of Johnny's family and his life support both ideas, for his network of kin and friends spreads to both sides of the line, while its center moves ever northward.

WHEN GEORGE CHAFFEY and the California Development Company brought a brigade of canal-builders to the heart of the desert in the winter of 1900–1901, Mexicali sprang up across the invisible border as a second workers' camp, mirroring Calexico in every way, even including its spliced-together name. Although both towns nearly washed away in the Great Diversion, the desert toehold of town life held on, ultimately to be surrounded by nearly a million acres of cultivated land, north and south. With additional spurts of growth fueled by prohibition or, more recently, industrial opportunity, these borderlands have never ceased to be a magnet for emigrants from Mexico's interior.

Johnny's grandmother, Amparo, and his father, José, crossed the border at Nogales in a horse-drawn wagon in 1923. They had journeyed the length of Sonora and most of Sinaloa simply to stay alive. When Villistas attacked Culiacán during the Mexican Revolution, Amparo took refuge with friends in a house with thick adobe walls. She hid José in a chimney and took cover under mattresses as soldiers clashed in the street outside and bullets pocked the adobe. After the fighting, malaria nearly accomplished what rifle fire and cannonades had not. By chance, a doctor visited the house. He recognized her pallor, her fever and chills, and said that if she wished to live she must move elsewhere. Two brothers had already trekked north to pick crops in California, and she determined to follow them. At Nogales, after hundreds of miles, someone gave her a glass of water with ice—the first ice she had tasted in her life. As she drank the water, she felt the welcome, liquid coolness spread through her core, and for the rest of her long life she marked her recovery from that miraculous moment—"El agua me hizo cambiar."

Amparo died in 1991 at the age of ninety-nine. She was both mother and grandmother to Johnny, the sixth of José's eight children. When Johnny was five, he moved with his family, full of hope, into a new, small bungalow that his father, now known as Joe R., had built near the corner of First and Brighton in El Centro. Johnny's mother was then pregnant with her eighth baby, but she would not live to nurse him. Complications of delivery took her life.

Within days, Amparo, then sixty-three, moved from her own house, a block away, to raise her son's family. A tiny woman who did not break five feet, she handled the brood of eight almost singlehandedly, for Joe R. was absent more than he was home, planting, thinning, and harvesting crops in distant farming districts or working long days in the fields of Imperial Valley. Today, the diaspora of the family continues. While Johnny and two sisters live in El Centro, other siblings have moved to Indiana, Los Angeles, San Diego, and Riverside.

Johnny's wife, Maria, tells a similar story. She grew up in Mexicali, although for part of her childhood she lived on the U.S. side in Seeley, not far from El Centro, where her father, without benefit of documents, worked on a ranch. "When my grandmother died, we lost the house," she says, "and so we had to move back." She finished school in Mexicali, but she and all her brothers and sisters (save one who drowned in an irrigation ditch) have subsequently made their way to permanent residence on the U.S. side. Only her mother, Angelina, still lives in the old house in Mexicali. Johnny and I have crossed the border to visit her.

OUR TAXI DROPS us a few blocks away. The houses, packed one against another, match the color of their dusty yards. Graffiti is scrawled on nearly every wall in sight. Johnny explains that Angelina's street, just off Avenida de Baja California, runs through Mexicali's toughest neighborhood. From Angelina's house you could almost throw a rock into the United States.

We jump a waist-high masonry wall into Angelina's yard. She does not answer when we bang on the padlocked back door, but her neighbor Juanita appears and gives us the key. Inside, the house is dark, stifling. The linoleum on the floor is worn through to the slab. There is no food in the kitchen. Johnny tries the faucet and a little water dribbles out, then nothing. He opens the refrigerator, gasps at what he sees, and slams it shut. We go to a small side room where two tiny beds with white coverlets are crammed into a corner. Johnny switches on an air conditioner, set in a boarded window. Soon after it clatters to life, we hear Angelina's steps at the door.

She is a small woman, with thin gray hair pulled back in a bun. She walks in a shuffle, so downcast she seems to be shrinking. We go to the cramped bedroom, where we sit before the wheezing wall unit. She tells us that thieves have plagued her without a break. Days ago, she turned her back for only a moment, and one of them dashed in and stole her purse from the kitchen table. It held her money, plus important papers and identity cards for herself and others. Later they took her broom. They have taken nearly everything that can be carried off. When nothing else remained, they stole the circuit breakers from the electrical box on the outside of the house. After that, she had no power and no light; the refrigerator warmed and everything spoiled. She fans herself with a scrap of paper and speaks of the thieves in a flat, emotionless voice, as though they were as obligate as weather. The frayed edge of a tear in her blouse flutters in the wall unit's exhalation. A neighbor has helped her get replacement breakers, and, she explains, now her best hope is to fix up one of the rooms and rent it out. With someone around, she says, maybe things will be safer, but her voice has no conviction.

Soon we leave, and when have gone a block down Avenida Baja California, Johnny vents his anger. "I don't know what's the matter with Maria's brothers. They could fix that place up; they could help her." But the answer comes in the next sentence: "The trouble is Richie. She won't give up on him." Richie is Maria's youngest brother, and he is a junkie. As badly as Angelina needs help from her children, her children see no use in giving more than they already have. They figure everything passes straight through to Richie's arm.

Richie, who has been in and out of recovery homes and on and off both methadone and probation, cannot cross back into Mexico, but his presence remains strong. Johnny believes the thieves besetting Angelina are junkies Richie brought to the house. Johnny says, "We're gonna have to do something, but I don't know what." Then he muses, not for the first time, "One of these days if we start some kind of import-export business, we could use part of that house for a base, put some money into it, keep an eye on things. That could work pretty well, don't you think?" Head down, hands in pockets, Johnny walks in silence.

Johnny plans and replans, all the time. Otherwise he could not manage the moving parts of his life: husband and father; brother, son, and uncle in a large family; career bureaucrat; active member of his church, member of the school board, of the Sunrise Optimist Club, and of the Mexican-American Political Association. Johnny recently became a master Mason in the Imperial Valley lodge, a heavy commitment. And he's a budding businessman: much of his spare time and extra money go into a corner store he opened in his old home neighborhood in El Centro. Everyone's life is a balancing act, but Johnny is like an acrobat who walks a wire with chairs and tables on his head. What seems to keep his load tied together are the twin goals of personal gain and service, joined like border towns, in everything he does. The job, the store, and the clubs enable him to lend a hand to others. And through income or contacts, they also help him get ahead. It isn't easy, he'll say, as his restless eyes search for your assent or disbelief.

EVERY FAMILY has core stories. The northward trek of Amparo and her devotion in old age to motherless grandchildren, in Johnny's words, "tells me who I am." No less important is the story of his father's life, about a path not taken. Except for his devotion to Amparo, Joe R. might have run with Jesse Owens in the Nuremberg Olympics.

One afternoon in El Centro, Joe R. explains: "I broke the color line at the high school. Me and a couple of other Mexicans were the first ones to go in there. We didn't have too much trouble because the city boys knew us from the streets. Only the country boys gave trouble, but pretty soon they found out what they were in for." Joe R. also won respect because he could run as no one at the high school had run before. He clocked a mile in 4:24; the 1320, or three-quarter mile, in 3:15; and the half mile in 1:59. In those days, the world record for the half was 1:54, and he was a mere kid.

Johnny and I are sitting with Joe R. in the shade of a sagging tarp in the old man's yard. The day is warming past 100, but Joe R. wears a dark, snap-button shirt with the long sleeves down. Occasionally he removes his baseball cap and runs a hand through gray, crewcut hair. "In the spring of 1935, we ran against the University of California over at USC stadium in L.A. I was running the 1320 against their best man. In the backstretch their coach, Dean Caldwell, yelled to him, 'You got him, take off.' But it was me who took off, and I beat him." Dean Caldwell soon offered Joe R. a full scholarship at Berkeley—a life-changing opportunity for a Mexican kid from Imperial Valley. Tryouts for the 1936 Olympics loomed not far ahead.

But in El Centro soon afterwards, Joe R. went into the fields with Amparo and his stepfather Juan Gonzales, a big man known as Juanón. It was harvest time, just before the blistering peak of desert summer, and the fields were as hot as ovens. It was also the midst of the Great Depression, and Joe R.'s mother and stepfather had hired out to pick corn for twelve and a half cents an hour. Joe R. watched them go down the long rows between the stalks, knives in hand, dragging bags they slowly filled with ears. Finishing one pair of rows, they turned and labored up the next, at the end of which they stopped for water, big Juanón and tiny Amparo, already gasping with heat and exertion.

"I seen her face," Joe R. says. "I seen the way she looked when she drank. I couldn't stand to see it. I said, 'Gimme your sack.' She gave it. 'Gimme your knife.' She gave it. My stepfather said, 'You do your row, I'll do mine.' We go down, then we turn and come back. When we get to her, she wants her bag, but I won't give it. I keep on all day. That was Sunday. Next day, Monday, I was supposed to go to school. I won't. I'm goin' back to work in her place, I tell them. They sure get mad. The coach from the high school came over, but he can't talk me out of it. Even Dean Caldwell came down. But he couldn't talk me out of it. He understood. If I had taken that scholarship, my mother would never have lasted me as long as she did. I never went back even to graduate from high school."

BACK AMONG the wide avenues and arcaded storefronts of central Mexicali, Johnny takes me to one of Joe R.'s old haunts, a vast, windowless betting hall, powerfully air conditioned. Silent men nurse coffees and beers and stare at overhead displays of race cards for Aqueduct, Hialeah, and Bowie, Maryland, any track where the ponies are running. "My dad used to park us kids at a barbershop for a haircut and come here to bet on horses or dogs. Sometimes he'd settle in at a bar down the street and give a kid a few pesos to run back and forth and place his bets for him." Outside, we round a corner and face the faded sign of Joe R.'s former hangout, El Imperio. When his children weren't in tow, Joe R. would have a few drinks, then go upstairs in the adjacent hotel with one of the women who gave El Imperio its reason for being. The bar closed its doors years ago, but Joe R., now on the far side of eighty, still visits Mexicali regularly. He drinks at a different bar but calls at the same hotel in the same old way—"Still taking care of his needs," says Johnny proudly.

Prostitution and gambling were Mexicali's growth industries in the 1920s. Always a negative mirror for the straitlaced, Protestant north, the border country boomed during Prohibition when thirsty gringos flocked to unrepressed pleasures south of the line. Then was born the margarita, as well as many a new variation on the myth of romantic, carnal Mexico, land of smiling and compliant señoritas.

Northern events continued to shape southern realities in the thirties, when Mexican farmworkers who lost their livelihoods in California—and feared to lose their lives—retreated back to Mexicali. Their fears were justified, notably in Imperial Valley, where growers and police combined to snuff out unrest with a heavy and brutal hand, painting every strike as the work of communists and agitators. When the authorities dispersed a band of protesters, they were not slow to swing their truncheons. On January 12, 1933, for example, a force of local police and deputized growers surrounded a meeting of several hundred lettuce pickers and their families at Azteca Hall in Brawley. The authorities sealed the doors from the outside and lobbed tear gas into the hall. The people within, who included many women and children, scrambled through the poisonous air and frantically broke every window they could reach. They tumbled out through the broken glass, gasped for a moment in the fresh air, and ran into the darkness bearing no one knows how many injuries. Whereupon the authorities entered the hall and destroyed everything there—typewriters, field kitchens, pamphlets. There was no need in Imperial Valley to preserve even a patina of legality; in this police action, no warrants were served nor arrests made.

Days later, a lawyer with the American Civil Liberties Union, A. L. Wirin, secured an order from federal court in San Diego barring local police from interfering with a subsequent gathering scheduled for January 23. Even a federal court, however, could not chasten the muscle men of Imperial Valley. Minutes before the meeting was to have begun, vigilantes seized Wirin at his Imperial Valley hotel, dragged him into a car, beat and robbed him, and later dumped him in the desert, barefoot, eleven miles from the nearest town. In the months and years that followed, neither federal inquiries nor a special mediator sent out by President Roosevelt's secretary of labor could bring the growers to relax their iron grip or lessen their intransigence.[2]

THESE AND other matters, however, passed over Joe R. Hernandez like the circuits of unseen planets. Trim as a greyhound and with stamina to match, he kept his head down, literally and figuratively, as he bent over his short-handled hoe in the onion and melon and lettuce fields for a dollar a day. Three long decades passed before he escaped stoop labor and landed a job as a truck driver, delivering ammonium and other chemicals to tanks in the fields. By then it was the 1960s, and César Chávez and the United Farm Workers were on the march.

The most dramatic rally of the decade was a 1969 UFW march in brutal May heat from Coachella all the way to the border at Calexico. The purpose was to entreat Mexican workers to honor a strike against vineyards in the Coachella Valley. National media covered the nine-day trek, which helped assure appearances at the crescendo rally in Calexico by Ted Kennedy, Walter Mondale, Ralph Abernathy, and a busload of entertainment celebrities. The marchers first descended the east side of Salton Sea, camping twice on its shores. Then they trudged down past Niland and Brawley, through the heart of the Imperial Valley, with César Chávez and the other leaders of their column bearing the stars and stripes, the black eagle flag of the UFW, and a banner of the Virgin of Guadalupe. In El Centro, they mustered in the park across Brighton Street from the Hernandez home, and Father Victor Salandini, pastor of the Virgin of Guadalupe Church at the end of the block, said mass for them.

Not that day but on other occasions Johnny served Father Salandini as altar boy. Parts of some summers he had worked in the fields, thinning lettuce and other crops, but his education in the world of the Mexican farmworker did not really begin until he met Father Vic, the "Tortilla Priest."

Johnny remembers assisting Salandini at mass in a stifling, cramped labor camp at Danenberg and Fourth, on the east side of El Centro. While there, every breath he took conveyed a taste of squalor, fatigue, and resignation that until then he had not imagined might exist. "When I was little," he recalls, "whites and Mexicans up in the road in Brawley used the public pool on different days. First whites, then us, then they'd drain the pool, refill it, and start over with the whites. When you grow up in a system like that you just accept it as the way things are. I never had a Mexican or Anglo teacher in elementary school. All my teachers were blacks who'd come from the South because of segregation over there. What I grew up with was just the structure of things." But the labor camp at Danenberg and Fourth did not fit that structure. "That was something else. What those people were going through was beyond anything I had seen or known."

It was clear to Johnny, and to much of the rest of the world, that Father Vic was a man of conscience. No less a person than César Chávez called him the chaplain of the farmworkers' movement. Salandini said mass at migrants' camps and for strikers in the fields; he stood duty on picket lines and endured arrest, side by side with Chávez, in a Coachella vineyard. The bishop in San Diego took a dim view of reports that Salandini wore the black eagle of the UFW on his vestments and that he consecrated corn tortillas and served them, torn in pieces, as the Eucharist. When the bishop demanded explanation, Salandini protested that he'd simply run out of the customary wafers. Not for a moment did anyone believe him, least of all the farmworkers who embraced him as *their* priest.[3]

SOMEWHERE BETWEEN one-third and one-half of all winter vegetables consumed in the United States are grown in the Imperial Valley, and the value of the valley's annual production exceeds $1 billion. Mechanized harvesters gather in the cotton, and powerful loaders stack alfalfa hay in long boxy ridges beside the fields, but most of the vegetables that come out of the fields are picked by day laborers from both sides of the line. Men and women, they filter down the crop rows, picking and lifting, hauling boxes and buckets. Having no special clothes reserved for work, they dress in the bright and various garments they wear at home or on the street, with the addition of hats and bandannas to guard against the sun. You do not often see them pausing, leaning on tools, or gathering round the water tank. In the fields, they work.

Johnny's career calculations, and the influence of men like Salandini and his father, led him to take a job as an outreach worker with the California Employment Development Department, or EDD. His job allowed him to roam the vast fields of the Imperial Valley in a state-owned Jeep Cherokee, meeting laborers, hearing their stories, and advising them of their rights under California labor law. Several years later, he transferred to an indoor job as a complaints specialist in an air-conditioned office in Calexico—no small advantage when summer heat hits 115 degrees—where workers brought their problems to him: "It might be sexual harassment, lack of medical attention for work injuries, all sorts of things, but in the farming community it most often involved nonpayment of wages." He later moved on to yet another post, collecting labor and crop data.

Johnny's language, when he speaks of his job, is measured and detached, but the crow's-feet at the corners of his eyes deepen as his story advances. He leans toward his listener: "Basically, in an area like this you have a large surplus of labor and people are hungry and willing to work for the lowest wage. The workers know that they are dispensable and that it will be hard for them to find something else if they make a protest and get fired. So the problems are probably worse than what we see."

Johnny is not given to long explanations. He'll observe that sweeping reforms of United States immigration law in 1986 greatly loosened restrictions on the entry of agricultural laborers into the U.S., but he doesn't detail the mechanics of how an open market in temporary work permits caused the labor pool to swell, undercut the United Farm Workers union by diluting its presence in the work force, and ultimately drove down wages and erased benefits. Such dynamics are the cold business of supply and demand. Johnny would rather you hear the story from the people who live it.

With that in mind, he takes me to the Jack-in-the-Box in Calexico, hard by the port of entry. Labor contractors recruit crews in the alleys nearby, making the all-night fast-food joint a convenient meeting place. As we bring our coffees through the checkout line, Johnny hails an acquaintance, and we take a table together. Antonio Reséndiz is a compact, wiry man with much on his mind. He tells us in Spanish that he's joined the local union board, which risks getting him blackballed by labor contractors, and he also serves as treasurer of a committee attempting to bring power, water, and schools to his *colonia* on the south side of Mexicali. Last night Reséndiz returned home from two days of topping onions in the mountains west of the valley. The job involved pulling onions from a furrow that a tractor had loosened, scissoring off the roots and top, and dropping the onions in a bucket. Three buckets filled a seventy-pound sack, and each sack earned him sixty cents. Reséndiz pulls a yellow slip of paper from his wallet. It is dated yesterday and bears the name "Castillo Farm Labor" over an El Centro address. On it are rows of numbers up to 250. A hole is punched through number 57, marking the number of onion sacks Reséndiz filled. The space for gross pay shows $34.20. Deductions for worker's compensation and social security dropped the net to $31.12. Reséndiz allows that benefits were decidedly minimal. Both nights he was up there, he had to sleep on the ground in the clothes he was wearing, which is why he came home.

A friend of Reséndiz's has joined our table. He, too, has returned from sleeping on the ground in the mountains. He pulls out another yellow slip, which shows he filled only twenty sacks, earning $12.00 for eight hours of labor. It was a bad field, he says, and there were problems with the tractor and the buckets. Johnny reminds him that the labor contractor is obliged to pay piece rate or minimum wage, whichever is more. He then explains how to file a complaint. The minimum wage of $4.25 an hour would have netted him within cents of what Reséndiz got.

The cheated man mumbles acknowledgment, but it is clear he'll not pursue the matter. Then he says, "Hay otro problema muy fuerte." He describes the condition of privies provided for workers in the fields. The labor contractor hauls the privies from job to job in trucks that bounce down rutted roads. Frequently, the contents of the privies splash out and spatter the interiors, rendering them too foul for use. "Then all we have are the irrigation ditches," he says. "It shames the crew and is bad for the water, but what can we do?"

In fact, they can go to Johnny, and Johnny can alert the Department of Industrial Relations or the labor commissioner for Imperial Valley, who then can reprimand the labor contractor and eventually, perhaps, obtain compliance with laws and regulations protecting workers. But there is only one labor commissioner for all of Imperial Valley, and remedies, where they exist, proceed at a hobbled, bureaucratic pace. "It's satisfying when you can help somebody," says Johnny, "but just as often it's frustrating that you can't do more."

Later, at the UFW union hall in Calexico, Johnny and I sit on folding chairs in the assembly room under a bright mural depicting César Chávez and a local worker, Rufino Contreras, who was shot and killed near here during a strike in 1979. (No one was ever charged with Contreras's murder.) We join a conversation with several men, one of whom explains that he has worked for Azteca, a major grower, for eighteen years and earns $4.75 an hour, two dollars less per hour than he did five years ago. He says his benefits, which used to include health insurance, are now nil. Over the years, whenever he complained about the cutbacks, his foreman would remind him, "Allí está la línea"—there's the line, meaning the white stripe down the center of the highway.

Currently the man is harvesting lettuce for Azteca in Blythe, California, several hours to the northeast. He commutes daily from Mexicali, crossing the border at three A.M. and returning at eight in the evening. It is a schedule he can endure for only a few days before exhaustion lays him flat, but he plans to keep on with it "toda la corrida de la lechuga"—the entire run of the lettuce—before moving on to harvests in Arizona and New Mexico. He pays $7.00 for transportation each day, so that after deductions, he nets about $28.00 for seventeen hours of effort. He grieves that his wife and youngest child cannot accompany him into the United States to ease his loneliness and, not least, prepare meals and keep a semblance of a home. "Es muy dura la vida." Life is very hard, he says with no particular emphasis.

SOUTH OF THE BORDER, Mexican farmers cultivate an area equal to that of the Imperial Valley but make do with only one-third to one-half as much water. Yields are correspondingly low. Before the Mexican economic crisis that began in December 1994, an agricultural worker in the Mexicali Valley earned the equivalent of $50 to $60 a week for workdays running from sunup to sunset. Uneducated campesinos could not then and cannot now hope for industrial jobs in the *maquiladora* plants on Mexicali's outskirts; even so, such jobs paid only about $100 a week. With the economic crisis, the buying power of those wages plummeted, then slowly and painfully crept upward again. By 1998, entry level *maquila* wages were still only about half their previous U.S. equivalent, although competition for capable workers guaranteed many employees a rapid advance to higher pay.[4]

The crisis checked but by no means ended the boom along the border. A drive around Mexicali on new highways, their curbs and shoulders perpetually unfinished, will remind you that Mexicali, Calexico, and the entire border zone are gradually declaring their independence from agriculture, just as water brokers like U.S. Filter, on the other side of the line, are also beginning to do.[5] Mexicali's new highways are lined by equally new Dodge and Ford dealerships, a Price Club, the shell of a vast unfinished Wal-Mart, and industrial installations with signs that proclaim Megaplast, IVEMSA, Amerpack, Accuride, Baxter, Purina, Kenmex, and Rockwell. Eighteen-wheeler trucks belonging to the J. B. Hunt company crowd the highway, belching diesel, and Hunt trailers stand parked at every shipping dock. None of the factories, processing plants, or transshipment warehouses, however, can compare with the vast new Coca-Cola bottling plant, largest in Mexico, a fitting symbol of the decidedly international economy that Mexicali's frenzied growth is indentured to serve. "We can be another Hong Kong," say the boosters. "We are just getting started."

One of those boosters is Johnny's close friend and co-worker at EDD, Manuel Abundis, whose optimism and easy confidence lighten Johnny's darker realism. Manuel is rotund and indefatigable, as effusive as Johnny is reticent. In slimmer days, he played defensive back for UCLA and served in Vietnam as a navy SEAL along the Cambodian and Laotian borders. Now, as a Mason, he gives much of his energy to the Shriners' crippled children program. We tracked him down one morning in an orthopedic clinic in Calexico. He was unsnarling paperwork so that one of "his kids," a girl from Sinaloa with a facial bone tumor, might cross the border to see a surgeon. The arrangements made, he took us on a tour of Mexicali.

Free to live on either side, Manuel chooses Mexico and commutes daily across the line. He says his dollars go further in Mexicali and that people there are warmer and more helpful. He asserts that Mexicali's schools, with their emphasis on discipline and order, are superior to schools in Imperial Valley. Johnny disagrees but doesn't press the point, preferring to sit back, a half smile on his lips, as Manuel rambles on, leaping from subject to subject, as we cruise the broad new boulevards in his Oldsmobile, air conditioner blasting. Manuel gestures proudly toward each new industrial plant and sings the praises of economic development. "These factories are how we transition from the situation of the campesino," he asserts. "Maybe the campesino of today won't benefit, but his children will." Like many in Mexicali, Manuel dreams no small dreams. He envisions barge canals to the Sea of Cortez, a deepwater port, and a north-south trade network from Canada, down through the western United States to Mexicali, thence along Mexico's Pacific coast to the rest of Latin America.

But we pass more than factories. Crowded in the no-man's-land of canals and railroad rights-of-way stand clusters of plywood shacks, their walls aslant and roofs sagging under flapping tin and rolled asphalt. Windblown plastic bags and shards of styrofoam bedeck the dirt of every road margin and empty lot. The smell of burning trash and manure sours the air. Several times in our tour we cross the squalid gulch of the New River, which flows thick as paint and reflects the sky with a sickly bread-mold green.

The New River, which collects the effluent of the *colonias*, the discharge of Mexicali's chronically malfunctioning sewage plant, the irrigation runoff of much of the Mexicali Valley, and the raw, bandit wastes of scores of industrial installations and deposits them all in the placid Salton Sea, can be a sensitive subject for Mexicali's defenders. Not for Manuel. He says that the worst is over, that Mexicali's industries are cleaning up and the authorities have shut down the main polluters. Next question. The city's advancement interests him; its failures do not.

Johnny, more circumspect, expresses the frustration of most Imperial Valley residents: "Politicians have been talking about cleaning up that river for forty years. It's time they just got on with it." I tell him about a conversation I had with a Border Patrol agent who says he regularly picks up illegal aliens who swim the New. They float under the ragged border fence hiding in what they call the river's "icebergs"— the heads of foul-smelling foam, some a yard high, that spin down the surface of the river. When we spoke, the agent had two such swimmers locked inside his vehicle. He would soon transport them, still wet, back to Mexicali. By contrast, the rare agent who has the misfortune to fall in the river is rushed to a hospital for decontamination and given a "pincushion" of shots.

Now Johnny grows impatient. He warns against making too much of the agent's story: "First of all, the only illegal alien I know is ET, and he went home. Second, the majority of the people who come across the border illegally and undocumented are coming across right under the nose of the Border Patrol at the main routes of travel. We can go downtown right now and find somebody to sell you an immigration card or a social security card or a passport."

But Johnny overstates his case. Undocumented immigrants do in fact take extraordinary risks to enter the U.S. They swim the New and occasionally drown in it. They swim the All-American and other canals and drown at the rate of two or three a month. They stagger across the Yuha desert, west of Mexicali, in the heat of summer and perish of thirst and heatstroke. They careen down dark roads in airless vans and freight trailers and contribute more than their share to California's highway mayhem. They wait at isolated rendezvous in the desert badlands for guides—coyotes—who never come, and when the last of their water is gone and hyperthermia begins to claim them, they pray for capture.[6]

Still they keep coming. In the early 1990s, the Border Patrol arrested about 40,000 illegals a year in its El Centro sector, most of them, as Johnny says, at or near the main routes. The numbers, however, rose discouragingly as the decade advanced. In fiscal year 1996, Border Patrol apprehensions climbed to 66,860; in 1997, they jumped to 146,210, due in large part to tightened surveillance in the San Diego sector, which deflected the stream of illegals eastward to Mexicali and the deserts around it. In 1998, apprehensions in the El Centro sector topped 200,000 even before the fiscal year had closed, and the human flood showed no sign of abating.[7] Along with people, the Border Patrol also seizes drugs, principally cocaine, and throughout the 1990s the street value of impounded dope averaged well over half a billion dollars a year—not a bad haul considering that, according to indictments handed down in February 1995, a fair number of border inspectors were in cahoots with smugglers. When, on one rare occasion, a day passed without a seizure of drugs, the jaded Imperial Valley Press ran the headline, "No drug busts . . . It's news."[8]

Efforts to stop the traffic in drugs and to control immigration, although mighty, are like trying to push back the tide. Estimates of undocumented immigrants residing illegally in California range between 1.4 and 2 million, and resentment against them is high, as evidenced by the approval in November 1994 of Proposition 187, a referendum limiting the services the state provides them. Johnny is hardly alone in believing that the state has a moral obligation to serve these people as generously as it serves its rightful citizens. But neither is he in the majority. Widely published estimates that undocumented immigrants cost California $3 billion to $5 billion a year in schools, health care, and other services helped persuade 63 percent of voters in Imperial County and 59 percent statewide to approve the referendum.[9]

Johnny concedes that those who would seal off the border and withdraw services from immigrants "call the shots for now." Such people "want to take us back to the way Germany and Berlin used to be." But in the long run, with the electoral power of Mexican Americans steadily increasing and with bittersweet Proposition 187 causing more *mexicanos* to apply for citizenship, he is confident in the eventual victory of numbers. "Give it ten or fifteen or I don't know how many years, and the people who don't want to change and don't want to give in are going to come along or they are going to get left behind. The tide and the flow of this is too strong. There's no stopping it by a fence or a brick wall."

JOHNNY'S PRINCIPAL investment in the future is his *tiendita* in El Centro. He fitted it out mostly with equipment he scrounged at the vast swap meets in Calexico, which beggar conventional flea markets and testify to a staggering transborder market in used goods. The rest of the gear he snagged at yard sales or salvaged from other people's trash—wire racks to hold chips and *chicharrones;* shelves for bricks of lard and family-size flagons of mouthwash; another rack for herbal cures in plastic bags and seasonings like sage and cinnamon; counters and cases that display Mexican candies and quarter-dollar necklaces. A friend gave him a cooler for sodas; the 7-Up company loaned a refrigerator.

The corner store occupies a place at the center of Johnny's life. It stands a block and a half from the house where he grew up, and a block from the church where he and Maria worship and where Salandini once presided. Johnny bought the store and its two twenty-five-foot lots for $18,000, and the note he pays comes to $120 a month. The previous owner mainly sold beer. When Johnny took over, there were holes in the outside walls and holes in the roof; the wiring was shot, plumbing nonexistent. "Used to be, this was the worst drug corner in town—guys hanging out, dealing, drinking beer, and urinating on the walls." With $10,000 and the help of family and friends, he worked nights and weekends to patch, rebuild, and rerig. Nowadays, he opens the store, off and on, with a friend tending it while he's at work, selling sodas, snacks, and sweets.

In the store, Johnny beams. Every feature of the place conceals a story about a trade he made or a gadget he contrived to allow something else to work. This is home ground, his place, where he can shuck off his border toughness. For years he'd heard the neighborhood calling. This is his answer. At EDD, there are always bosses with different ideas, snarls of red tape, the sense of rolling a rock up a hill. But in the store, all possibilities belong to him, and they are nearly unlimited: "Maybe when the kitchen's ready, I'll sell like a deli: cold cuts, Mexican sandwiches, shrimp cocktails, and burritos." He plans and replans. On cool winter nights he might grill *carne asada* outside on the sidewalk and put tables around in back. Or maybe he'll buy the next lot over, then the one past that, then another if he can. With income from the store and a few rentals, he might retire from the state, dabble in politics, who knows.

Maybe it is only a dream, but the dream has heart: to build a real business on a rough corner in a tough neighborhood, and to make that corner an anchor for renewal. He won't sell alcohol or cigarettes, and people ask him how he hopes to succeed. Doubters don't bother him: "I can take food stamps, and gradually carry more groceries. There are no other stores within blocks and blocks of here."

Johnny's gamble is that he can operate a store where no one else would dare to open doors. He knows the residents of every house on the street, has known them all his life. He knows the gang kids wearing beepers. He tells them, "You mess me up and you are hurting your folks. They won't have any place to come for water or ice or a soda." Johnny's father, who lives a block away, tracks the neighborhood scuttlebutt every day. If anyone messes with the store, Johnny counts on knowing who it was.

When it came time to paint *"Johnny's" MKT.* and *"Mi Tiendita"* in big red letters above the door, Johnny paid a tough ex-con, fresh from state prison to do it. He was buying insurance as well as a paint job. "You think somebody like that's gonna let anybody mess with what he did?"

The idea behind the store is to have a place to be, a place to stand. The cash register is important, but so is the sense of standing behind it, of having a bit of a pulpit, a special authority. In a tiny office in the back of the store hangs a framed certificate from the El Centro Citizens Club that accompanied the $150 scholarship that sent him to college. It reminds Johnny of the breaks he's had. He likes to ask kids who come into the store to show him their schoolwork. When he spots an "A," he offers to buy it with a candy or a trinket. Some of those "A" papers get pinned on the front wall opposite the counters. Next to them is a poster, vintage 1960s, depicting a doctor, a lawyer, a pilot, and a scientist. At the bottom of the poster is the legend, "Sí se puede." Yes, you can.

ON A FRIDAY afternoon I am standing near the front of the store, waiting for Johnny to close up, when in marches a boy about five years old. His face bears deep scars from a windshield, a pit bull, or some other enemy of flesh. He is followed by two scruffy hench-tykes, one of them a girl. Faces unwashed since yesterday's popsicle, T-shirts smudged in multiple shades, and, except for the girl, rat's-nest hair that may never have seen a comb. The scarred leader steps to the candy counter, which he cannot see over. He hammers the service bell on the countertop like an irate mogul summoning a bellhop. "Johnny! Johnny, come on!" He's got three coins in his greasy grip and pauses from ringing to confer with his partners. They stand, noses pressed to the candy display, urgently calculating all that seventy-five cents might buy. I can hear Johnny moving boxes in the back. He does not hurry.

The boy, brash, unshrinking, resumes banging on the service bell. "Johnny! Johnny, get out here." Ringringringringringring.

Finally, I hear Johnny coming forward. Moments ago he said to me, "I've got my kids, my family, this store, the community; for me this is what is important." Now he appears, his face a scowling mask, and he barks at the children in mock fury, "Who are you and whadya want?" They jump back. Scar-face, uncertain now, sticks out a combative lip. Johnny breaks into a grin and leans toward him. In a stage whisper, as though they were old drinking buddies, he says to him, "I bet that pretty girl's your sister!"

The boy screams with delight, leaps again to the counter, and mashes the bell for all he's worth. "Johnny!" he shouts. "Johnny!"

Side Trip: Calipatria

I am at the foot of the World's Tallest Flagpole, looking straight up. I can't tell if a flag is up there or not. The top of the pole, invisible against a sky as white as a boiled egg, is 184 feet overhead, achieving an elevation—which is obviously the point—of precisely zero. It is the only thing that reaches sea level in Calipatria, California, a place that likes to call itself the "Lowest Down City in the Western Hemisphere."

Anyone who stands beside this gallant mast, full thirty fathoms lower than the whitecaps of all the seven seas, will want to know what keeps those whitecaps away and prevents their brine from pickling both the pole and its observers. The answer is, geologically speaking, *not much*.

If the nearby San Andreas fault suddenly convulsed, together with the Imperial fault, to which it here cedes primacy, and if a trench opened southward through the low deltaic plug of the Colorado River, then indeed the Gulf of California would rush into this desert sink to reclaim the territory it long ago occupied. After the waves subsided and the silt settled, we could stand here in our aqualungs, you and I, and look up past the bellies of mullet and sharks to see, far above, the green glow of sunlight playing on the wind-riffled surface. The only solid thing breaking the waters of the new sea would be the putative but as yet unconfirmed brass eagle on the top of Calipatria's mighty flagpole.

For now, the thought of region-wide inundation is mildly refreshing. Here in the overcooked heart of the Imperial Valley, Coke, Pepsi, and Gatorade are winter beverages; in July at the foot of the World's Tallest Flagpole you get a thirst only cataclysm can quench.

The great pole is not the only source of municipal pride in Calipat (as the town is known to its friends). Calipat is also home to the world's most lethal high-voltage fence, which encircles the new maximum-security state prison on the outskirts of town. The fence saves the cost of some thirty tower guards and sharpshooters and represents the latest advance in penal automation. Sandwiched between two companion fences that prevent "casual" contact, the thirteen-foot-high barrier packs 5,000 volts into each of its fifteen strands. That's enough juice to fry instantly anything that touches it: coyotes, jackrabbits, scissor-tailed flycatchers, or inmates. As long as Southern California Edison keeps the 500-amp circuits humming, the Calipat prison will remain the most cost-effective felon containment vessel yet devised.[1]

Only one major hitch emerged in the fence's first months: it toasted too much wildlife, especially birds. Animal protection groups protested that the slaughter was inhumane, and their pressure caused the Department of Corrections to install warning wires for rodents, anti-perching devices to keep gulls and finches aloft, and a network of tunnels to accommodate burrowing owls (but not burrowing convicts). Today, these adaptations make the death fence eco-friendly, but its humaneness remains a subject of debate. Four thousand inmates, including 10 percent of the convicted murderers in California, live warehoused within its arc. Two by two, they dwell in cells six feet wide and ten feet long. That's about the size of a parking space, which of course is what each cell is, except that the occupants are double parked, a third of them for life.

In 1992, when the first inmates began arriving and the Department of Corrections filled a thousand new jobs to staff the prison, Calipat found itself the fastest growing community in California. Its population jumped from less than three thousand to nearly seven, producing the first certifiable economic boom the town has experienced since its founding ninety years ago. The boom may yet have further to go: a high-security women's facility has been discussed.

Compared with the prison, the flagpole is a quiet offering and represents far calmer times. In 1958, storekeepers Harry and Helen Momita, with supplementary donations from many others, gave it to the town. In a little ceremony they dedicated it to "Good Neighborliness." It was the Momitas' way of saying thanks for the opportunities that Calipat and the United States had given them.

Still, the flagpole has its problems. According to my friend Larry, who is actually a friend of a friend and at whose house I stayed for a while, the column of welded aluminum is, more or less, a homemade job. No one, let alone an engineer, has calculated what stresses it might withstand. Larry advises that I not go near the park if the wind blows hard.

Larry lives safely several blocks away in a house with a pool green with slime, an unpredictable doberman, and a cat that takes too many privileges with the room I'm in. It is very much a bachelor environment. In the time I knew him, Larry looked after the real-estate interests of the geothermal plant a few miles away at the edge of the Salton Sea.

One night I went to a party at the home of one of Larry's co-workers, Danny. I got there at sundown, late, and missed the early evening entertainment, which had involved driving out into the desert in a small convoy of four-wheelers to drink beer and shoot guns. Danny, I learned, has a small collection of hinged and welded steel targets that sing when you hit them and rock back and forth. He likes to set them up on one gravel knoll and have everybody stand on another. Then the shooters get going, everyone blasting and plinking in turn.

Later at the house, the self-appointed armorers and weapons specialists of the group speak glowingly of Danny's beautiful wife, Reba, whose voice is like honey. The shots she made, long range, with her .22 caliber Mountain Eagle were simply astonishing. Reba basks in the praise, crosses her long legs, and modestly says nothing.

I've been to the spot out on the knoll where they blazed away. The empty cartridges make it easy to find. Late in the day, when the wind is not too hot, it is a good place to be. You can watch the shimmer of the dying sun's reflection on the Salton Sea, and sometimes you can halfway hear or mostly feel the perfect stillness of the desert. Other times you hear the heavy *chuk chuk* of fifty calibers over at the SEAL base just across the Coachella Canal, or you watch the fighter-bombers out of Yuma streak toward the gunnery range. As night falls, the flash of their rockets and tracers against the Chocolates is as bright as fireworks on the Fourth of July.

You can't find food better than Danny and Reba's. Their barbecue was the best I ever ate, and their beans, corn on the cob, and fresh green salad were superb. As a stranger, I felt the grace of their open-hearted hospitality. Their house, set back in a thicket of mesquite, had a broad outdoor deck where people lounged and chatted, with a fire pit off to one side. The temperature stayed in the nineties, but with the air so dry and the beer cold, it was a welcome heat. Mostly people talked about their guns, which they passed around for comparison and admiration, and about how good it was to be living so far from L.A. and all its crime and confinement. The harshness of the desert, they said, was nothing compared with the danger and harshness of the city, and it was clear that these people had been drawn to guns not just because of a fondness for target shooting. When I left, a half moon had risen, shining strong enough to cast shadows. Back at Larry's, I slept like a stone.

But now the mercy of the night is gone. Not a breath of air stirs, and the World's Tallest Flagpole is silent. No ropes slap against the aluminum; no clips or buckles rattle. Old Glory is limp, unflapping. The sun pours down in a rain of heat, a hail of glare, a solar wind. Every curtain in Calipat is drawn against the dazzle. Every shutter is closed. The charged air feels stiff as gum, plastic and gluey. I lean against it, expecting its support, as I retreat from the flagpole, then move back still more, to improve my angle so that I can see the top—and its brass eagle, which I pray is there. Vertigo pulls at me; heat presses down. Mentally I want to fuse myself to the image of the jubilantly erect flagpole—else, I may topple. I want to know what is up there at sea level, what kind of beacon the pole raises atop this low-down world.

13 | HOME BY THE RANGE

Could this be a dream?

A man, neither young nor old, looks to the sky above his native Vermont and sees a great and marvelous shape floating softly above him. It is as large as a whale but lighter than air, as colorful as a circus, as quiet as prayer. It is a great silent inverted teardrop with a basket hanging underneath: a hot air balloon, the very contraption that carried the Wizard of Oz back to Kansas. The man can see one or two people leaning from the basket, as the wizard once leaned, waving to the rooftops, and the people below, waving to the wizard. Emblazoned in big bright letters on the enormous bulb of the balloon is the name *Budweiser*—or some other name of some other product, the identity of which is immaterial.

What matters is the thought that name provokes: what if the message on the balloon were one that mattered? What if it said something valuable to people, and said it clearly and simply? What better way might there be to say it than to float softly overhead, with the message in bright letters on a beautiful vessel silhouetted against a blue sky? What better way might there be to sow happiness and to reassure the world?

This was no dream. Tall, handsome in a rail-thin, wild-eyed and beakish way, Leonard Knight looked up into the Vermont sky many years ago and saw the advertising balloon, more or less as I have described it, except for the business about the Wizard of Oz. The sight of it both charmed and galvanized him, changing his life. I threw in mention of the wizard because, for me, Leonard is as improbable as that strange coot. You will remember that the wizard, in the end, proved to be a charlatan, and I mean none of that where Leonard is concerned. But you will also remember that the wizard was capable of such genuine and strong belief that he could make the impossible seem real and dispense courage, heart, and brains as though they were dispensable. If only briefly, he could awaken similar belief in others, even the cynical. Right there, in the simplicity of the wizard's convictions, you have Leonard in a nutshell.

Leonard's path across the continent was like the path of a balloon. Events blew him here, then there. He was not the kind of man to fight a headwind. In the early 1980s he fetched up in Nebraska, where he earned a living driving grain trucks by day. By night, he sewed. You will gain some sense of Leonard's capacity for patience and persistence by knowing that he sewed on the same project for six years. The object of his piecing and stitching, as you will have guessed, was a hot air balloon. But it was not just any hot air balloon. Most flying balloons are 60 to 90 feet high. Leonard sewed a balloon that stretched 230 feet from base to crown. The immensity of the thing, as Leonard saw it, derived from its function. On the side of the balloon he stitched a message in letters 10 feet high: *God Is Love.*

In such a craft with such a message this wizard of evangelism proposed to float above the people of America. The Nebraska heartland, unfortunately, posed an obstacle: it was too windy, and Leonard could not get launched.

So he loaded his balloon onto his truck, a small flatbed with a dual axle. In this vehicle, which he had decorated with hand-painted Biblical messages, clouds, flowers, trees, and fanciful dots and dashes of color, he drove halfway across the continent to a little-known spot in the California desert near the Salton Sea.

In the early haste of World War II, the Department of War ordained that Camp Dunlop be built east of Niland on a few hundred acres of oven-hot desert, the better to prepare Marines for the environmental pleasures of battling Nazis in the Sahara. Engineers poured a score or two of concrete slabs and quickly erected Quonset huts. Marines came, trained, and went away to war, and after the peace was won, the engineers returned and removed the huts. The slabs remained behind.

It was not long before people began to pass the word that the "Slabs" or "Slab City," as its growing population came to call it, was a good place to camp. You couldn't ask for a cleaner or more level place to park your truck or camper. Or you could share a slab with others as a miniature village square or even a dance floor. From the time of John C. Van Dyke to the present, the wastelands of the Salton basin have beckoned to those in search of warmth, solitude, and freedom from interference. They were few at first, but their numbers grew as the gear supporting comfortable nomadism became more available. They came mainly in winter, escaping the snow and sleet of more northerly places, and they collected around the edges of the basin at the hot springs and oases that had formed where geologic faults permitted the liquids of Hades to leak upward to the surface.[1] Sometimes their seasonal squatter villages became a nuisance or simply attracted too much attention, in which case the authorities ran them off. But they were like crows in a farmer's field. Shooed from one place, they lighted in another. But no one ever shooed them successfully from the Slabs.

When Leonard Knight brought his balloon there, the Slabs provided haven for several thousand winter refugees—snowbirds, as they are called. No one counts them; estimates of their number range as high as seven thousand or more. They park their trucks, recreational vehicles, and travel trailers not only at the actual slabs of Camp Dunlop but anywhere nearby where they can find a nook between dunes, the shade of a mesquite, or the merest space beside a set of ruts to pull off and kill the engine. With thousands of snowbirds jamming the Slabs from November to March, the grail of solitude seems thoroughly forgotten, but warmth and fundamental freedom are much on everyone's mind: the sun shines every day, and no one is in charge.

To this unruled but not unruly place, Leonard Knight brought himself, his gaily painted *God Is Love* Bible truck, and his *God Is Love* balloon. Days passed, and people from the Slabs helped him finish the balloon and prepare to launch it. They helped unroll the giant thing in expectation of lift-off. It was probably too big and unwieldy ever to have been inflated, but that is not what finally dashed Leonard's plans. He looked it over carefully and was horrified by what he saw. Deep within the fabric's labyrinth of folds, moisture and heat had been at work; the sun's rays on exposed sections of synthetic nylon did their part too: the balloon had rotted.

You might ascribe the disaster to galloping incompetence, and be right. Surely Leonard should have taken greater care in storing the balloon. You might also ascribe the disaster to divine intervention, as many do. If Leonard had ever gotten airborne, he almost certainly would have killed himself. And God, or so it seems, had other plans for Leonard.

Leonard was despondent at the demise of his balloon, but not for long. The edge of a little mesa near the entrance to the Slabs caught his eye. He conceived the idea of displaying his *God Is Love* message in the face of the mesa, which was composed of a semi-hard caliche. He could shape the edge of the mesa by digging. He could raise up letters by sculpting with adobe. He could give it contrast, color, and ornament, and seal it from wind and rain, by lathering it with paint. And so he set to work on what he began to call the Bible Mountain Sculpt.

Leonard's efforts did not proceed without setback. His first three years of scraping, sculpting, and lathering came to naught when the unanchored adobe and housepaint skin he had pasted on the face of the mesa slid loose and collapsed in a gummy heap. But a man who sews a balloon for six years does not give up easily. He revised his plans and resumed sculpting the mesa almost immediately. "When I get excited about God," he says, "I get a little crazy, but I can make 101 mistakes and correct them 101 times and get on to the next and keep on working."

This time he cut back the mesa more steeply, using a small front-end loader that had been given him—and which, of course, he soon adorned with dots and splashes of color, trees and birds, and Biblical messages. He quarried scrap from a broken-down truck and used the metal as foundation pins to anchor a new facade into the hill. And then he started shaping his adobe and mixing his paints.

By his accounting, he lacked for nothing. He lived in the funky camper on the back of his old flatbed, now defunct, which he had parked under a mesquite in front of the mesa. He paid no rent, no taxes, no fees. People brought him paint in half-used cans: gallons, quarts, and pints by the score. One fellow brought him a couple of fifty-gallon drums full of half-pint cans of miscellaneous paint. It took Leonard most of a day to open all those little cans, the contents of which he poured together and mixed in one goopy slosh. People gave him money, too, enough to keep him in food (though he often ate for free at the Hi-Line Cafe in Niland), and if he had cash left over, he would buy a can of paint. One person gave him a moped, which he rode several times a day into town for coffee and, not least, an air-conditioned break from the heat. Someone else troubled to position a old bathtub in the trickle of a nearby hot spring, and so he bathed there as often as he pleased. Overall, needing little, he had all he needed, and so by day or, when the brutality of summer heat was greatest, by night, he built his mountain.

First he brushed the native caliche with water and built up a skin of adobe, mixing in loamier soil, but not too much sand, which he suspected had partly caused the first collapse. He added shredded palm fronds or grass to give it fiber, and when the adobe dried, he painted directly on it. *God Is Love* became, of course, the main message at top-center of his creation. Below it he fashioned a big heart in bas relief enclosing these words in raised white letters: JESUS I'M A SINNER PLEASE COME UPON MY BODY AND INTO MY HEART. Elsewhere he added John 3:16: "For God so loved the world that he gave his only begotten son. . . ." And lest any visitor forget, he put "God Loves You" emphatically up high near the cross that crowned everything. Lightning struck the railroad-tie-and-telephone-pole cross once, blasting it into four large pieces and an infinite number of splinters. People asked Leonard if he thought it was a sign of anything. He said, well yes, probably, but he didn't know what. He added, "I think I better get a hearing aid and listen to God better." Evidently God gave him the go-ahead, because Leonard soon scrounged more timbers, which he painted, assembled as a cross, and erected as quickly as he could.[2]

Leonard converted a natural cleft in the face of the mesa into a kind of painted grotto. He wrote the Lord's Prayer on one side of it in a slightly recessed panel that seems almost like a door. He built up the words by applying ordinary window caulk to the panel with a caulking gun. The letters, painted red, seemed to harden up just fine. Above and to the left of the prayer, he painted an American flag. Leonard also terraced part of the face of the mesa into a succession of ledges. You walk along them and feel the paint spongy underfoot; the air smells of latex and linseed. The stairstep of ledges seems to invite the waterfall that he painted tripping colorfully down it. He bordered the waterfall with bright gardens of dabbed-on flowers and little forests of fanciful trees that suggest a gingerbread, folktale landscape. Nothing could be more different from the surrounding Salton basin. You look out from the top of the Mountain Bible Sculpt into the bleak flat windswept distance and see a panorama as sere and monotone as an enlarged sheet of sandpaper. Behind it lies the drearily serene mirror of the Salton Sea. Leonard says proudly that his bright mountain stands out so much from the desert that pilots navigate by it. Still, he finds no fault in any of his surroundings.

We were standing by the *God Is Love* Bible truck, in the shade of the lone mesquite. "I think God must have put me in the garden of Eden," Leonard said. "These have been the best years of my life. I work on my mountain; sometimes I get to preach a little. I get lots of friendly visitors, and people are so nice and bring me things."

I said, "Leonard, it's not often one meets a genuinely happy man."

And he replied, without a trace of self-consciousness, "Well you know, I think that's probably true."

LEONARD HAS become the most famous of the hundred or so year-round residents of the Slabs, but by no means are all of them as mild and harmless as he. Nor are they as open and conformist as the thousands of snowbirds who settle about the place when winter drives them south.

Just as a real river, the Colorado, ends in the Salton Sink, so does a river of myth and dream. Leonard Knight, the snowbirds, and the other residents of the Slabs, no matter how dark and various their histories, find themselves there for similar reasons—foremost among them being freedom from economic pressure and from rules in general. Theirs is one of the oldest American stories.

Generation after generation, Americans have kept pushing: California or bust. New lands, new hopes—or bust. Initially they fled the drab, gray nightmare of Europe where "misguided religion, tyranny, and absurd laws everywhere depress and afflict mankind."[3] Later they fled the structures and strictures their neo-European parents and brethren brought with them. If one spur prodding the horse of westward movement was economic and political, the other was psychological. No doubt a tortured graduate student will one day write a thesis detailing the importance of family dysfunction as an impetus for white folks' moving west. Huck Finn, you will remember, took up residence in a hogshead after Pap beat him half to death, then opted for roomier separation by way of floating down the Mississippi on a raft. When he later "lit out for the territories," he was simply extending the original trip. Many other Americans, before and since, have done the same.

J. Hector St. John de Crèvecoeur, one of eighteenth-century America's most ardent admirers, was well aware of this. In his "Letters from an American Farmer," he gushed about the dreamy promise of new lands: "We have in some measure regained the ancient dignity of our species," he crowed. But he also noted that the edges of settled America were untidy:

> He who would wish to see America in its proper light, and have a true idea of its feeble beginnings and barbarous rudiments, must visit our extended line of frontiers where the last settlers dwell . . . where men are wholly left dependent on their native tempers, and on the spur of uncertain industry, which often fails when not sanctified by the efficacy of a few moral rules. There, remote from the power of example and check of shame, many families exhibit the most hideous parts of our society.[4]

Crèvecoeur's final sentence may be a little harsh as a description of the fellow I'll call Mad Jack, but his general theme applies. Mad Jack dwells in a concrete bunker. We met amid the flea market that periodically manifests along the main thoroughfare of the Slabs. I had been browsing Mike's Junk Mall and Patty's Palace, pipe-frame affairs under awnings of tattered camouflage netting where for next to nothing you can pick up an inoperable fuel pump or a dozen heads of week-old cauliflower. You can buy an entire library of water-damaged pulp novels with their pages stuck together, although anything by Louis L'Amour costs extra. You can paw through mounds of unmatched socks or purchase a half dozen bent forks with kernels of brown rice glued between the tines. The possibilities are nearly limitless, and everything wears a film of windblown silt.

Mad Jack gestures expansively at the tables and the racks of worthless goods, and at the travel trailers and defunct school buses behind them, which stand parked upon the slabs where Quonset huts once stood. "In other words," he begins, "this place is the last outpost of freedom."

Jack prefaces nearly everything he says with the phrase "in other words," no matter whether any words have previously been spoken. One senses that his inner ear has a lot to listen to and that the words that leave his mouth are simply the portion of his internal monologue that happens to intersect the air. "In other words," he continues, "I wrote a letter to my family, who is very religious, and I told them that they have always been telling me that if I do right like I am told that then I will go to heaven, but I told them why the hell would I want to do that since I am in heaven right now exactly here because nobody hassles me, I can do just like I please, and it don't cost nothing. Do you think they get it? Yee-ha! Hell no, they don't get it. Do you?"

In other words, Jack professed not just to be a free man but a happy one. He characterized his situation in much the same terms as Leonard, if somewhat less convincingly. At his side was a young female of the species whom I at first took to be Jack's daughter, but as our conversation progressed, I concluded that Jack was unlikely to have a daughter, or if he did, such a daughter would be unlikely to live with such a father. In any event, Jack's blond and evidently teenaged companion spoke rarely and then in fragmented monosyllables. Mostly she gazed into the middle distance with a satisfied, chemically oriented grin. Jack's eyes, on the other hand, never paused. He looked me up and down, examining every feature of my clothing and bearing, just as I was examining his. While Jack may finally have taken satisfaction in concluding that I was not a cop, I am certain I had the more interesting part of the exchange.

Jack explained that he chose to live at the Slabs from among a range of options. He said he also had a place in the Central Valley, another in Florida, and still another at an undisclosed location in Central America. He could have chosen to live in any of those places, but preferred the Slabs.

It struck me that he could also have elected to run his belt through belt loops, but chose not to. Jack wore a vast pair of trousers pulled up high on his body and maintained them there by means of a narrow belt, which he fastened around himself six or seven inches below the trousers' neglected waistband. Loose above the belt, the pants flowered out like the fused petals of some weird desert plant, and from that scruffy and malodorous blossom erupted Jack himself, the human stamen, the ultimate anther of which was a dirty wool watchcap that stood at maximum extension from his head. Jack's discolored teeth were long, his whiskers also long but not amounting to a beard, and his right hand, always gesturing, held a beer can from which he never drank.

"In other words, what more do you want? I've got a bunker, a generator, air conditioner, TV, anything I need. And what I don't need ain't here: taxes, utility bills, an address, or a boss. Yee-ha!"

Jack, you might say, was self-employed. He called himself a "range runner," a name calculated to conjure images of Mad Max and other post-apocalypse heros, but to the rest of the world Jack and his colleagues are better known as "scrappers." In the dark of night or anytime jets whistle overhead toward the Chocolate Mountain Aerial Gunnery Range, Jack and a handful of others scramble behind the wheels of their big-tire, chopped-down, welder-on-mescaline desert buggies and roar off across the Coachella Canal and into the dark, wild desert beyond. Wherever they see the flash of exploding ordnance, that's where they head.

Here, as in many other times and places, the frontier edge of what passes for American civilization is dominated by the U.S. military. The Chocolate Mountain Aerial Gunnery Range, consisting of 400,000 acres northeast of Salton Sea, has been regularly bombed and strafed since at least 1942. Notwithstanding that the majority of the bombs falling on the Chocolates nowadays contain cement instead of TNT and that most of the aerial cannon rounds are nonexplosive, the hail of high-velocity heavy missiles can nonetheless be hazardous to one's health. For this reason, the range is off-limits to all but a few credentialed wards of the defense budget. Unofficial wards like Jack, however, routinely ignore the closure because it is the very hail of angry metal that they would harvest and sell in order to maintain themselves in the style to which they have become accustomed—and addicted. They comprise a kind of cargo cult who pray for aluminum bomb and missile fins, high-alloy casings, machine gun clips, and other scrap from the sky. In their souped-up dune buggies and rattletrap pickups, they race across the creosote wastes to be the first to reach a smoldering target, where they hustle and sometimes quarrel to grab the high-dollar items, load up on bulk metals, and clear out before the next flight comes barreling and blazing in.[5]

"In other words, I can make six hundred dollars in three hours, like I did yesterday," says Jack, "and then I take it easy." The scrap gets sold down the line to local junk kings who in turn sell to small Mexican foundries, at one of which some years ago a live antitank round entered the melting mix undetected and exploded, killing a worker. The range runners, too, sometimes blow themselves up. Besides selling scrap, they also feed a separate, more exclusive market for actual explosives. It takes steady hands to dismantle a shell or bomb, but steadiness of hand is not a widespread characteristic among the range runners of Slab City. More often, their hands are shaking when they most need an infusion of income, which a pound or two of high-grade explosive can supply. The Imperial County Sheriff's Department estimates that "thefts of government property," as they term the scavenging of the gunnery range, produce two serious injuries per year and a death every eighteen months.[6]

PERIODICALLY, the authorities respond to the mayhem by cracking down on the scrappers, making a few arrests, and sending someone to jail. Periodically, they also announce intentions to clean up Slab City. The place has no water system, no sewer, and no garbage collection, and it can be argued that when seven thousand or more snowbirds and their vehicles pack the place fender to fender, a threat to public health exists. More to the point, regulators abhor a vacuum, and where rules are concerned, the Slabs are as empty as outer space.[7]

The oft-proposed solution is to make the place a public park and impose "modest" fees, the income from which would pay the cost of providing services. One day, such a plan may come to pass, especially if something particularly dreadful occurs at the Slabs to stir public opinion in favor of control. But meantime, the voters and merchants of Niland, where Slab City's denizens spend their money, continue to defend their satellite town and its anarchy. They fear, probably rightly, that if fees were charged at the Slabs—if, in effect, it became the same as any of several score of other snowbird waystations in the basin or nearby, at Winterhaven, Lake Havasu, or points beyond—then people would abandon it. Its physical attractions, after all, are hardly prepossessing.

"There's rumors every year that they're gonna hit us with rent or permits," says Mad Jack. "In other words, let em try. It's nothin but government land all around here, and we can move somewheres else." Jack gestures, beer can in hand. He tells me once again that he absolutely will not under any circumstances go on camera, notwithstanding that I have no camera, nor is any in sight. His ragged girlfriend rocks on her heels beside him, blank as a sand dune. "Doing what we've been doing," he says, "in other words, it's working."

"In other words," continues Jack, "we're misfits. Some of us are plain crazy, like that guy." Jack points to a truck going by, a kind of jiggered olive-drab bus with pipe-fitting tools and a host of other weird gear sticking out the back. The words "Free American" are scrawled in yellow on its door.

"No, he's not crazy," says the girlfriend, as though waking from a nap. "He's nice."

"No, that guy is crazy. He's just another misfit. I'm a misfit too. I am the most misfit of them all because I can live in society or I can live out of it. I can go back and forth. So that's okay with me."

UP WHERE the truck of the free American disappeared, past the torpid flea market, past the last of the actual slabs, past the final broken chunk of the Slabs' paved road, dozens of truck trails snake away through the sands. You'd be surprised by the abundance of trees, mainly mesquite and salt cedar, plus palo verde and smoke trees in the washes. They grow, not thick enough to form thickets, but they grow, nourished by seepage from the Coachella Canal, which divides the Slabs from the gunnery range. Nearly all the trees are doomed by the lining of the canal, as the Slabs may also be doomed by "improvements" in other spheres, but for as long as they grow, each one that rises high enough casts its shade on an old Airstream trailer or an engineless school bus or a ring of camper shells, circled like pioneer wagons. Nearly every camp has a big wooden wire-spool, lying on its side, to serve as a table. A swirl of shade netting stretches between tree and trailer, and nearby a few solar panels lean at drunken angles, their wires all a-hoo. Off to one side lies a jumble of water drums and gas cans. A pair of scabby mongrels stare balefully from beneath the trailer or the bus. Here lies a disassembled carburetor on a scrap of plywood, over there a stack of egg cartons, filling with sand, and a doll's left arm. You recall that the human services people say child abuse is rife here. You drive on, mindful of the ruts and the washboards; each camp is different in detail, each the same in loneliness and desuetude. Up ahead, a coyote sprints across the road with a rabbit or a sixpack in its jaws. From a distance it is hard to tell. It could have been a baby.

LEONARD KNIGHT has had problems with the authorities, too, but not with all of them. Soon after he got started on the Mountain Bible Sculpt, two sheriff's deputies came out to see what he was up to, heard his story, and came back later with a gift of paint. "I think we've got the best police anywhere," says Leonard. But others have questioned whether the presence of "Leonard's Mountain" on public land violated the principle of separation of church and state. Not to worry, said one

Imperial County commissioner. He ordered toxicology tests of the painted hill, which as expected, found evidence of lead in the paint. The county announced its intention to condemn the Bible Sculpt as a toxic site and bulldoze it down.

Friends rushed to Leonard's aid. Besides local allies in Niland and the Slabs, folk-art curators from as far away as the Carolinas launched a letter-writing campaign asserting the significance of Leonard's creation. A film-maker from Los Angeles shot a documentary about Leonard and his painted hill, which gradually assumed the more grammatical and high-sounding name of Salvation Mountain. Ultimately, the county yielded, the aggrieved commissioner hollowly explaining that public opinion had not swayed him, only that a lack of funds to finance the bulldozing had caused the project to be tabled.[8]

THE HARSH natural world of the Salton Sink nurtures eccentrics like Leonard, misfits like Mad Jack, and migratory refugees like the thousands of snowbirds who seek to escape winter, not just at the Slabs but at Bombay Beach, Desert Shores, and a dozen other low-dollar hideaways. The natural environment also contains their analog.

It is the desert pupfish, *Cyprinodon macularius*, a chunky and quite ordinary-looking creature not three inches long. The fish is officially listed as endangered.[9] Its adaptations are its most impressive feature, for it can tolerate and even flourish in water that is so hot, salty, or deficient in oxygen that fish of any other species would turn belly-up. It inhabits the salty hot springs of the basin and the withered creeks that descend from them. It also finds a home in the weedy agricultural drains that carry water from the farm fields of the Imperial and Coachella valleys into the Salton Sea. Formerly, the pupfish also made a home in the sea itself, but the introduction of tilapia, sailfin mollies, and other predators banished it from its old home, just as similar exotic introductions helped drive it from its native waters in the Gila and Colorado rivers.

The great weakness of the pupfish is that it cannot tolerate competition. Nearly any fishy predator can gobble up its eggs or young or even adult fish to the point that no member of the species remains. Its one successful evolutionary strategem, its stock in trade, so to speak, is to withdraw where it will not be followed, to pull away into aquatic redoubts twice as salty as the ocean or so hot as to risk a slow poaching. What a pupfish needs, more than anything else, is protection from molestation. As *Cyprinodon* survives in the fetid tributaries of the Salton Sea, so do colonies of human pupfish survive in the eddies and backwaters of America's river of real-estate dreams, not a few of which have drained into that same final sump.

14 | HAVE WE GOT A DEAL FOR YOU

Perhaps the event was ordained before it took place. Perhaps it was no less fated than the flood of the real river that Rockwood and the others set loose in 1905. One need not forgive the Colorado River Development Company its greed and incompetence to say that it did not *cause* the Great Diversion; it simply released a flood whose hydrologic causes were embedded in the sediments of millennia.

So, too, it might have been only a matter of time before the real-estate dreams of the American mainstream spilled into the Salton Sink, drawn there by the accidental sea and the images of shorefront luxury and frolic that its waters made possible. The dreams that ended in that unlikely place descended from others that had flowed west in a human current all the way from Europe. That current of dream had promised landownership for countless thousands who might never have imagined owning land except on the broad American continent. No one checked the progress of that river of dreams for long, not for the sake of the land's previous Indian owners, though the attempt was repeatedly made, nor for the sake of wise settlement and the rational use of land, though that attempt was also repeatedly made.[1] The current of people's hope and hunger for land flowed over or around every obstacle it encountered on the long trip west across the continent. And then in California, it hit the limit of the land. With nowhere else to go, it pooled against the coast and poured into the Los Angeles basin the way the real Colorado once poured into the Salton Sink. It filled the basin higher and higher, mounting into the lower reaches of its smoggy atmosphere.

Thus impounded, the dream steeped and darkened into a tea of ever-stronger desires. In time, the dream waters rose to the height of San Gorgonio Pass and spilled eastward to places they earlier had bypassed. Palm Springs bathed in those waters from early on, and together with its satellite communities, it bathes in them still. In 1958, while Palm Springs swelled, a freshet of dream water spurted farther down the pass and kept flowing all the way to an expanse of thirsty sand beside the Salton Sea. The land that drank up that brief flood of optimism was perfect for the playing out of dream, for it was empty and isolated, a blank geographic canvas that imagination, always more colorful than reality, might paint as it wished.

M. PENN PHILLIPS made the first bold strokes on that canvas, and he made them as no one else could have. In the growth-mad, dirt-moving California of the fifties, Phillips was Mr. Big: he had a big smile, a big cigar, big money, big ideas. Over a span of thirty-six years, he had built the M. Penn Phillips Company into what his brochures claimed, perhaps truthfully, to be "the world's largest land development and building organization." The Phillips company had caused whole communities, like Azuza and Vista, to spring forth where only "raw ground" had previously existed. It had tied itself to the fortunes of existing towns like Palm Springs and Compton and boosted their growth to new heights. Its successes ran the length of the West Coast, from Ensenada, Mexico, to Coos Bay, Oregon. It was active in Baldwin Park, Arrowhead, and Big Bear. It had recently begun selling lots in the fully planned city-to-be of Hesperia. And now M. Penn Phillips was bringing his Midas touch to 19,600 acres on the shores of Salton Sea. There he would build Salton City, crown jewel of "the Salton Riviera."[2]

People believed that he would do exactly as he said, for he had never failed, at least not publicly. Like a celebrity who is famous for being famous, Phillips inhabited a wonderful money-making tautology: he was successful *because* he was successful. His company bragged that 125,000 people owned land that Phillips had developed; 8,000 children attended schools in towns he had brought into being; factories and business with millions of dollars in sales prospered on land to which he had brought roads, electricity, and water. His sales literature displayed the motto "Building California," and he had truly built a lot of it. Now he would build a "superspectacular" and "ultra-modern" resort city beside the Salton Sea, and he promised that he would put every resource he had behind the effort: "Without any question, Salton Riviera is the most important development in our history. It is what we have been building toward for 35 years. In the days ahead, we shall see the fruits of all our labors grow: the city, the family of communities we have dreamed of so long and so fervently."

Sincerity, confidence, professionalism, and vision: these were the qualities that Phillips and his coterie strove to project. Trust us, they said; they would provide Salton City with everything that experienced city planners and "two electronic brains" (*computer* was not yet a familiar word) could conjure up. They would spare no expense in equipping the new city with marinas for sailing, power boating, and water-skiing, a country club, golf courses, a recreational complex, shopping centers, a private airstrip (fly right in to your own house at Salton City Airpark Estates!), resort hotels, and more. Some $20 million (in 1958 dollars) would be invested in roads, sewers, water mains, telephone lines, and other infrastructure, ensuring an immediate upward leap in the value of every lot, a value that in turn would be protected and enhanced by the imposition of farsighted, carefully crafted restrictive covenants.

So much for the bells and whistles. At its core, the Phillips sales effort would pursue two main themes: vanity and greed. It would appeal to vanity by democratizing luxury. At Salton City, the high life of Palm Springs would become available to people of ordinary means. Caressed by the "cool sea breezes" of this new desert oasis, any Joe from Hoboken or Alice from Anaheim might bask in the same carefree opulence that cushioned the lives of movie stars and moguls. And, if they had the foresight to invest early at rock bottom prices, who was to say they did not deserve the life of princes and princesses? The Phillips company's promotion of Salton City sang a love song to the ego of every Joe and Alice in America: they were important, and they deserved the best.

But if appeals to vanity wore down their good judgment, it was greed that moved them to pick up the pen and sign where the salesman pointed. The price of entry to this incipient miracle in the desert was as little as $250 down and $29 a month (of course, that was for the smallest lot in the loneliest corner—wouldn't you and Alice like to see something nicer? It only costs a little more . . .). In many cases, if not most, sales were structured under a deferred-payment real-estate contract that granted the buyer no equity in the property until the contract was entirely paid off. Nevertheless, the curious and the gullible were made to understand that prospects for investment return were unlimited: "It may well be that this entirely new resort area—bigger than Capri and Monaco and Palm Beach combined—will become the most valuable piece of resort property on earth." The message was clear: by investing in Salton City, you could not only live like a wealthy man, you could become one.

THERE IS AN art to selling worthless land, and the essence of that art is to create the perception of value where none has existed before. When Penn Phillips bought

19,600 acres on the western shore of the Salton Sea, the land's chief worth, in the jargon of rural land appraisal, was merely "to keep Hell from shining through." According to a 1970 inquiry by the California attorney general's office, Phillips paid between five and ten dollars an acre. Eventually his company would retail thousands of lots, one-half to one acre in size, for an average of $3,500 per lot.[3] He would also wholesale chunks of his holdings—still at logarithmic markups—to other developers, who would promote mini-resorts of their own. Still other developers would buy land adjacent to Phillips's until the circus of Salton Sea promotion, illusion, and salesmanship sprawled over 30,000 more or less contiguous acres. The challenge underlying the nascent value-creating crusade was to impart motion to a flywheel of publicity and to keep that wheel spinning so that its hum changed the background noise in everyone's ears. No longer would people hear the stillness of the desert. Instead they would hear a buzz of great activity. They would hear the growl of bulldozers, the rat-tat of hammers, the jingle of coins. The source of the hubbub would seem to lie just over the horizon, but—hey, did you hear that?—you would swear it was coming this way.

Everyone heard it, M. Penn Phillips included. The easiest way to misunderstand a great scam, whether in real estate, politics, religion, or any field built upon a foundation of belief, is to assume a sharp distinction between the shearers and the sheared. The bunco or boiler-room artist, the siding salesman, the grizzled old fart with a rigged wheel at the county fair, these fellows know just what they are doing; they are entirely cynical, but their scams are small-scale. A massive fleecing, a shearing of thousands, on the other hand, demands both moral and cognitive ambiguity. The general of the sales force must convince his captains that their product is good for people. To do so, the general must have belief. The captains, for their part, must transmit that message with an intensity that galvanizes the troops. The fog of this particular kind of war is the fog of believing that you can prey upon suckers and do them a favor at the same time. Officers and soldiers alike must believe that everyone can benefit and that making indecent amounts of money by playing on the gullibility of the guileless hurts no one because *in the long run the promises will be made good.* In this fog of war, it becomes all but impossible to distinguish between combatants and noncombatants, between suckers and suckees.

The Phillips Company rallied its sales force first. In March 1958, more than three hundred sales representatives gathered for a strategy briefing, pep talk, and barbecue dinner under tents at Salton City. Phillips and his executives extolled the wonders of the city they would create and the munificence of the income that would flow from it. They asked—no, told—their sales reps to put their address lists to the service of a direct mail campaign, to host dinners, make calls, get on the radio, make the papers. And to come back two months later for the real work of opening day: May 21.

The flywheel began to spin: the people came. On the morning of May 21, 1958, long lines of bulbous sedans with whitewall tires and incipient tailfins breezed down the highway from L.A. and Palm Springs. They came down from the mountains through Borrego Springs. They crossed eastward from San Diego and westward from Yuma, converged in El Centro, and motored north. They parked on the level, gravel-skinned desert, wherever they could, little noting that ecologically and topographically, the landscape of Salton City was very like a parking lot.

There were enough tents to house a circus now, flying pennants from their masts as if for a coronation or a festival. Beside the tents stood a few wooden shacks, their facades as spruce as a new-built midway. Wherever the eye roamed, it encountered billboards depicting the wonders to come, the vision of M. Penn Phillips—homes with swimming pools and emerald lawns, streets, clubs, esplanades—all rendered in bright colors at grand scale. When the breeze stirred, you could smell the billboards' fresh paint and glue. You could also scent the charcoal promise of several field kitchens, still being organized beside a sea of tables covered with homey, checkered cloths. But it was tough to think about what you saw or smelled—in fact, it was tough to think at all on that day, in that place, because several sets of loudspeakers, the most powerful the rental folks in L.A. could provide, blared without a break, cheering you on, chanting the buzz: "Ladies and gentlemen, there's a grand opening special marina presentation in Tent 2. Tent 6 has seaside lots on sale, insurance protection at no extra charge; this offer is available for a limited time only." Meanwhile, a Hawaiian band, with amps turned as high as they would go, ukes and yammers from somewhere behind you. And the ear-splitting messages continue: "M. Penn Phillips is scheduled to arrive very soon, don't miss his talk in Tent 1. Corner lots are going fast in Tent . . ."—but gees, didya see those hula dancers? Were the corner lots in Tent 6 or 4?

When Phillips arrives, the TV cameras and radio microphones are ready. The reporters have their pads out. Jack Dempsey, the heavyweight champ, is with him. And what's-her-name—the sleek brunette from the beach movie—is beaming right behind him. The loudspeaker blares a welcome. The VIP entourage and the human tide around it edge toward the podium. Suddenly, celebrities are as ubiquitous as bad coffee. People point and crane. Somebody says, "Over there's that quiz show guy with a buncha Hollywood girls." You look, but instead you see a hard-eyed man in a thousand-dollar suit smoking a cigarette. He looks like the heavy in the new gangster movie. Yeah, that's right. That's him. Salesmen are sliding through the crowd like barracudas, their hands like fins before them, ready to shake yours. They pat everybody's back; they act like they've known you forever. They ask, "Are you seeing everything you want to see, have you heard about the lots in section 9?" They don't let on that their face muscles ache from smiling.

And now Penn Phillips is speaking:

"Ladies and gentlemen, suppose someone handed you four million dollars and said, 'There's plenty more where that came from, but take this as a starter and plan the world's perfect city.'" That, said Phillips, was the assignment he had given to his staff some months earlier. What you see around you, he continued, "is like the tip of an iceberg. We see only one-seventh above the surface, but the broad base of men, money, and imagination is there." As Phillips described his vision, his audience saw it too. He said nothing they had not already heard, but in hearing it from him, they felt they were hearing it from the very source, from the oracle himself. They felt the way the first farmers of Imperial Valley must have felt when they listened to Charles Rockwood or George Chaffey. M. Penn Phillips was a man of equal vision. He dressed like a senator and possessed the confidence of a king. He stood before the booming microphone and painted a verbal picture of the Shangri-la that was to be:

> Think about the picture you have in mind of the perfect place, and the ideal setting. Wouldn't it be much like this? A place ringed by snow-capped mountains and bathed in warm sunshine winter and summer, and cooled by sea breezes. A place where you could go swimming in warm smooth salt water the year round. Or boating. Or water skiing. Or just loaf on the beach under the clearest of blue skies, breathing air so clean you can see for fifty miles. A place where you can ride horseback. Or hunt. Or just sip a tall, cool drink. There is such a place coming to life on the shores of our largest inland body of water: Salton Riviera on the Salton Sea. A place where you can buy now for enjoyment and hold for income and a share of the profits in what may become the most fabulous resort city in the world. I have never been able to stand on that rise of land above the Salton Sea without seeing a great resort city. Now our dream is coming to life."[4]

Sales that day topped four and a quarter million dollars, and they purred profitably along for the next year and a half. Bringing new technology to the art of flimflam, salesmen took their prospects on aerial cruises of the area while an assistant on the ground marked key lots with giant *X*s. On the not-infrequent occasion when the airborne salesmen's rhapsodies caused the impressionable buyers, usually a man and his wife, to sign a sales contract right there in the plane, the assistant, alerted by radio, scurried back to the lot in question and flipped over the plywood sign bearing the *X* to reveal the word *Sold*. Then the plane circled while the new owners admired their possession, and the salesmen grinned.

Few homes were built in Salton City at that time, but a motel and clubhouse went up, providing the basic equipment with which the Phillips Company wined and dined its prospective, land-buying "guests." Construction crews laid miles of sewer and water lines. Scraps of irrigated landscaping came to life, and nearly nine thousand fan palms, icons of the California good life, were planted and watered along the empty bulldozed roads. The palms, along with several score of street signs emblazoned with meaningless names, were all that stood higher from the ground than a boot top. They stippled the landscape as improbably as if a cloud of fronded darts had fallen from heaven. Such was the imaginary "city" into which thousands of middle-class dupes poured their money.

And then, in November 1960, the inconceivable occurred. M. Penn Phillips, who had called the Salton Riviera the crowning glory of his luminous career, abruptly bailed out of the project.

Phillips sold his interests in Salton City, neighboring Desert Shores, and far-off Hesperia. At seventy-three, he had reached an age when many men retire, but Phillips, by his own account, remained at the height of his powers and was still deeply involved in projects for which he expressed the profoundest commitment. He offered no explanation for his abandonment of the Salton Riviera.[5] If the enterprise, without Phillips, should collapse, everyone connected with it—lot owners, investors, builders, suppliers, other developers—stood to lose what they had committed. Nervously they watched as ownership of Salton City passed through a pair of holding companies before landing finally, in October 1961, with the Holly Corporation, a Dallas company with interests in oil and gas and other fields. Almost immediately, real-estate development became the company's obsession. The assets Holly acquired at Salton City included real-estate contracts with a gross value of $25 million (reflecting the lots the Phillips Company and its immediate successors had sold) and an inventory of unsold lots that possessed, in the eye of an extremely optimistic beholder, a retail value of $70 million.[6] All this "wealth" derived from the 19,600 acres Phillips had bought three years earlier for about $150,000 and into which he had invested a few million dollars on promotion and infrastructure. Though Phillips never made public the price at which sold his Salton City stake, one may assume that he did not suffer from the transaction.

In the manner of developers everywhere, the Holly Corporation promised to accelerate the pace of construction and to provide the long-awaited infrastructure and amenities that would at last make the Salton City dream come true. At the top of its list was the Salton City Golf Course, and soon a nine-hole course did indeed emerge from the desert. It was a dim, pale shade of green, with an overabundance of natural sand traps at the edges of the fairways. Holly's sales reaped the benefit of this demonstration of viability, and money poured in.

Holly's greatest investment, however, was in promotion. Where Phillips had dazzled buyers with plane rides, Holly organized boat races. Beginning in 1961, Salton City hosted the "Boat World's Indy"—a five-hundred-mile speed race, round and round the Salton Sea, for inboards and outboards of every description. Thousands of people came to sit in the sun and gawk at the distant wakes of scores of motorboats angrily droning in circles out on the sea. It must have been as exciting as watching bees commute across a hayfield. No doubt many enjoyed it. CBS was said to have sent a camera crew. As the beer and cocktails flowed, spectators and participants alike discovered that the throng of fellow boating enthusiasts included many who also happened to be real-estate salesmen. The race weekend, they discovered, afforded unlimited opportunities for learning more about the advantages of lot ownership along the Salton Riviera.

IT IS HARD to identify the moment when air first began leaking from the Salton City balloon, but by 1963, Holly realized that Salton City could not succeed solely as a resort. It began to court industry. In 1964, it exultantly hooked a prospect, Atlas Plastics Corporation, which commenced to build a plant there. The developers' elation, however, was short-lived. Atlas was the only industrial client Salton City ever captured, and because of shoddy construction, Atlas's machinery repeatedly tore loose from its cement-slab mountings. After several years of headaches, Atlas left town. By 1968, the Salton balloon had thoroughly deflated.

As many as 22,000 buyers had paid (or owed under contract) more than $80 million for Salton City real estate, but not more than 200 homes had been built on the 23,000 platted lots.[7] The sewer system didn't work because there was not enough flow to move the waste. There was no school and no commerce to speak of, except for a struggling motel and restaurant on the highway. Medical services were distant. Property values were nil. Everyone hungered for someone—or something—to blame.

Clearly, what the place needed was more people, enough to generate the critical mass of economic activity that would pay for services and amenities, as well as prop up land values. Although investors and residents were rarely happy with the Holly Corporation, they could not criticize the company's development vision without impugning their own judgment. And so they pointed accusing fingers at the newspapers and other media that had begun to report the story of flimflam by the sea. Outraged, they argued that Salton City's main problem was the negativity of the press, which exaggerated the area's problems and scared away buyers.[8]

Another malefactor, in the view of many, was the Salton Sea itself, long the center attraction of the development dream and now a leading villain in the nightmare the dream had become. Whenever the Salton Sea rose and drowned waterfront property, whenever it suffered a foul-smelling algae bloom or an even more odiferous fish die-off, the boosters of the Salton Riviera complained that the sea, the liquid mirror of all their fantasies, had turned against them. Of course, the sea was not so much at fault as those who ill used it. The people of the Salton Riviera accused the farmers of the Imperial and Coachella valleys, the irrigation districts, and any number of state agencies of negligent and, on occasion, criminal mismanagement.

The constituents of the once "superspectacular" but now stillborn resort city expected the stewards of the sea to keep it as Penn Phillips had described it: clean, clear and vibrantly healthy.

As people cast about for explanations of Salton City's demise, no one who had earlier bought into the dream wished to admit that at Salton City the Theory of the Greater Fool had simply run its course: suckers had sold to other suckers, who drew in more suckers, until the supply of suckers ran out. Finally, when the mesmerizing din of hype and promises stopped fooling people, only fools remained. They were the last to realize (if they ever did) what was plain to the rest of the world: the shoreline of the Salton Sea lacked the attributes of a good place for a city, let alone a resort. It was too hot, too foul, too isolated, too vulnerable, too hostile. For all but a handful of human pupfish, it was not a good place to live.

BUT DON'T tell that to Norm Niver.

If you were to hire someone as chief citizen advocate for the Salton Sea, you might include the following in the job description: "Must be imaginative and articulate, willing to work at all hours; must be unconcerned with compensation; must have a facility for recasting liabilities as assets; must be a person for whom the glass is never just half full but who finds the contents downright abundant, thirst-quenching, and delicious; must be as persistent as the passage of time itself, and immune to discouragement."

You might advertise widely and interview thousands, but you would still pick Norm. His house is on the waterfront. He has caught twenty-pound corvina by casting off his dock. The meat is perfect, he says, pure and white. The warnings about selenium are "way overblown." He has scooped up whole washtubs of tilapia, chopped off the heads, and frozen the bodies for his cat, Charles. Charles lived to the ripe age of nineteen, a prodigy of feline good health. Norm loves the sea. He knows why it smells, and so the smell does not bother him. He knows that the best fishing is always when the water is as dark as root beer and even slightly fizzy. "People get the wrong idea," he says. "They think the water is brown from New River sewage, but it's just algae and plankton. Every body of saltwater in world has them. Nothing could be more natural. You never catch fish when the water is clear."

Norm Niver believes that the enormous population and continued growth of southern California require that the Salton Sea become once again the recreational mecca that it was in the mid-1960s, when thousands came there to fish and water-ski, weekend after weekend. "It's not just property values," he says. "It's *what are we gonna do with this region?*"

He speaks in a rapid-fire volley of urgent predictions: "Twenty-four million people, *twenty-four million,* will be in the Coachella Valley and roughly in this area and they will see the Salton Sea as an asset." Inwardly, you pause, as he rushes on, realizing that Coachella Valley couldn't hold all those people. Norm must have been citing a population projection for all of southern California, but he's going so fast, with such enthusiasm, that the edges of things begin to blur. That's all right, you say to yourself, you can sort out the details later:

> It's common knowledge that if this thing is done and done right, it will rival if not surpass farming as far as the multi-billions of dollars it would bring. It could be the world's largest water park. It's water-skiing, fishing, swimming, and even though the hype that you have heard about the sea, and it is a bunch of hype about being a sewer and polluted—I don't care, I get upset up there in Sacramento and I've been trying to get the task force to wake up: quit letting people put the sea down! If you sell a car, you wash it up, you vacuum it out, you detail it out—you sell a house, you clean it up, you paint it and stuff, and I just want them—these agencies, the original water quality . . . in fact, at that meeting and I'll say it again, we did . . . if you want to clean it . . . every time I hear "clean up the sea": What are we talking about? What are you talking about when you say clean up the sea? What ARE you talking about? People see the brown, they think "the sewer." It's not the case. The brown is simply plankton. . . . It is the life of the sea. It's brown. You want us to take the brown out? The green, the algae, the green-blue algae we have here, these things are worldwide, they are in every body of water, they are just more concentrated here. We have the green-blue algae here. You want to take the algae out, take the brown out? You want to vacuum to get the selenium?"

Norm pauses for a question. He is a genial man, well into middle age, perhaps a little heavier than he would like to be. He's a modern California jack-of-all-trades, and he still fervently believes the old Penn Phillips dream that the Salton Sink can be an waterfront Eden-in-the-desert.

It has been years since the developers stopped paying to water the original Salton City golf course. The greens, of course, dried out and blew away. It has been years since the nine thousand palms enjoyed their last drink of drip irrigation, and all but a few have long since let fall their up-reaching fronds and either blown over or become hotels for nesting starlings. It has been years since the Holly Corporation unloaded its last holdings into the custody of bottom-fishing real-estate scavengers in gold chains and tassel loafers. It has been years since anyone kept much track of this group's or the next one's hype; mostly people remember that each of Holly's successors came with promises and left without a forwarding address. It has been years since the yacht club was abandoned to vandals and their spray paint. It has been years since the high water of the seventies drowned the marina and everything else on the shoreline and then went down again, and years since the even higher water of the eighties rose and fell but lasted much longer. It has even been years since the rumors started about a new rash of "investors" buying up lots for

$500 a pop and marketing them, with the usual promotional malarkey, to inner-city blacks, Hispanics, and Asians in an attempt to recruit new volunteers to the armies of the bilked.

Norm Niver has lived through nearly all of that, and he still believes. There is an agility to him, a mental and emotional adaptability that seems to permit the emergence of happy outlooks, if not outcomes, from the direst circumstances. You see his agility in the way he earns his living. To say that he works as a musician, electrician, and newspaper columnist is a gross oversimplification. When I first met him, he wrote for *three* newspapers—mostly about the sea. He played the tuba *and* the bass. You might have seen him in lederhosen, oompahing around the state fair one day and playing the low notes with a band of Dixieland strollers the next. He did Chinese funerals, wedding dances, hotel bars and restaurants. The big places from Indio up through Palm Springs were his bread and butter. On bass—acoustic or Fender jazz—he has backed up some of the best. He can play the beat, lie low, or push it, any kind of music—it doesn't matter, he's been there, done that, for the better part of forty years. He backed Ray Price on the road in the sixties when Ray was a big name, and he has played with lots more top talent at the Stouffer, La Quinta, and other shiny places up and down the valley.

Years ago he wanted to be a park ranger and studied zoology, but he found out that nobody could raise four kids on a park ranger's salary. So he stayed with music and dabbled in electronics on the side. Partly self-taught, partly trained in night school, he has serviced forty-eight hundred televisions in the Coachella Valley, including Gene Autry's. He keeps thousands of dollars' worth of oscilloscopes, voltmeters, and service computers in the shop at his house, and he's got a complete analysis system no bigger than a suitcase that he takes on the road: "You plug it in, and it'll walk through any TV top to bottom."

When I met Norm, he was doing fewer TVs and getting more into what he called Christmas crafts and neon art. His wife, an excellent seamstress, whipped out little Christmas-motif pillows with ruffled edges. She decorated them with Christmas-tree covers or patterns showing Santa and a stick-reindeer. Norm then equipped them with a circuit that ran four months on a single AA battery, so you could turn on the pillow the day after Thanksgiving, throw it on the sofa, and from then until long past Christmas, Santa's green-diode eyes would keep blinking and the reindeer's nose, of course, would flash red. The next year, you wouldn't even have to change the battery: you'd just plug in the little clip in the back and there you'd go, all season long.

The neon art was a new area altogether. A lot of it was custom work: a sign, say, for a man whose wife is named Harriet, that says *Harriet's Cocina* in crimson, scripted letters. Or a big pink martini glass with a green olive in it. "These rich people," said Norm, "will buy anything, whatever you want." The Coachella Valley, in his view, "is a valley of service. It is what's so neat about here. . . . You can babysit cats; you can shop for people; they will pay you to go get their groceries. People across the street from me came out here and they set things up with a house-cleaning business; they've got so much work they just have to get out of town. They make good money. It's a valley of service out here and that's why I don't mind the rich getting richer."

For a man of Norm Niver's energy and imagination, the possibilities in any area of endeavor are rarely less than endless. So it is with the sea. Norm, like many, views salinity as the main problem. For reasons that will be detailed in the next chapter, the salinity of the sea has risen to the point that Salton Sea is now over 25 percent more saline than the Pacific. While adult fish may survive such saltiness, their eggs and young become less and less viable. This, of course, spells doom for the fish and the sport fishery, which once boasted the unheard-of success rate of 8.5 fish per fisherman per outing. It also bodes ill for many other creatures, including water birds that feed on fish or other animals the salt will kill.

Norm is thoroughly versed in every scheme that has been devised over the past thirty years to control the salinity of the Salton Sea: a pipeline to the gulf, a pump-out to evaporation ponds, desalinization plants, and diking schemes. His house is full of the impact statements and feasibility studies that three decades of study and deliberation have produced. Norm has actually read them. Almost anything would work, he says. The issue is to pick a plan and get on with it. Forget about the money. The growth of the area, the recreational activity, the yearning of those twenty-four million people for places to go and have fun, the multi-billions of dollars from all those people and all that energy will pay, he says, for saving the sea. For years Norm has waged a crusade to convey one message, and one message only, to county administrators, irrigation districts, state water authorities, and federal officials. He has delivered it to everyone who would listen and to many who wouldn't. He tells them: *The sea is worth it. Now stop studying. Do something. Act.*

IN SALTON CITY and Desert Shores, no one disagrees about saving the sea, but there is plenty of disagreement about nearly everything else. Struggles within the Salton Community Services District over the provision of infrastructure, over budget priorities, over the authority and status that come with holding a leadership position—these struggles have been famously divisive and all but sanguinary.[8] The situation is a little like that of academic departments at most universities, where feuds last for decades and disputes over the postage budget elicit all the eloquence and venom their gifted participants can muster. People with little to fight over seem to fight with particular viciousness. So it is within the depressed communities of the former Salton Riviera, among those whose lives are embittered by failed dreams and who believe that fragments of their dream might still be rescued, if only so-and-so weren't blocking the way.

The path toward rescue is, of course, the path toward growth, even if that growth may now depend more on trailer parks than yacht clubs. But while the most vocal members of the community loudly debate the fine points of how growth should be attracted, others quietly abide in silence. Some of the silent ones believe they have found what they wanted, and they think it's all right if things don't change. In their view, the failed subdivisions beside the Salton Sea possess a pleasing quality of refuge and quiet, just as they are.

You sense it among the snowbirds of Desert Shores: old folks in lawn chairs, early in the morning, checking the paper to see how their stocks did, playing endless games of bridge or rummy, telling the same jokes, finishing each other's sentences. Even their chihuahuas and poodles seem sedate; they scratch at their fleas with more diffidence than conviction. A bumper sticker on a travel trailer has the ring of truth: "Used to be Wine, Women and Song and now it's Beer, the Old Lady and TV." So be it. The occupants of the trailer have nothing to prove. They say, "Show us another place where we can get by on $300 a month."

WHETHER THE outcome is tragic or comic, life is always a puzzle: you try to assemble its pieces so that you live as much as possible on your own terms, satisfied and secure. The Declaration of Independence, after all, does not promise happiness, only the pursuit of it, and the pursuit is the effort to solve the puzzle. Some people shoot the moon and play for the big score; others, like the snowbirds of Desert Shores, make a thousand small adjustments and compromises. Either way, a person tries to fit the pieces of the puzzle together. For plenty of folks, finding a way to get by on $300 a month is a big victory. They get gleeful about it, as if they've won a jackpot or outsmarted the Furies. Plenty of snowbirds have that twinkle in their eye, and in spite of the ill temper infusing the Salton Community Services District, so do a fair number of permanent residents.

Lester and Patty Murrah solved part of their puzzle by buying a lot in Vista del Mar, a subset of the Penn Phillips dream, where every lot was adorned with a palm tree, corner lots with three. Lester jokes that the developers sold land by the tree, icon by icon. Long after prices had crashed but before all the palms had withered, the Murrahs bought a corner lot, 75 feet by 125, where they built a house, watered their palms back to health, and made a home. To avoid heavy fees, they bought a lot that had escaped the jurisdiction of the homeowners' association, leaving them with few fixed costs save for annual taxes of less than $500. To the best of their knowledge, the Salton City area offers the cheapest land in California that lies within an easy drive of places that sell what a person might need, be it in Coachella or Indio or Brawley or El Centro. Lester and Patty have some money, but not a lot of it, and this is a place that suits them well.

They like having friends, not close neighbors. Lester and Patty have plenty of the former, none of the latter. They've even bought some of the lots surrounding theirs, and they wish they could buy more, the better to insulate themselves from future development. Lester likes being by himself, likes puttering around his place. Out here, he said, "you can dance on your patio with your wife under the stars, or you can talk to your yard bare-assed naked. And nobody can complain."

Lester hails from Alabama and has the gift of easy and eloquent profanity that seems to flourish in the South. He can look you straight in the eye and suddenly and sincerely proffer the most vivid details about giraffe genitalia. He does so with a matter-of-fact candor that makes you feel he must have said it a thousand times or never before. Lester joined the navy in 1943 and spent most of his tour in San Diego, where he settled after the war. He made his way as an electrician, married twice, and raised a pair of daughters. When his chance at retirement came, he grabbed it, and one of his top criteria in choosing a new place to live was that it had to have a low density of idiots.

Because of the idiots who crowd around traffic wrecks and other calamities, blocking the efforts of emergency workers, it would take you as long to get to the hospital in San Diego, according to Lester, as it would to ride the thirty-five miles to Indio from Vista del Mar. He admits, though, that the Salton City area is not idiot free: "There are plenty of fools on dirt bikes and three- and four-wheelers running all over the place, tearing things up." But their presence is sporadic.

This is one of the reasons Lester is content with the area's scant progress. More development means more people, which is a guarantee of more idiots. Like many who have gravitated to Salton City, Lester comes from a rural background. He points out that people who were raised in urban situations tend to be heavily concerned with prestige and the ownership of things, whereas he and most of his desert neighbors are practical people. They want their money to go far, they don't care about ornament or stupid things like lawns. Lester used to have a lawn in San Diego, but he got so fed up with all the trouble of taking care of it and so tired of idiots and their dogs coming across it and leaving their presence behind that one day he pulled the whole thing up, and the irrigation too.

Coming over the mountains from the coast, he and Patty left a lot of things behind, including all that pretense. Here they tend a trim, quirky-looking cactus garden that includes the grave of "Cactus Jack," who, according to Lester, "came out trying to steal my cactus, so I shot him and buried him right here." He points to the ground. "Ole Jack couldn't get away cause his boots were on the wrong feet." And, indeed, a right boot protrudes from a gravelike pile of rocks where the left should be, and vice versa. Traces of a tattered straw hat project from the head of the pile.

Lester's garden includes an upside-down palm trunk wreathed in baling wire, several tangles of walking-stick cacti, and, not least, five or six golden barrel cacti, their double-hued thorns almost luminescent. "People warned us not to put them in here 'cause folks will come and steal 'em. They're worth a lot of money. But hell," Lester says, "you can't go through your life worrying about what other people are gonna do." He planted them anyway and has never had any trouble. "People here look out for each other," he says, proud that Vista del Mar, in spite of its pretentious name, has never become so busy that he and his neighbors lose track of what is happening around them.

The Murrahs don't worry about leaving things out, and they've never had their house bothered during the long stretches when they've been away during the summer. They pull a travel trailer behind Lester's stretch-cab F-250. Not long ago, they paid a visit to Oroville. Soon they will run up the road to Laughlin, Nevada, and see what "good sport" they can find there. "Don't mind the bright lights," says Lester. "I just don't want to live around them all the time."

They always like it when they get back. They go dancing with friends at various places around Salton City. They rise early to drink their coffee in the loneliness of bright, clear mornings and savor the illusion that they can just reach out and touch the mountains. Yes, they are concerned about the smokey, smoggy haze that clouds the view at times, but they love the open space, the long vistas, the moonlight at night, the stars overhead, and it's not hard to tell that Lester, at least, gets a little romantic when he's looking up at those stars.

Sometimes the stench of the sea carries up from the shore to their house, "but I can live with that," says Lester. It's an acceptable trade-off. But the sea's deeper illness is not. Lester does not have a boat, nor does he fish, but he is troubled by what he hears about sewage from Mexico, about the die-offs of birds and fish, about the predictions of what the sea may become. He asks, "Will we keep our sea?" By which he seems to mean, Will we keep it alive, so that when we see the shimmering reflection of dawn or moonrise, we can feel that it is good? Lester is no caricature. He fits no stereotype and is no one's fool. He has solved part of his life's puzzle by finding a spot beside the Salton Sea, and he likes that spot for what it is. Never having fallen for the fakery of the early years, he carries no bitterness about what might have been. He is attuned and adjusted to how things are. But he wonders, and the answer is important to him, "Will we keep our sea?"

Dimensions: The Sea

Highest level reached in historic time (February 1907): 197.5 feet below sea level.[1]

Approximate 1998 level: 227 feet below sea level.

Maximum depth at −227 feet elevation: 51 feet; average depth: 31 feet.

Desired level, as expressed in the Sonny Bono Memorial Salton Sea Restoration Act (introduced in 1998): 230–240 feet below sea level.

Approximate 1998 salinity: 44,000 to 45,000 parts per million (ppm).

Desired salinity, as expressed in the Sonny Bono Memorial Salton Sea Restoration Act: 35,000 to 40,000 ppm.

Salinity of seawater: 35,000 ppm.

Approximate 1907 surface area: 500 square miles.

Surface area in 1998: 243,718 acres (381 square miles).

Average annual evaporation: 5.8 feet.

Approximate 1998 inflow: 1.3 million acre-feet of water, including 4 million tons of dissolved salt.

Greatest 1998 length (NW-SE): 36 miles.

Greatest 1998 width (SW-NE): 15.5 miles.

Average annual precipitation within basin: 2.5 inches.

Number of days annually when temperature exceeds 100 degrees Fahrenheit: more than 110.

15 | A SEA OF TROUBLES

In low places consequences collect, and Clark Bloom, the grizzled manager of the Salton Sea National Wildlife Refuge, has collected a lot of them.

As Bloom explains, perhaps for the thousandth time, the fundamental problem of the Salton Sea, being lowest of the low, is that it cannot pass along its troubles to someone or someplace farther downstream. The greater part of the American West stands above it and discharges a fair portion of its waters into it. Without an outlet to the ocean, the Salton Sea must stew in its own juices. Because infiltration to the underlying geology is negligible, only vapor leaves the sea's tepid precincts. Everything else—salts, minerals, pesticides, organic compounds, toxins—stays behind, and the resulting witches' brew, in recent years, has proved lethal for a substantial portion of the wild creatures unfortunate enough to depend upon it.

Bloom, on this early March morning, is in his office giving yet another interview. The day outside is warming toward a high in the mid-nineties—mild by local standards. The sky is windless and cloudless, and the unruffled sea, reflecting the emptiness above it, presents an image of peacefulness, or at least of torpor. Elsewhere, water may symbolize restlessness, but at the Salton Sea it more often stands for inertia. Nothing moves in the seascape, not the statuesque cormorants in the dead, flooded trees at shoreline, not the glassy, somewhat oily surface of the water, not the flat hazy horizon. The fierce weight of downpouring heat and light seems to hold everything immobile.

It is good to be indoors, away from the sun, and Bloom's cool office, like so many hot-country buildings, is almost windowless. Its concrete block walls give a cave like, protective feel, somewhat like a bunker, which, under the circumstances, is an appropriate analog for the headquarters of the Salton Sea National Wildlife Refuge. "It's not like I am managing a refuge now," says Bloom. "It's more like a battlefield." His voice is gravelly, his tone resigned. It is not hard to hear the inflection of fatigue and discouragement in what he says. Bloom, whose pressed uniform, trim moustache, and close-cropped hair give him a slightly military air, is a veteran of nearly three decades with the U.S. Fish and Wildlife Service. He has had abundant experience in the management of budgets, visitors, and habitat and in the operation of wildlife refuges that function essentially as migratory bird farms. Before coming to the Salton Sea refuge in 1992, he had almost always been able to set his own and his staff's day-to-day work priorities according to goals and plans that he had helped determine. Not at Salton Sea, where circumstances far beyond his control hold sway. "What we do here," he explains, "goes by what's dying and how fast it's dying."[1]

The dying began a few months before Bloom came to the refuge. On December 16, 1991, as the full moon approached, a refuge equipment operator working east of the Alamo River delta reported seeing a group of eared grebes that appeared sick. Given the sea's huge population of wintering birds, especially grebes, a handful of dying birds was hardly unusual. Nevertheless, refuge staff collected several carcasses and sent them to the National Wildlife Health Research Center in Madison, Wisconsin, for necropsy. They expected the lab to determine that the birds had died of avian cholera, a common affliction of large, crowded populations. To everyone's surprise, the report came back with no finding of bacterial or viral infection. In fact, the lab failed to identify any cause of death.

A month later, in the third week of January 1992, refuge managers discovered nearly two hundred dead and dying grebes on the barnacle beach west of the refuge headquarters. Some of the birds were partly decomposed, but others, still alive, sought shelter among the few rocks and larger shoreline flotsam in order to escape the herring gulls that preyed upon them. The grebes appeared unable to dodge or flee as the gulls stalked among them, attacking with thick bills, hammering at the grebes' heads as they would at a mussel until the skulls of the smaller birds split open. The gulls feasted on the warm brains. The living grebes appeared too dazed and disoriented to resist. Nor did they resist the refuge workers who picked them up and wrung their necks as a farmer might kill a chicken. Eight more specimens were sent to the Madison lab, and again, a suite of tests proved inconclusive. Soon, however, the moon began to wane, and repeated patrols of the refuge shoreline turned up only three more dead grebes. The crisis, whatever it was, appeared to have passed.

Then came the third week of February. Waves of thousands of dead grebes began to wash ashore, and not just at the refuge. Phones rang incessantly as birders called in with accounts of strange grebe behavior and residents of Salton City reported large numbers of carcasses windrowed on their beach. On February 24, refuge biologist William Radke (who got his start in the Fish and Wildlife Service working for Clark Bloom in northern California) chartered a plane and made a circuit of the sea. From the air, he estimated the sea's population of eared grebes at over a million birds, an unsurprising number for the Salton Sea. But the rest of what he saw was hardly usual. Many grebes, he later wrote,

> were concentrated at the mouths of virtually every freshwater source flowing into the Sea, including the New River, Salt Creek, San Felipe Creek, the Whitewater River, and most agricultural drains. . . . From the air it was obvious that many grebes were dead at sea and had not yet reached the shoreline, though some of those washed ashore were being dashed apart by wave action and rocky shores, or were buried by sand and barnacles. Predation on sick and dead grebes was also widespread, with large numbers of coyotes and tremendous numbers of gulls observed feeding on the birds.[2]

Interestingly, few of the predators showed signs of illness as the die-off progressed, indicating little, if any, secondary infection or poisoning. The grebes, however, fell to the as yet unidentified epidemic in astonishing numbers, and as they succumbed, they exhibited behavior that, for grebes, could not have been more bizarre.

The eared grebe is a red-eyed, chunky bird that rides high in the water. No more than a foot long from tip of beak to end of tail, it dives for its food, which consists mainly of invertebrates—pile worms and copepods in the case of the Salton Sea. A sharp, thin bill conveys the impression of pursed lips, and tufts of golden feathers on the sides of the heads of breeding males give the birds a quizzical expression, as well as their name. Few birds are more thoroughly aquatic than the eared grebe, which under normal circumstances may spend its entire life on water. They even shun the land when they nest, laying their eggs instead on mounds of vegetable material that they pile up in shallow water. Few birds are better adapted to the high salinity of inland basin lakes. At various points in their annual migratory cycle, eared grebes also congregate in massive numbers on Mono Lake, at the foot of the east slope of the California Sierras, and on the Great Salt Lake of Utah. Compared with these locations, which register salinities of 100,000 and 230,000 parts per million (ppm), respectively, the water of the Salton Sea, at between 44,000 and 45,000 ppm, is almost sweet.[3]

But as tens of thousands of grebes fell victim to the mysterious plague of the Salton Sea, they left the lake's familiar waters for the unwelcoming and inhospitable beaches, where they lolled and staggered. They preened constantly, as though their skin were irritated or they could no longer produce the oils that prevented their feathers from waterlogging and drowning them. They arrived on land like aliens on a hostile planet, and they yielded to the voracious gulls as hopelessly and passively as sheep at a shearing. Radke and his colleagues watched them gulp freshwater, as though in a losing battle against dehydration. With every wave and gust of wind, more carcasses washed in.

Refuge personnel collected specimen birds by the dozens, both live and dead, sick and healthy. They also collected samples of the pile worms and other invertebrates on which the stricken grebes had been feeding. They sent the carcasses to Madison wildlife lab and the invertebrates to Patuxent National Wildlife Refuge in Maryland. A team of biologists from the U.S. Fish and Wildlife Service and California Department of Fish and Game assembled at the refuge. They were acutely aware that a mass mortality of wild, aquatic birds constituted a rare event and that such a die-off in the vicinity of a national wildlife refuge, with resources and researchers available to record the outbreak, was rarer still. They felt the weight of a considerable responsibility, and the weight only grew heavier as the laboratories failed to answer their questions.

Consistent with earlier tests, the results of necropsies continued to rule out avian cholera and botulism as primary causes of the grebe die-off. Cholera, however, proved to have killed most of the relatively small number of other birds, mainly ruddy ducks, whose carcasses were collected along with the grebes. Lest a secondary cholera epidemic or another infectious disease break out, the refuge organized a large-scale carcass cleanup, which continued through March and most of April, during which time the die-off mercifully abated. Ultimately, volunteers and agency personnel picked up some 46,040 dead birds from forty miles of shoreline.

Extrapolating those results to the rest of the sea, and factoring in data from supplemental aerial and shoreline surveys, researchers concluded that approximately 150,000 eared grebes had perished. This colossal toll, one of the largest recorded bird die-offs in North American history, represented nearly 8 percent of the estimated continental population of the species. Additionally, about 5,500 shorebirds, gulls, ducks, and other species died of avian cholera. Fortunately, a larger outbreak failed to materialize.

The laboratory results continued to trickle in. Try as they might, researchers found no evidence that the grebes had died of bacterial or viral infection. Indeed, no cause of death could be determined at all, yet the necropsies still yielded plenty of chilling data. Compared with samples collected years before the die-off, the livers of the grebes showed elevated levels of selenium, mercury, DDE (a derivative of the pesticide DDT), and chromium. Average selenium values in both grebes and pile worms appeared to have doubled in three years, and all grebe liver samples showed selenium concentrations higher than the known threshold for adverse health impacts. For a long time, biologists had been monitoring the concentration of selenium in the ecosystems of the Salton Sea, and for just as long they had been expecting a dire manifestation of its toxicity. Now, although proof of such a manifestation seemed close at hand, the levels of selenium and other contaminants in the grebe carcasses, though dangerously high, were insufficient to have caused death. Indeed, none of the contaminants, acting alone, appeared to have triggered mortality among the grebes. Still, no one familiar with the contaminants' potential for toxicity could avoid the conclusion that they might have compromised the grebes' immune systems and made them vulnerable to the mysterious agent that ultimately killed them.

A further conclusion was also obvious: the Salton Sea was tainted and sick, and it posed considerable danger to the wildlife that innocently depended on it. The danger seemed no less grave for being only partly understood. Contrary to their hopes, few observers expected that the worst was over.

FROM AN AVIAN point of view, the Salton Sea is one of the great oases of the West. Only the Texas Gulf Coast has more recorded species, and as the temporary home for huge populations of birds, the Salton Sea seems a fecund and protective haven. On a given February day, over a million and a half eared grebes, some three-quarters of the bird's North American population, may float upon its surface.

Over millennia, the repeated manifestations of Lake Cahuilla offered a similar place of rest and nourishment for the hundreds of species and millions of individual birds that migrated along the Pacific flyway. When the Colorado flowed into the sink and the delta was starved for water, the lake promised alternate habitat. As the lake grew salty and shrank, the delta bloomed. In the mutual history of lake and delta, the habitats they made available covered an enormous range of possibilities: the lake wholly fresh or hypersaline, the delta and its lagoons running the gamut of brackishness.

Migration was rarely an easy passage. Storms took their toll. Oases that offered plenty one year might yield little the next, or the bloom of needed food plants might be early or late. Some birds stayed; others flew on, and the only mercy was that the stepping stones of the flyway lay so close together they formed a kind of bridge along which the birds streamed in constant pursuit of day length, food, and safety. Geese, ducks, bent-legged shorebirds, pelicans, gulls, the musical passerines, the extravagant plume-bearers of the heron family, the drab coots and tufted grebes all made their way while the hard-eyed predators and scavengers followed and watched them. The flyway became a river of bird life flowing up the continent in spring, down in fall, sometimes in sky-darkening flocks, more often in small groups invisible against the magnitude of the sky. Every lake, playa, and desert oasis became a beacon to part of the kinetic throng, and one place in particular—the vast, wet, lush delta of the Colorado—became a haven to all.

Today, the line of migratory stepping stones no longer makes a bridge. Key stopping places such as the lakes and riparian gallery of the Owens Valley are simply gone. Others, such as the wetlands of the Klamath basin or Mono Lake, have shrunk from grand islands amid the desert to mere atolls, from Bermudas to Bikinis. Worst of all, the shriveled delta of the Colorado waits for death or rescue, its fate uncertain, its health at best precarious, and its carrying capacity a shadow of what it once sustained.

Which leaves the Salton Sea more vital than ever it was in the past.

The rapidity with which birds responded to the new habitats of the Great Diversion and the early sea amazed observers. George Wharton James, riding the floodwaters in 1906, joyfully recorded passing through great squawking concentrations of waterfowl, pelicans, and other birds.[4] When Joseph Grinnell surveyed the freshwater sea two years later, he was awed to find breeding colonies of guillemots, cormorants, puffins, white pelicans, and other birds already established on the bars and buttes the waters had failed to engulf.[5]

In 1930, the federal government recognized the sea's value to migratory wildlife and designated 23,425 acres at its southern end as the Salton Sea National Wildlife Refuge. In 1940, it took the further step of leasing 24,000 acres from the Imperial Irrigation District, most of which it assigned to the refuge, so that the refuge now consisted of 37,576 acres of lake, marsh, scrub flats, and fields. Unfortunately, the government proved as bereft of clairvoyance in founding the refuge as it had been in reserving land for the Torres-Martinez Cahuilla. Today, more than 90 percent of the refuge lies beneath the waters of the sea; only about a thousand acres near the mouth of the Alamo River and another slightly larger tract near the mouth of the New remain dry or shallow enough for a tall heron to stand on.[6]

Despite its diminished size, the Salton Sea refuge remains one of the most surprising and impressive components of the national wildlife refuge system. Some 384 different bird species have been identified there, more than at any other refuge. It hosts hundreds of thousands of shorebirds, both winter and fall, plus thirty thousand geese and twice as many ducks from November through February.[7] Although prior to the die-offs of the 1990s the refuge remained little understood and largely ignored by the people of Imperial Valley, it had become a mecca for birders from every other corner of the globe. People intent on adding species like the wood stork, the yellow-footed gull, and the brown booby to their life lists made a pilgrimage to the Salton Sea. Many came to glimpse the endangered species of the refuge: the California brown pelican, the Yuma clapper rail, the least tern, the Aleutian Canada goose, and the more familiar bald eagle and peregrine falcon.

Refuge managers were proud to tell their visitors that, given space enough and clean water to grow bulrushes and cattail reeds, they could raise reclusive Yuma clapper rails "like carrots" in marshes at the edge of the sea. With the refuge and surrounding lands supporting about two-fifths of the thousand-bird population of Yuma clapper rails found in the United States (many of the rest are along the lower Colorado River), this success might lead to eventual removal of the rail from the federal list of endangered species.[8]

Unfortunately, the bad news emanating from the refuge since 1992 has by far overshadowed the good. Clark Bloom and his refuge staff have found themselves thrust to the forefront for reasons that have nothing to do with species recovery. They have found, to their dismay, that the refuge is ideally situated for the study of environmentally induced avian disease.

Not that the U.S. Fish and Wildlife Service has made the most of this unfortunate situation. In the half dozen years following the great grebe die-off of 1992, the service initiated little meaningful research on the sea and its problems. People impatient with the lack of government action to remedy the ailments of the Salton Sea complain that the sea has "been studied to death," that the time to act came long ago. Unfortunately, they miss a vital point. The Salton Sea has been the object of much speculation—you could say it has been "talked to death." But it has still not received the serious, long-term study it needs and deserves.

The U.S. Forest Service is no paragon of organizational efficiency, but at least it possesses the capacity to respond to emergencies—witness its fire-fighting prowess. The U.S. Fish and Wildlife Service, by contrast, has no similar capability where wildlife disasters are concerned. After 150,000 grebes perished on the Salton Sea—tens of thousands of them within the boundaries of the agency's so-called refuge, neither the Fish and Wildlife Service, the Department of the Interior in which it is housed, nor the Congressional committees with authority over its budget and operations took meaningful action, let alone invested in the kind of science that might have provided an explanation.

Perhaps the most tangible outcome of the die-off was provided by Clark Bloom, who took over management of the refuge shortly after the epidemic had abated. "I could not believe how we treated dead things around here, which was pretty much to do nothing—just let them lay and rot. That bothered me. At the first financial opportunity we bought an incinerator and had it installed so that we could dispose of things properly." Even so, more than two years passed before the incinerator was purchased. When, during yet another die-off, it was fired up and run day and night, day after day for weeks, Bloom and his exhausted staff found to their dismay that it was not big enough to dispose of the harvest of death that surrounded them.

TO UNDERSTAND the ecology of the Salton Sea one must first appreciate that in the true sense of the word, it is an *extraordinary* system. Viewed from one angle, the

Salton Sea is an intriguing, if disorderly, biology experiment, conducted on a grand scale. The sea is a young and therefore unstable ecosystem. Left alone, it would become a hypersaline lake, valuable to wildlife but vulnerable to serious perturbations. But it has not been left alone. All the life in and around the sea depends on its hugely productive food chain, the key elements of which are species humans have introduced. The development of this linked sequence of creatures eating creatures is a story partly of happy accidents and partly of accidents waiting to happen.

Like the prehistoric manifestations of Lake Cahuilla, the historical Salton Sea was at first a freshwater lake. The floodwaters flushed bonytail chub, humpback suckers, and other Colorado River fishes into the sea, even including rainbow trout, which biologists hypothesized might have been washed down from the uplands of Arizona. For a time, these freshwater creatures survived and perhaps even prospered. But the environment on which they depended soon changed. The sea shrank rapidly in its first years, and its water grew brackish. By 1929 a few trout were still present, and mullet fry, spawned in the gulf, migrated up the river and down the Imperial Canal to mature in the sea, but aside from the redoubtable desert pupfish, other freshwater fishes had vanished. In that year and the next, California officials attempted to establish populations of striped bass in the sea, but the fish did not survive. While gross salinity may have been acceptable, the composition of salts or, more likely, elements of the food chain were not. A 1934 planting of fifteen thousand silver salmon similarly swam into oblivion.[9]

More important to the future biology of the sea was the introduction in 1930 of several buckets of pile worms, *Neanthes succinea*. The pile worm, at its largest, is too small to grace a fish hook, but it soon played an enormous role in the ecological dynamics of the sea. Viewed closely, the pile worm looks like a reddish centipede. It possesses two rows of tiny limbs that, by waving in a loose approximation of unison, move the worm through the water. Salton Sea pile worms spend most of their lives burrowed in the sediment of the sea bottom, where they feed on the detritus of plankton and other organic matter that gently rains from the water above. Drawn by who knows what annelid urge, they periodically forsake their burrows to swim to the surface and reproduce.

In the mid-1940s, the pile worms witnessed the arrival of barnacles, which probably hitched a ride to the sea on pilings or the hulls of boats or seaplanes associated with the navy's Salton Sea Test Base. The barnacles found the sea less than ideal, for its soft, muddy bottom offered few hard structures to which they might attach. But wherever barnacles encountered rocks, submerged brush, rusting car hulks, or any other solid object, they soon covered it entirely, and then covered each other, creating colonial encrustations that swelled with waves of new arrivals. Pile worms soon wriggled among the interstices of the barnacle shells and thrived with the expansion of their habitat.

In 1950, the California Department of Fish and Game adopted a new approach to planting fish in the sea. In May, a team of department biologists took to the waters of the Gulf of California off San Felipe and threw out their nets. They placed the whole mass of what they caught in large tanks, hauled the miscellaneous creatures, milling and slowly suffocating, northward across the border, and dumped them in the Salton Sea. In December and again the next year they repeated the process. In this way, wrasse, halibut, totuaba, bairdiella, bonefish, anchoveta, roosterfish, two species of corvina, and a half dozen other species were invited to colonize a rather empty sea. Some made it; most did not; but those that did reproduced with explosive rapidity.

The robust but narrowly based food chain on which the survivors depended began with phytoplankton, which harnessed solar energy to metabolize organic compounds dissolved in the sea. Various herbivorous algae fed on the plankton. Unlike the more complex food chain of the ocean, that of the Salton Sea lacked fish or other higher animals that might consume the algae and plankton directly. Instead, the next operators in the system were bacteria, which decomposed the remains of defunct algae, and the pile worm, which dined on the detritus of the bacteria and their meals. Next came the pan-sized bairdiella, which preyed upon the pile worm when it left its burrow to swim and reproduce. At the top of the chain was the orangemouth corvina, a prodigious gobbler of bairdiella, and a fisherman's dream.[10]

One more introduction deserves mention. In the 1960s two species of tilapia, a small, omnivorous grazing fish from Africa, were released in nearby irrigation drains to control aquatic weeds. The tilapia soon found the open waters of the sea much to their liking. If anything had limited the sea's productivity as a corvina farm, it had been the bairdiella's dependence upon *swimming* pile worms. Tilapia remedied that. They grazed along the bottom of the sea, nipping off such pile worm parts as the worms allowed to protrude from their burrows. On such easy pickings, the tilapia population grew to untold millions. Everybody who could hold a fishing rod caught them, residents and visitors, kids and seniors, rank beginners and crusty old salts. The only way a fisherman could be sure of peace, quiet, and perfect indolence was to throw a line with an unbaited hook into the water, or better yet, a line with no hook at all, for the tilapia sometimes seemed willing to bite anything.

The corvina, like the fishermen, flourished. The orangemouth—a big, hard-fighting, predatory torpedo—gobbled tilapia as it had previously gobbled bairdiella. It multiplied until its numbers crowded the tepid waters of the sea to a degree rarely duplicated by top predators elsewhere on earth. Twenty-pound corvina became commonplace. Thirty-pounders were not rare. They fought as hard as bonefish or tarpon. Only one creature preyed on them: *Homo pescans*. People reeled in the corvina from boats and docks. Some fishermen took to wading out into the warm sea, hundreds of yards from shore, towing an inner tube with a peach basket to hold their beer and gear. They would wade out into a glimmerglass stillness of water and sky, and there they would cast and cast, ripples spreading where the bait plopped, but in the good old days they would not have cast many times before *bang!*, the rod bent, the line ran, and their reels began to smoke.[11]

EVEN IN THE heyday of the fishery, people knew the party had to come to an end. The problem comes down to this: if you take a liter of sea water and boil it carefully away, your boiling pan, in the end, will contain about thirty-five grams of white powder. You might select your sample from any ocean, and the result will be the same, give or take a gram. The powder, of course, is salt—not one salt but a range of compounds derived mainly from chlorine, sulphur, sodium, calcium, magnesium, and potassium. Corvina and other ocean fish evolved in an environment of remarkably consistent salinity. In an environment where salinity relentlessly increases, their eggs and fry will at some point cease to survive; at still higher levels, adults will lose vigor. Eventually the fish die out.

The saltwater fishery of the Salton Sea thrived as long as the salinity of its waters did not far exceed that of the oceans in which its constituent fish had evolved—about 35,000 ppm. But the basin that contains the Salton Sea is essentially a grand, terrestrial boiling pan, toward which the salts of the Colorado River watershed slowly migrate and which they cannot leave.

The snowmelt headwaters of the river begin their downhill journey perfectly fresh—with no salt. Springs with dissolved minerals and the leaching of native soils add flavor along the way, but the greatest load of salt comes from irrigation, which repeats the leaching process again and again. Every time water is spread upon the land, about three-quarters of it is taken up by plants or simply evaporates, and a fair portion of what is left percolates through the soil, dissolving salts and carrying them back to the mainstream via seepage and drainwater. In the long journey downriver a given gallon of water may be repeatedly diverted, ending as a sharp-tasting pint. Evaporation from the river, and especially from its vast, quiescent reservoirs, accelerates the concentration of salts. Evaporation from Lakes Powell and Mead accounts for the loss of nearly one-tenth of the Colorado's annual flow, an amount roughly equal to Mexico's entire share. It also accounts for a commensurate increase in salinity. Ultimately, the water diverted into the All-American Canal at Imperial Dam has a salinity of about 900 ppm, and every acre-foot of it carries a ton of salt.[12] Even so, the augmentation of salinity has scarcely begun.

Since time immemorial, the soils of Imperial Valley have collected salts. Through all the prehistoric comings and goings of Lake Cahuilla, through every inland flood of the river and every application of irrigation water, they have retained what evaporation and flowing water could not carry away. As a result the valley's agricultural soils tend to be two to five times more saline than the water used to irrigate them. Farmers must therefore flush these soils regularly—usually during the hot summer season when fields are sometimes fallowed—to drive salts below the root zones of their crops. Toward this end they use about 15 percent more water than their crops actually need for biological growth. And the crops they grow are thirsty ones: an acre of alfalfa, one of the principal crops of the valley, requires about six acre-feet of water. This water would accumulate underground and gradually waterlog the fields of the valley, pushing salts back up into the root zone, were it not for drainpipes buried under virtually every field. These pipes collect wastewater, together with large amounts of salts, and deliver it to the canals that ultimately empty into the Salton Sea.[13]

The result of all this flushing and draining, this fight against the salty buildup that doomed the civilizations of Mesopotamia and other ancient lands, is to deliver water to the Salton Sea that is three times as salty as the "sweet" Colorado River water that teases life from Imperial Valley fields. Four million tons of salt—three tons in every acre-foot of inflow—enter the Salton Sea each year. Without water to do the work of transportation, one would need forty thousand railcars to move such a quantity of salt.

Salt and water enter the sea, but only water vapor leaves. Inexorably, salinity increases. The threat this poses to aquatic life has long been understood. In 1961 a state-sponsored study of the ecology of the sea forecast destruction of the existing food chain by 1980 or 1990.[14] The days of gloom actually arrived later than predicted because of the huge dilutive inflows of the 1970s and 1980s, but the study was essentially on target.

By 1985 the salinity of the sea had passed 40,000 ppm, and a decade later, as it approached 45,000 ppm, it had entered the fateful zone in which the reproduction of its coveted fishery began to falter. Adult fish appeared able to endure the caustic waters, which were now a quarter again as salty as the Pacific, but their eggs and fry survived less frequently. Fishing declined.[15] The halcyon days of endless "luck" and thick corvina filets receded into memory. People searched for someone to blame. Without fishing, there would be no recreation. Without recreation, there would be no resurgence of land values and no revival of depressed communities by the water's edge. The sea, everyone said, must not die.

SAVING THE sea means different things to different people. To charter-boat captains and recreational fishermen, it means saving the corvina fishery. To property owners and investors it means arresting and even reversing the sea's erosive effect on land values. To Clark Bloom and the staff of the Salton Sea National Wildlife Refuge, it means having a refuge again and not a battlefield littered with corpses. But in the mid-1990s, no one got his wish.

In the early months of 1994, eared grebes began dying again, succumbing to another visitation of the same mysterious ailment that had killed 150,000 of them two years earlier. Bloom and his people again collected carcasses. In this instance, the toll was less apocalyptic: only 20,000 birds, a number roughly equal to the human population of Brawley or Calexico. Looking with compulsive good cheer on the "bright side," one might have observed that the entire continental throng of eared grebes declined by only 1 percent, losing the equivalent of a town, not, as before, a state. But such losses could not be sustained indefinitely, and no one yet had identified the cause of the mortality.[16]

If biologists had possessed the capacity to "round up the usual suspects," they would have quickly arrested selenium, a naturally occurring mineral that the waters of the Colorado have collected from vast areas of the West and concentrated in the sediments of the Salton Sea. Of all the substances on earth necessary for life and health, none is subject to narrower tolerances. An essential nutrient, selenium becomes toxic to humans when ingestion rises by just a few ten-thousandths of a gram per day. A healthy male, for instance, needs about 0.000076 grams daily, but that male will become unhealthy—suffering decline of the immune and nervous systems, plus a host of skin, digestive, and motor disorders—if his daily intake rises to 0.0005 grams. These small amounts grow still smaller for the very young and very old. Given a choice between consuming equal amounts of arsenic and selenium, one should not hesitate to take the arsenic, for selenium is five times more poisonous.[17]

Selenium is ubiquitous in the American West, for it is a common constituent of the marine shale formations that overlie thousands of square miles from Montana to New Mexico. Although mineral selenium is not readily soluble, in the process of weathering it oxidizes into water-soluble selenate. Additionally, certain plants are well adapted to seleniferous soils and can extract mineral selenium from deep within their root zones, recombining it in organic compounds which they build into their tissues. As the plants grow, shed leaves and blossoms, and ultimately die, they deposit a halo of selenate on the ground around them. Rainwater and snowmelt dissolve these compounds and carry them away. Wherever western rivers end in an evaporative sump, the water's freight of selenium, like its salt, stays behind, relentlessly increasing.[18]

Prior to the die-offs at the Salton Sea, an outbreak of avian birth defects at Kesterson National Wildlife Refuge in California's central valley had alerted the world to selenium's unhappy relationship to irrigation agriculture. Kesterson, like the Salton Sea, was a sump for drain waters, and in the early summer of 1983 biologists became aware that Kesterson's ponds and reedy marshes had grown unnaturally silent. Investigation revealed the cause. In nest after nest, they found dead chicks with stumps where legs or wings should have been, external stomachs, bulging brains, eyes missing, corkscrewed beaks, all manner of deformity. Nearly all species present were affected: gadwalls, mallards, teal, stilts, avocets, and killdeer. Hardest hit were the embryos of the lowly coot and, as at Salton Sea, the eared grebe, of which 64 percent suffered death or deformity.[19]

Eventually, Kesterson's ponds and wetlands were drained and its incoming wastewater diverted elsewhere. The revelations at Kesterson also spawned programs to monitor selenium levels at similar sites throughout the West, including the Salton Sea. The California Water Resources Control Board began testing fish from the Salton Sea in 1984. The high levels it detected in corvina, tilapia, sargo, and croakers prompted the Department of Health Services, in 1986, to warn that pregnant women and children under fifteen should avoid eating fish from the sea, and healthy adults should limit their consumption to no more than eight ounces a month.[20] In theory, the advisory was to have been posted at places around the sea where fishermen were likely to go. In practice, notices are scarce, and most anglers believe the selenium scare to be overblown. Said Mac Prentice of the Red Hill Marina in 1993, "I've been eating the fish from the sea all my life, and I know lots of others who have been, and I don't know anyone who glows in the dark yet."

Indeed, not even the birds were glowing. Selenium levels in grebes that perished in the die-off of 1991–92 were high, but not fatally high. Although periodically selenium might be blamed for a bad hatch of eggs, nothing like the Kesterson outbreak seemed to be developing—yet.[21] One theory for the sea's reprieve is that a selenium-respiring bacterium may live in its salty water and at the brackish mouths of its tributaries. The bacteria appear to remove selenate from solution, possibly releasing some of it to the atmosphere and carrying the rest to the bottom when they die. Although pile worms then consume the detritus of the bacteria and continue the transmission of the mineral up the food chain, concentrations remain low enough that widespread selenium poisoning has not yet occurred.[22] Managers like Clark Bloom and researchers at places like the Madison wildlife lab have therefore focused their energies and limited resources on other problems immediately at hand, not promised for the future.

IN 1995, a shaft of light penetrated the obscurity of the grebe mystery. Researchers at the National Wildlife Health Research Center in Madison had sent samples of grebe livers to Wayne W. Carmichael, a professor of aquatic biology and toxicology at Wright State University in Dayton, Ohio. Carmichael detected in the livers potentially fatal amounts of a toxin produced by the misnamed blue-green algae, in which the Salton Sea abounds. These organisms, which are among the earliest lifeforms to have evolved on earth, are properly described as planktonic cyanobacteria, and they form a large and diverse family that includes the familiar "pond scum" of stock tanks, stagnant pools, and quiet fishing holes. Cyanobacteria are generally benign enough that certain forms are used as food (including the popular "spirulina" dietary supplement), yet toxic strains are not at all uncommon, and their toxicity persists even after the death of the organism. For well over a century cyanobacteria have been implicated in the deaths of farm animals exposed to water tainted with pond scum. On several occasions they have also caused the die-offs of waterfowl on the Mississippi flyway, and today they are suspected of contributing to high rates of liver cancer in parts of China, where water supplies are none too pure.[23]

Although no one has proved that an outbreak of toxic cyanobacteria triggered the grebe die-offs, no one has disproved the hypothesis either, for the necessary research remains lacking—and the opportunity for it may have passed with the passing of the toxic bloom itself. In any event, the conditions that favor growth of cyanobacteria are conditions that prevail in the Salton Sea with increasing frequency and strength.

Cyanobacteria prosper when large quantities of nutrients enter a body of water and fail to be flushed away. These days, the nutrients usually consist of nitrogen and phosphorus compounds derived from sewage, fertilizers, and detergents. In a warm, calm, and neutral to slightly alkaline aquatic environment, cyanobacteria multiply rapidly, accumulating as a slimy presence that may extend for miles across the water and reach many meters deep.

You might say that the Salton Sea is a pond-scum paradise. The water is alkaline and warm—even in winter its temperature rarely drops below 50 degrees Fahrenheit, and in summer it may reach into the 90s. Calm days, with no wind or chop to disturb the bacterial colonies, are common. As for nutrients, few bodies of water receive such a wealth. No small portion of the fertilizers laid upon the hundreds of thousands of acres of the Imperial and Coachella valleys dissolves under irrigation and is carried by tailwaters to the tributaries of the sea.

Nutrients also constitute the greater part of the freight of the New River, which flows from Mexicali north through Imperial Valley and discharges into the Salton Sea a few miles west of the wildlife refuge headquarters. Wherever the river drops over a fall or speeds through a narrow channel, you can see, and usually smell, a foam of detergent suds scudding down its surface. At the international border, mounds of such suds may grow large enough to conceal the occasional international swimmer. Within the river's flow, large quantities of human and animal wastes also journey seaward.

It has been said that "one of everything" has been found in the waters of the New: all three types of polio virus, several of hepatitis, and the agents for a full gamut of gastrointestinal illnesses. The bacteria responsible for infectious diseases such as cholera, tuberculosis, and typhoid have also been found in the New River, as have salmonella bacteria and a medley of carcinogens. Thanks to the chronic inadequacy of Mexicali's municipal sewage system, however, no other contaminant or pollutant rivals the river's profusion of fecal coliform. Health officials have repeatedly measured coliform in the river at levels several thousand times that which triggers beach closings in San Diego.[24]

Fears concerning Mexican waters go back to the earliest days of the Imperial Valley, when typhoid headed the list of potential scourges, but the present character of the New River, repeatedly identified as the continent's most polluted, beggars all comparison. In April 1994, for instance, the *Imperial Valley Press* reported that a truck careened off a bridge and crashed into the river. Twenty-four hours later, when a wrecker pulled it out, the "water" of the New River had stripped the truck of its paint. The vehicle was carrying a load of San Diego sewage, but officials ruled, with no overt attempt at irony, that its cargo need not be recovered, as it posed no threat to the riverine environment.[25]

Water quality officials for the state of California acknowledge that industrial toxins originating in Mexicali's tanneries and manufacturing plants remain a subject of concern and that the same may be said for the New River's cargo of mercury, other heavy metals, and pesticides, most of which, like its sister river the Alamo, it picks up in the Imperial Valley. Nevertheless, they emphasize that the disease-carrying viruses and bacteria that swirl in the river expire before reaching the Salton Sea.[26] Anyone greedy for consolation on the subject of the New River will cling hard to data such as this.

The great irony of the New River is that the same qualities that cause people to detest it have also—and equally—provided a foundation for the once-fabled Salton Sea fishery. The huge volume of fertilizers and municipal sewage that the New disgorges into the sea helped charge the food chain with nutrients enough to fuel its frenetic, hothouse growth. In its heyday, the fishery of the Salton Sea was a marvelous thing, but it was a fishery on steroids.

The Alamo and Whitewater rivers and other drains connecting to the sea give generously of nitrates and phosphates too. Without rivers of food flowing into it, the Salton Sea could never have supported the hyperactive overachievement of its plankton, pile worms, pan fish, and corvina. At one time, the sea was like a champion sumo wrestler, eating voraciously, heavy but strong, and capable of bursts of enormous power. But time is kind neither to wrestlers nor to self-contained seas. The aging wrestler loses strength and quickness, while fat remains, and the aging sea cannot abate—or escape—the inflows of its high-energy diet. Fish may so proliferate that no matter where you go on the sea, they skitter in ripples from the approach of your boat. They may so proliferate that scaled bodies, as much as water, seem to compose the volume on which you float. But all that burgeoning life places still more demands on other, more limited elements of the ecosystem.

In the best of times the Salton Sea was subject to "algal blooms" (more often cyanobacteria than true algae). Even when such blooms consisted of entirely non-toxic strains, the effects could be exceedingly unpleasant as huge quantities of formerly living cells died and decayed—which is to say, they oxidized, drawing large quantities of dissolved oxygen from surrounding waters. A vigorous algal or cyanobacterial bloom, feeding upon a superabundance of nutriment, can generate such vast quantities of organic material that its decay consumes all available oxygen in the area of the bloom. Several conditions result, one of which is merely unpleasant. The continuing and now *anoxic* process of decomposition becomes more fetid and malodorous, producing ever greater quantities of the "swamp" gases that foul the air at the edge of the sea.

Worse for the sea and its creatures is the second effect: the intersection of fish with a bloom must necessarily turn out badly for the fish. Some may be caught in a bloom as it develops—for a bloom may be miles long and wide. Others may wander into lifeless waters and fail to find escape, or they may find themselves trapped by such waters against a shallows or some other barrier. In any case, the fish, which depend on oxygen in the water as we depend on its presence in the air, must necessarily suffocate, and the toll of death can swiftly grow colossal.[27]

And there is a third result: even without mass fatality, fish in oxygen-depleted waters become stressed and vulnerable to disease, which, in turn, can cause them to become vectors for the transmission of disease to other animals. Such a condition of cumulative beleaguerment seems to have sparked the cycle of disaster that finally brought the Salton Sea and its problems to national attention.

FOR CLARK BLOOM, 1995 was basically uneventful. Sporadic fish kills and a brief outbreak of avian botulism, which killed fewer than three thousand ducks, punctuated the year, but overall it was a peaceful time for the Salton Sea National Wildlife Refuge. And 1996 began with promise. For the first time, brown pelicans were found to be nesting at the Salton Sea, a cause for rejoicing among refuge staff. The brown pelican is a bird of ocean shores and estuaries, while its white cousin, a common year-round resident at the Salton Sea, breeds mainly on inland lakes. Because the western North American population of brown pelicans had been unfortunate enough to earn a place on the federal endangered species list, the discovery of three successful nests at Salton Sea seemed a kind of vindication of the refuge and of the sea itself. As Bloom observed, "It appeared that the ecological health of the Salton Sea was better than previously believed."[28]

The seeming remission of long-term illness, however, was lamentably brief. On August 15, a neighbor called the refuge to report seeing dead pelicans along the southeastern shore. Bloom was on his way to a meeting when he got the news. His route paralleled the shoreline where the birds had been reported, and so he drove over to have a look. Over a distance of two miles he saw the carcasses of six white pelicans, enough to raise concern. After the meeting, he asked biologist Ken Sturm to take one of the refuge airboats and reconnoiter the area. In little more than an hour, Sturm was back. The news was bad. He had seen sick pelicans scattered over a large expanse of sea, pelicans so sick they made no effort to get away when he brought the boat up next to them, many of them in a strange state of collapse, weakly paddling to shore. There was no doubt that something dreadful had begun. And from that point on, says Bloom, "things went downhill fast."

The victims of the emergent epidemic were fish-eating birds, mainly pelicans, which lolled in the water unable to lift their great, ungainly heads, and thereby risked drowning. The birds that could make it to shore crawled up on the beaches to die, their heads dangling as though their necks had neither strength nor structure. In some quarters, avian botulism goes by the name "limber neck," an apt description for what Bloom and his people were seeing. In due time the National Wildlife Health Laboratory would confirm that the sick and dying birds were infected with avian type "C" botulism, but on the afternoon of August 15 at the Salton Sea National Wildlife Refuge, none of the staff needed a formal diagnosis to know that a wildlife disaster of frightening breadth was unfolding around them—again. They scrambled to respond.

From the outset, Bloom and his staff resolved to save as many sick birds as they could, an effort that brought them quickly into new and unfamiliar territory. They improvised pens and rigged canvas awnings to shield the birds they brought in from the 115-degree heat. They dispatched boats to collect the sick and the dead, but no one had much experience wrestling wild pelicans, big birds with plenty of muscle and a hook at the tip of their bills that can inflict a nasty wound. In the first days of the die-off, rescuers simply stowed the birds they caught in their boats, and some of the healthier pelicans responded naturally enough by flailing and lunging at their handlers. Several rescuers suffered severe bites, and one was badly raked across the face. A solution was soon forthcoming. The rescuers gathered all the pillow cases they could lay their hands on. They cut off a corner from the closed end, and presto: a pelican straitjacket. Birds were plucked from the water, their bills and heads quickly thrust through the opened corner, and the rest of the pillow case pulled down over the wings and body. Thus restrained, the birds waited calmly through their boat ride to shore.

The initial chaos of the first days yielded slowly to a semblance of order. Boats patrolled established routes to collect birds. Periodically they put in at transfer points on shore, and their live birds were lifted out and placed in pickup trucks for transport to the triage center at headquarters. At day's end, or sooner if necessary, the boats returned to the dock at headquarters and unloaded their gruesome cargo of dead. In these ways, birds both alive and dead arrived at the refuge by the score, then by hundreds, soon by thousands.

This was August, two hundred feet below sea level, in one of the hottest deserts on earth. Conditions were miserable: the heat was debilitating and the work heartbreaking; the reek of death surpassed description; prospects for relief were bleak. Soon it was September, and the crisis did not abate. October came and the Salton Sink was still an oven, and birds kept dying. Refuge staff worked from dawn to dark, seven days a week. Allies from the California Department of Fish and Game and from other units of the Fish and Wildlife Service, together with volunteers and private wildlife specialists, joined the rescue force until it numbered eighty or more. Even so, they were as continuously overwhelmed as they were exhausted. From the first hours of the emergency, Bloom had contacted wildlife rehabilitators to come and help nurse the stricken birds, but there were far more of the sick than could be handled. The order went out: ignore the dying white pelicans, bring in only the ailing—and endangered—browns. "That was the shame of it," says Bloom. "We could have saved a lot more birds, if only we had had the means to take care of them."

The handling of the dead was especially stressful. The incinerator Bloom had purchased in the aftermath of the grebe die-offs was at last operational, but it was undersized, designed to burn one hundred pounds an hour, which is a lot of grebes, but only four or five pelicans. The furnace roared twenty-four hours a day for weeks on end, only slowly eating away at the pile of waiting carcasses, from which ooze and nastiness seeped in damp trails through the dust, the flies swarming, the stench overpowering.

As hard as they labored, as exhausted as they worked, as many injuries from fatigue and strain as they suffered (one rescuer fell from his boat, which then ran over him and broke his arm), the rescue force, or "Joint Incident Command," as it was officially named, never controlled the outbreak—it remained for cooler weather in late October and early November to break the biological chain of transmission by which the botulism spread. But without the labors of the rescuers things could have been much worse: the collection and disposal of carcasses probably thwarted secondary disease outbreaks and may have damped the amplitude of the botulism.

The strenuous efforts of Bloom's large team also saved the lives of hundreds of birds and probably constituted the largest and most successful rescue ever of an endangered species. More than eight hundred brown pelicans were picked up and shipped to the Pacific Wildlife Project, a rehabilitation group, which somehow gained the use of a wing of the Orange County dog pound, soon renamed "Pelican Hall." More than two hundred white pelicans also made the trip to the hall, and altogether over 60 percent of the treated birds survived, most of which were subsequently released to the wild.

Still, the outbreak was devastating. The rescue force handled some 14,000 birds representing 66 species. Most of them were dead or so close to death that they were euthanized. The 8,539 white pelicans that perished in the outbreak represented approximately 15 percent of the western North American population of the species, while the 1,129 brown pelicans that died represented close to 10 percent of a population that prior to the die-off numbered no more than 5,000 breeding pairs. Actual mortality in both species was somewhat greater still, as rescuers were unable to pick up all the birds felled by botulism. The total death toll for the year, including other species, was finally estimated at 20,000.

Although dramatic, the outbreak was in no way exceptional, at least not in the purest sense of the word. It resulted not from the sudden visitation of a foreign scourge but from the growing debility of the local ecosystem. Researchers ultimately concluded that the spread of botulism had depended on a succession of organisms, beginning with tilapia and other small fish. Stressed, presumably, by high salinity, low oxygen, a smorgasbord of contaminants, and assorted other miseries, the fish fell victim to a bacterium, *Vibrio alginolyticus*, which is widely present throughout the seas and oceans of the world but prospers best in hot, very salty water. Under assault from the vibrio, the fish's alimentary canals became blocked, allowing botulism spores that were already present to multiply. The sick fish bloated and rose to the surface, where they swam about slowly, stupidly, and—from the point of view of a pelican or cormorant—enticingly. The birds feasted. Botulism then infected the birds, which, as they died, were preyed upon by gulls, as grebes in the earlier die-offs had been. The birds' carcasses soon swarmed with maggots, a favored food of many ducks. The maggots brimmed with botulism spores, and so the ducks died and bred more maggots, and the cycle wore on.[29]

The die-off of 1996 differed from those of 1992 and 1994 in many ways: the greater human effort made in response to it, the rescue and rehabilitation of sick birds, the greater understanding of causes. Greatest of all, however, was the public response. Clark Bloom estimates that after the first week of the 1996 crisis, fully half of his time was given to interviews with the media. Every TV station in southern California sent a crew to the Salton Sea. Every newspaper, large or small, sent a reporter. The national dailies and networks provided extensive coverage. And each one made demands. "I wasn't a very nice person," says Bloom. His real concerns were elsewhere, and the twelve-hour days eroded his concern for niceties, but he gave the interviews; they were part of the job.

The reason for so much interest was that pelicans were involved. Birds, generally, are not easily portrayed as warm and cuddly. They are, after all, bird-brained and facially unexpressive; nothing that wears a beak can smile. But if there is one bird that can compete with puppies, koala bears, and Bambi for human empathy, it is the pelican. The unthreatening Pleistocene strangeness of the creature makes it beloved of animators, illustrators, writers of light verse, and, quite naturally, children. Where earlier the deaths of a much larger number of grebes had caused people to scratch their heads, now the agonies of dying pelicans caused them heartfelt pain. Bureaucrats and politicians took note. Slowly the audience of listeners willing to consider the plight of the Salton Sea began to grow.

Grow as that audience might, it could only watch as the problems of the sea persisted. In the early months of 1997, the refuge collected 2,500 sick and dead grebes, felled again by the mysterious agent. Periodic fish kills continued. As spring bled into summer, refuge personnel collected about 200 dead grebes a week. Biologist Ken Sturm kept a lookout for sick pelicans and sent several dozen to Pelican Hall in Orange County. In April, 10,000 tilapia washed up dead; they were heavily infected with *Vibrio alginolyticus*. In May, refuge staff found well over 2,000 fish-eating birds dead, in association with an ongoing fish die-off. In addition, 1,600 double-crested cormorants on Mullet Island near the mouth of the Alamo River, most of them nestlings, perished of Newcastle disease, a viral disease that causes nervous system breakdown and paralysis. In the same area, 100 newly hatched Caspian terns and various mature birds succumbed to ailments that remain unknown. In August, a three-mile-long raft of dead tilapia suddenly appeared. Someone estimated that it consisted of a million fish. It was a good, round number, and no one particularly disagreed. On this occasion the grim reaper appeared in a new guise: *Amyloodinium ocellatum,* a dinoflagellate gill parasite that proliferates in warm waters. Most of the fish sank, adding tons of decaying flesh to the sea's high nutrient load. By the end of the year, more than 6,000 birds of 51 species had been picked up and cremated at the refuge, and 300 sick pelicans had been delivered to the care of rehabilitators.[30]

The dance of disease and death spun on—the 1998 body count had already reached 17,600 by late July, when a still larger raft of dead tilapia, estimated to consist of more than 2 million fish, washed up on the south shore of the sea between the mouth of the New River and the refuge headquarters.[31] The heat of August was still approaching. No one knew which element of the sea's beleaguered ecosystem might lead the next turn, or how the elements might join together, accumulating woes until some new threshold of fatal beleaguerment was crossed. A watery world that once seemed the province of trophy corvina and wheeling flights of waterfowl now heaved with tides of epidemic disease. Its most powerful actors possessed neither eyes nor feathers nor even scales. They included selenium-saturated pile worms, cyanobacteria, *Vibrio* bacteria, botulism spores, and gill parasites of the tongue-twisting genus *Amyloodinium.* No one could escape the fear that these faceless agents might recruit yet another co-conspirator, some unnamed creature smaller than a pinhead that only bided its time until the rancid soup of the Salton Sea should spoil some more.

In the summer of 1997, the Fish and Wildlife Service opened a wildlife hospital at the refuge for the treatment and recovery of stricken birds, especially pelicans. In August the service convened a workshop in Palm Springs attended by more than a hundred scientists, agency managers, and others. The purpose was to identify research priorities that might lead to an understanding of the sea's ecology sufficient to provide a basis for remedial action. Participants identified thirty-one major projects, spread over three years, the cost of which they estimated at $36 million. More than six years had elapsed since 150,000 grebes died, and at last the Fish and Wildlife Service had a wish list. Meanwhile, back at the refuge, Clark Bloom, thanks to repeated emergencies, found himself severely over budget (and therefore less than popular within the agency) yet eager for "the next available financial opportunity" when he might order a second and more capacious incinerator, this one gauged to pelicans.[32]

16 | PIPE DREAMS

If the Salton Sea could be heaved up on a gurney and wheeled into the most efficient emergency room on the planet, treatment would still be slow. The sea's medical history would take a long time to record, there being so much of it. And the sea's girth and ponderousness, combined with its diversity of ailments, would require the examining physician to employ the utmost ingenuity and patience.

Suppose a capable doctor might be recruited to the task, and suppose the doctor, having examined the patient and sifted the available data, asks the sea's friends to hear her out as she presents her findings. What might she say?

She would likely begin by saying that the patient's vital signs are bad: salinity is much too high; the accumulation of wastes threatens a condition of fatal uremia; the interplay of salt and waste with pesticide metabolites, selenium, viruses, and other agents remains largely unknown and cannot be assumed benign. The doctor would further explain that no matter how severe and expensive the treatment, a favorable outcome can in no way be assured.

Nevertheless, the doctor gravely proceeds with a review of available therapies, of which there are principally three.

First, *dialysis*. One might ring the sea with desalinization plants, continuously treating its water to remove salts while also filtering out organic wastes, undesired minerals, and anything else deemed harmful. Trouble is, the cost in plants and in the energy to run them would be staggering. With a sigh, the doctor explains that desalinization of seawater is generally considered too expensive even as a means to provide water for human consumption, let alone to recycle the two and a quarter *cubic miles* of Salton Sea. She pauses. There is also, she says, the nettlesome problem of what to do with the large volume of waste the plants would produce, a caustic slurry of salt and sewage. The doctor clears her throat, as doctors do upon touching a subject not easily resolved, and moves on to the next alternative.

Second, *transfusion*. The idea here is to connect the patient to another body of water with healthier fluids, and to exchange them. The donor of healthy fluids would have to be very large in order to have fluid enough to spare—and still larger in order to absorb the fluids of the Salton Sea without ill effect. Only the ocean fits that description. Fortunately, the Gulf of California lies not far away. One might pump ocean water into the Salton Sea, and Salton Sea water into the ocean.

At this point, the doctor's auditors begin to smile; they have been waiting for a hopeful sign. But the doctor shakes her head and solemnly continues. The concept has an appealing simplicity, she explains, but the numbers underlying it are obdurate and discouraging. The transfusion of one body whose salinity measures 45,000 parts per million with 35,000-ppm water from a second body cannot quickly produce substantial change. Even to reach a salinity of 40,000 ppm over the span of a decade would require whole rivers of water to flow in both directions. Such a transfusion would need a pair of pipelines, each sixteen feet in diameter, with more than a million acre-feet of water surging through each one every year.[1] In order to achieve salinity in the treated body equal to that in the donor—to restore the Salton Sea to the salinity of ocean water—the transfusion would have to be even more massive, effectively replacing the contents of the inland sea with water from the ocean. A pipeline, even a gang of pipelines, would be insufficient to move such volumes. A canal with locks might suffice, but as with dialysis, the cost of construction, operation, and maintenance would mount to the stratosphere. There is also, the doctor adds, the question of securing the cooperation of the Republic of Mexico, a nation that owes the United States no gratitude where the Colorado River is concerned. Again, she clears her throat.

The third treatment alternative is *amputation and prosthesis*. If the sea were smaller, the doctor explains, the dilutive effect of its existing inflows would freshen it more. If a quarter or a third of the sea were cut off, she says, we might save the rest for a time. A pall falls over the small group of listeners. Dismemberment of their beloved lake was not what they had in mind. One of them asks, What becomes of the part that is cut off?

It will serve as an evaporative basin, the doctor answers. A prosthesis, in the form of a dike, will separate the surviving sea from the amputated basin. Gates in the dike will allow inflows of healthy water to the remaining sea to push saltier water into the basin. The listeners do not find this answer satisfactory. One asks, Won't the surviving sea eventually become as salty as the whole sea is now? Won't more amputations be required until the living part is entirely whittled away?

Essentially, yes, the doctor admits, but only over the span of generations. The amputation buys time and holds promise of enabling the greater portion of the sea to remain productive far into the foreseeable future.

But the doubter persists: What happens in the evaporative basin? Doesn't it become a rather large and dangerous waste dump?

The doctor clears her throat.

A FOURTH alternative goes undiscussed. It is the alternative of inaction, of no treatment, of letting the patient die. Every previous recurrence of Lake Cahuilla succeeded eventually to hypersalinity. Why not this one? Unaided and unaltered, the consequences that have been collecting in the Salton Sea since the Great Diversion of 1905–1907 will cause the sea to become a hypersaline lake devoid of fish early in the twenty-first century.

From the longest and broadest perspective, the sea's current agony of bird die-offs and fish kills might be viewed as the expected, though certainly regrettable, death throes of the marine environment as it "flips" from saline to hypersaline conditions. Certainly, fish must die. The same was true in the day of Lake Cahuilla, when the suckers and other fish that the natives caught in their fish traps died out, leaving only pupfish behind. Many of the birds that preyed on the fish moved elsewhere, but others did not. A hypersaline sea may still teem with life—witness Mono and Great Salt lakes.

Hypersalinity would cause the Salton Sea's introduced marine fish to die away; its pupfish would retreat to salty tributaries like San Felipe Creek. Eventually, except at the mouths of freshwater inflows, pile worms would also succumb, but life in the broad deep sea would not cease. Brine shrimp, the staple of countless aquaria and not a few lakes throughout the world, are already present in the sea, and they would multiply in the absence of aquatic predators and become the foundation of a new food chain. Grebes, which relish brine shrimp, would thrive, while pelicans would not. Long-legged waders would still stalk the fresh-to-salty shallows of the deltas and drain mouths, probing the mud for invertebrates or snatching minnows from the eddies. The effect of hypersalinity on the constituents of the ecosystem would vary, species by species.[2]

In centuries past, the same thing repeatedly occurred as previous lakes evaporated: creatures that could neither travel nor adapt died out, while displaced birds flew on. And in those previous metamorphoses, there were a number of other places, notably the Colorado delta, for them to fly to.

Today, those alternative oases, if they exist at all, are much reduced and beleaguered. The dried-out sloughs of the delta hardly offer haven for the tilapia-gobbling pelicans of the Salton Sea; indeed, not even the undamaged delta would have provided such birds with hunting grounds as rich as the overpopulated sea. To say that displaced birds would move on to other habitats ignores the fact that comparable habitat may not exist. A fair portion of the displaced birds, like many human refugees of war or famine, will die outright or suffer a slow decline.

The consequences of a hypersaline sea are potentially worrisome. Will the sea's high nutrient load cause brine shrimp populations to boom and burst as the tilapia's did? Will such die-offs result in new cycles of disease? Without pile worms to gather selenium from lake sediments, one might assume that less of the toxic element might enter the food chain, but would the bacteria that now remove selenium from the system also die out? Would selenium then begin to enter the tissues of wildlife and humans by some new and unmapped route? The uncertainties that attach to future scenarios for the Salton Sea are virtually innumerable, and some may lead to outcomes dangerous to human health. If, for instance, inflows were diverted from the sea (to be purified for agricultural or human use), the level of the sea would drop, exposing the selenium-rich lake bed to sun and wind. The same forces that built the Algodones dunes from ancient sandy lake shores would lift those unwelcome sediments into the air and carry them far and wide, to the likely misfortune of people downwind.[3]

NONE OF THE alternatives for addressing the woes of the Salton Sea can be considered attractive—and this is one reason why, after decades of discussion, nothing has yet been done. The Salton Sea confounds the region's and the nation's traditional confidence that physical problems must inevitably yield to engineered solutions. William Smythe was right when he foresaw that no land on earth, save perhaps the dike-ringed lowlands of Holland, would be more transformed by engineering than the old Colorado Desert. Yet even he could not foresee the eventual, total control of the wild and mighty Río Colorado, converted now into a chain of lukewarm lakes. Nor could he or anyone foresee the emergence of a sixteen-million-inhabitant megalopolis on the dry southern coast of California.

The story of southern California in the twentieth century is fundamentally a story about the control of nature. And yet the Salton Sea is out of control. It is the physical contradiction of a tenet of American life. No good and easy solution to its problems exists, and its problems grow steadily worse. In increasing numbers, people have decried the resulting injury to wildlife, to property values, to the area's economy. They have lamented the loss of recreation and development potential. They have demanded action, and yet action has not come.

The California Department of Fish and Game accurately and publicly predicted the death-by-salt of the Salton Sea in 1961. A plan for saving most of the sea by diking off part of it as an evaporative sump—the alternative of amputation and prosthesis—sparked widespread public interest and discussion as early as 1974.[4] Yet despite continuing complaint and agitation from activists like Norm Niver, nothing was done.

The year 1988 brought a modest breakthrough: a hodgepodge of government and private interests formed a Salton Sea Task Force. Although it did little but meet and talk for the next five years, the task force helped focus attention on the sea. Its very existence became a reminder of the somnolence of public agencies and irrigation districts, which should have been developing a plan of response. Ultimately, in 1993, the task force metamorphosed into the Salton Sea Authority.

The new entity had substance. It was brought into being by a joint powers agreement executed by the two counties, Riverside and Imperial, in whose territory the sea lies, and by the two irrigation districts, IID and the Coachella Valley Water District, whose drainwaters maintain it. The agreement empowered the Salton Sea Authority to receive and spend public money and to enter into binding contracts. Together with the U.S. Bureau of Reclamation, it soon set about a long-neglected task: the systematic evaluation of alternatives for saving the sea. By the authority's definition, "saving" the sea meant restoring its salinity to a level near that of ocean water—between 35,000 and 40,000 ppm. It also meant lowering the elevation of the sea to its approximate elevation in the go-go days of Penn Phillips, between 230 and 235 feet below sea level.[5] With a pragmatism rare in Salton Sea affairs, the authority further specified that a program to save the sea should entail "O&M" (operation and maintenance) costs of no more than $10 million a year. The authority assumed that while the state of California and the federal government might fund the initial cost of a dike or pipeline, local entities would have to bear the year-to-year expense of tending the device.

The authority evaluated fifty-five alternatives embracing a range of schemes for dialysis, transfusion, and amputation and prosthesis, the last of which included proposals for onshore evaporation ponds independent of a dike. The authority completed its evaluation in the summer of 1996, as large numbers of pelicans began to die. It soon selected the alternative that fit its criteria best: the construction of a dike in a shallow portion of the sea. Further study would determine the exact location of the dike and the size of the evaporative basin it would enclose. The project was expected to cost between $100 million and $200 million—an enormous sum but still less than pipelines costing an order of magnitude more. Moreover, annual O&M costs would not exceed the $10 million limit the authority had set. The authority did not exclude the possibility of pursuing other strategies in the future, including building a pipeline for discharge of Salton Sea water into Mexico's Laguna Salada, but for the time being, the recommended dike was the only affordable alternative. It was a first step—enough, it was hoped, to get things started.[6]

But events intervened. While the authority deliberated, pelicans kept dying.

Dozens of camera crews came and went. Journalists filed scores of stories. In the words of one member of the Salton Sea Authority, "The pelicans changed everything. Nobody knew about us before then."[7]

THE PELICANS changed people's assumptions about the limits of the possible. Elected officials who had scarcely been aware that the sea had problems now began to call it an "issue." One of these was the Republican congressman from California's forty-fourth district, the former mayor of Palm Springs, Sonny Bono.

This was the same Sonny Bono who, wearing bellbottom jeans and a goofy moustache, had attained celebrity in the 1960s by playing the clown to his statuesque wife and singing partner, Cher. Upon his improbable election to Congress, Bono's name instantly appeared on every list of the "Ten Dumbest" or "Least Likely to Succeed" congressmen, but to his credit (or, depending on point of view, to the discredit of his colleagues), Bono proved to be a relatively effective member of the House of Representatives and an energetic advocate for his district.

Pelicans may not have been much on his mind in the fall of 1996 as he won election to a second term, but early in 1997 Bono experienced a kind of epiphany. He was flying with his chief of staff, Brian Nestande, from Palm Springs back to Washington when he looked down from his airplane window and meditatively studied the broad, placid Salton Sea far below. After some minutes of silence, Bono turned to Nestande and remarked the beauty of the sea and his fondness for the days of his youth when he had boated and water-skied on it. As Nestande later recalled, by the time their conversation ended Bono had resolved to champion the sea's restoration. "He started rattling off a hundred things for me to do when we landed, and the next day he formed the Congressional Task Force on the Salton Sea."[8]

The task force consisted of Bono and four other southern California congressmen with whom he met regularly. Their purpose was to write legislation to guide the rescue of the sea and to persuade both congressional and administration leaders to support their bill. They were sufficiently energetic that the dynamics of Salton Sea politics immediately shifted. Task force members such as Representative George Brown, a native of Holtville and a powerful Democrat, openly favored a pump-in/pump-out transfusion scheme, and no one on the task force blinked when told that the necessary pipelines might cost a billion dollars.[9] The prudent deliberations of the Salton Sea Authority were soon forgotten. Prospects for lavish funding suddenly seemed high, and the universe of possible projects again loomed large.

Proponents of dialysis, such as Richard Heckmann of US Filter, took heart. Heckmann had never liked the limit of $10 million in O&M costs set by the Salton Sea Authority. As head of the world's largest water purification corporation, he hoped for big, profitable projects with giant budgets. Heckmann commissioned a study of the economic impact a cleaned-up sea might have. As one of his lieutenants put it, if a rehabilitation plan "generates $9 million, $10 million [in O&M] is too much. If it generates $300 million, $10 million is not very much." The study, to no one's surprise, buttressed US Filter's position by claiming billions of dollars in economic benefit from a cleaned-up sea.[10]

Even the proponents of diking felt encouraged. If political leaders were willing to spend a billion dollars, they reasoned, the dike could be more than an earthen berm in shallow water. It could be an engineered dam in the deepest part of the sea, capable of withstanding a "head" of water—that is, a difference in elevation and pressure between the water outside the dike and the water inside. With a deep-water dike, the surviving Salton Sea would become an irregular doughnut, and the evaporative sump would be its hole. Placing the dike in deep water had several advantages. More water might be displaced—that is, greater amputation achieved—with less loss of surface area. No one's shoreline would be lost. The dike would remain effective even if the elevation of the sea substantially dropped—and there was every likelihood that it would, because of both water transfers and cleanup of the New River.[11] In fact, if inflows declined, the prosthesis of the dike might be used to maintain the sea's present elevation and, by extension, a pleasing shoreline. Not least, a deep-water dike would keep the unpleasantness of the evaporative sump farther out of sight, and out of mind.

Advocates of diking, including Richard Thiery, an environmental scientist with the Coachella Valley Water District, acknowledged that "the ocean pump-in/pump-out idea is attractive; it has a sound-bite quality. Understanding that it won't work requires doing a little arithmetic, which is not something people frequently do or like doing." Plus, he points out, there remains the nettlesome business of securing agreements with Mexico, which means that the pipeline concept would "eventually have to face reality," leaving diking the only reasonable alternative.[12]

But advocates of other strategies feel similarly about dikes and impoundments: "Inside the dike is gonna be a mess, folks, a mess," argued a senior vice president of US Filter at a 1998 symposium on the future of the Salton Sea.[13] No one disagreed. Environmentalists dislike the amputation strategy because they fear the resulting sump will poison wildlife and the general environment. Those who beat the drum of economic development reject it just as emphatically, for they understandably believe that the public may have qualms about the wholesomeness of a recreational lake that encircles a dead-zone of toxic sludge.

THE MORE THE congressional task force reopened possibilities for the sea, the more the meaning of "saving the Salton Sea" began to shift. Bono, clearly, was not interested in merely rescuing a fishery or preventing birds from dying. He wanted the sea to be a busy playground for coastal Californians, as it had been in his teenage years. US Filter's Heckmann, a master of persuasion, echoed him. They and others argued that the Salton Sea should be saved not just for the birds and fish or for the good of those who already lived beside it. They argued that the sea could become an engine for economic growth for the entire region, attracting the development of a new generation of marinas, towns, and resorts. The sky, again, was the limit.

In January 1998, these shifts of meaning and political momentum received a boost in a most unfortunate way when Sonny Bono died in a snow skiing accident. Suddenly, the campaign to save the Salton Sea featured a fallen hero (and an open congressional seat, which Republicans badly wanted to keep). Politicians of national prominence spoke of their obligation to "honor Sonny's memory" by enacting legislation to save the sea. Every speech about the sea—and they now came thick and fast—called for speaker and auditors to "do everything we can to make Sonny's dreams of the Salton Sea become a reality."[14]

Where the Salton Sea was concerned, Sonny Bono dreamed of nothing less than the rebirth of the "Salton Riviera." It was as if he borrowed images straight from the Penn Phillips scrapbook. According to his press secretary, Bono thought the sea should become "a premier destination resort and residence opportunity for the people of this region. He envisioned marinas, perhaps floating islands for marinas, incredible development along the shoreline, great economic opportunity for the residents and the tribes in the region, and really putting a first-class resort on the map that would energize this whole region economically."[15]

In American politics, the line that separates cheerleading from leadership blurs easily. Salesmen, promoters, entertainers, and other "communicators" attain popularity and sometimes high office because they tell people what the people want to hear. They become the apostles of grand visions and do not allow themselves to be troubled by the facts, if the facts promise to be troubling. They rally and enliven their followers, and once in a great while, they render society the service of redefining the limits of the possible and inspiring people to accomplish more and better things than they might otherwise have done. More often, however, they take their believers on a voyage to nowhere. The landfalls conjured in their rhetoric turn into banks of clouds or melt into a sea of partisanship and impracticality. The energy they unleash goes to waste.

Sonny Bono's immediate legacy was to energize the spiritual heirs of M. Penn Phillips, the promoters of an unlimited future for the Salton Sea. Like their antecedents, these visionaries were baptized in the continent's ever-flowing river of real-estate dreams, and also like Phillips and his contemporaries, they believed they could divert that river down from the golf meccas of Palm Springs, Palm Desert, and Indian Wells to the desert at the edge of the sea. Magician that he was, Penn Phillips harnessed the dream-river long enough to irrigate the Salton desert and start a crop of subdivisions, but nothing he sowed ultimately ripened. His heirs are poised to try again. The obstacle they face is not a mountain, a desert, or a rocky shore; it is another river, a real one with actual water, that reached the basin first.

The threshold question is whether the Salton Sea, bereft of outlet, can be made to serve the river of dreams, as it already serves agriculture by accepting its waste.

If it were possible to rule nature by decree, there would be no contest. Earthly powers would repeal the law of evaporation, order the Salton Sea to find an outlet to the ocean, and command the Colorado River to double its flow. But with such deeds out of reach, the legislators of the Congressional Task Force on the Salton Sea attempted the next best thing, which was to order that a solution to the sea's problems be put in place with all haste, leaving the details to others. This was the strategy embodied in the "Sonny Bono Memorial Salton Sea Reclamation Act," which the task force introduced to the House of Representatives early in 1998. The bill called for the secretary of interior to evaluate the feasibility of alternatives to "reclaim" the sea and, in consultation with the Salton Sea Authority and others, to select the best alternative, design a plan for its physical implementation, develop an estimate of its cost, ensure its environmental compliance, and report back to Congress—all this within a year's time. The bill also suspended the public's right to object to the secretary's decisions and provided a whopping $350 million with which to commence the actual construction of an unidentified project of unknown cost. It further commanded that construction begin within two months after submission of the report. The sense of the bill and the opinion of its authors was that the time for talk had ended; it was time to "move dirt."[16]

For a while, it seemed that dirt might indeed move, no matter how errant or precipitate the order to move it. In midsummer, the Bono bill narrowly passed the House by a vote of 221 to 220, mainly along party lines. Sonny Bono had been a Republican; his widow now held his seat, and the Republican leadership give the bill its full support. In the Senate, however, cooler heads prevailed, and the House's bill was entirely redrafted. Ultimately, the 105th Congress provided $5 million for wildlife studies at the Salton Sea National Wildlife Refuge and $3 million for experimental wetlands to help clean up the New and Alamo rivers; it also added $13.4 million to the budget of the Environmental Protection Agency, with most of the money earmarked for assistance to the Salton Sea Authority. Without requiring or funding the moving of dirt, it further ordered the secretary of interior to proceed with the feasibility study outlined in the earlier bill, and in a touching bit of political theater, it renamed the Salton Sea Wildlife Refuge in honor of Sonny Bono.[17]

NO ISSUE stays long in the news or in the minds of politicians or the public, neither selenium scares, nor prophesies of fishery collapse, nor flood and property losses, nor even the mass death of pelicans or the passing of a popular congressman. Advocates for the Salton Sea have watched public attention to their cause rise and fall like a penny stock. But the long-term trend of attention has favored them. Each time the spotlight finds the sea, it shines brighter. And now a new factor will help to keep it focused there.

The new factor is California's obligation to adopt a meaningful "4.4 plan"—that is, to reduce its withdrawals of Colorado River water to a level that at least approaches the annual limit of 4.4 million acre-feet set by the Colorado River Compact.[18] It will be impossible for California to produce a credible 4.4 plan that does not include large-scale transfers of water from the Imperial Valley to the coast. Present versions of the plan assume a transfer of 200,000 acre-feet from Imperial Valley to San Diego, the general terms of which have been approved by both IID and the San Diego County Water Authority. Nevertheless, many obstacles remain. If implemented, the transfer may ultimately involve as few as 130,000 acre-feet or as many as 300,000, but it cannot proceed at any level until it achieves environmental compliance at a cost IID and San Diego are willing to bear.[19]

The IID-San Diego transfer may substantially alter both the quantity and the quality of water entering the Salton Sea. To the degree that those effects are deemed material to the sea's long-term viability, the fate of the transfer will be tied to that of the sea, and pressure to complete the transfer will increase pressure to aid the sea. If, as seems likely, a federal project is launched to remedy the problems of the sea, then the project will indirectly constitute a subsidy of the water transfer and, by extension, of California's movement toward 4.4 compliance. More than the fate of pelicans or the hopes of real-estate developers, the water politics of the Colorado basin and of one of the thirstiest, most densely populated desert regions in the world will undergird the effort to "save" the Salton Sea.

If something is done, how likely is it to be the right thing? Nearly every plan for the sea has emphasized the control of salinity and elevation. Certainly both are important, yet the vital issue of nutrient loading and resultant oxygen depletion lurks in the background, largely ignored. The dynamics of selenium and other potentially powerful biological agents, meanwhile, remain a cipher. Any plan of rescue needs to address the totality of the sea's many problems, but much of the $36 million, three-year research program identified for the sea in 1997 remains unfunded.[20]

In addition to the imperative need for good information, a rescue effort should obey two other imperatives. The first is to integrate the rescue effort—*deliberately*—with emergent environmental plans for the rest of the lower Colorado region, including the Colorado delta in Mexico. The second is to address the question of water rights for the Salton Sea.

At present, the Salton Sea contains too much salt and too many nutrients. If the sea is to be saved—if it *can* be saved—salt and nutrients must be removed. Irrespective of the means of extraction, the material removed must be safely disposed of, which means finding a way to move it "uphill but downstream" toward the ocean. This might mean building a pipeline for salt slurry along the Coachella and All-American canals to the mothballed Yuma desalinization plant, thence along the route followed by Wellton-Mohawk effluent to a saltwater outlet in Mexico. It might mean a pipeline to the Laguna Salada or to an outfall in deep water in the head of the gulf. No matter the destination, Mexico will have to bless the project, and Mexico's blessing will not and should not be forthcoming without the resolution of other border-region environmental problems, among them the survival of delta ecosystems. The forum for negotiating a balanced regional approach to the needs of the lower Colorado already exists: it is the International Boundary and Water Commission, which administers the 1944 Water Treaty between the United States and Mexico. In the half century since the treaty went into effect, diplomats have repeatedly adjusted and refined it through the formal adoption of "minutes," which function as administrative amendments to the treaty. To begin development of a new "environmental minute" would in no way strain the IBWC's authority. Indeed, it is hard to see how any serious rescue plan based on transfusing the Salton Sea can move far forward without such an agreement.[21]

An environmental minute to the treaty might address trade-offs between aid for the delta and a pipeline for the Salton Sea. Almost certainly, such an agreement would address the use (or partial use) of surplus Colorado River flows for environmental purposes—particularly in Mexico. The use of such flows to sweeten the sea, incidentally, is expressly disallowed under the Salton Sea Reclamation Act of 1998, as well it should be considering California's chronic violation of its 4.4 limit and the comparatively greater need of the Colorado delta for freshwater.[22] The Imperial and Coachella valleys, after all, consume more than one-fifth of the entire flow of the river. The Salton Sea and its problems exist because of the valleys' diversions, and the valleys' wealth exists, in part, thanks to their ability to dispose of wastewater freely and with scant regard for consequences. The idea of sending yet more water into a basin that has already consumed so much offends logic as well as justice.

In the end, the taxpayers of the United States will likely pay the bill for resolving—or at least attempting to resolve—the problems that water use in the Imperial Valley has created in the Salton Sea. Depending on the outcome of a suite of environmental decisions (some of which will surely be litigated), such a bailout will help further the IID-San Diego water transfer and California's nominal adherence to its 4.4 obligation. No one need be surprised. Where the West and especially the hydraulic Southwest are concerned, U.S. taxpayers have a long tradition of furnishing subsidies and bailouts. The region ought not be let entirely off the hook, however, which brings us to the idea of a water right for the Salton Sea.

The transfer of 200,000 acre-feet of water to San Diego, if unmitigated, will reduce inflow to the sea. (The amount of reduction will depend on the conservation methods employed.) With less inflow, the elevation of the sea will drop, and so will the rate at which it is diluted; with less dilution, the concentration of salts and nutrients will rise even faster than they are rising now.

By the grace of geography, the Imperial Valley is positioned so that a semi-accidental lake, instead of more farms and irrigators, lies downstream of it. If the reverse were true, the IID would be legally obliged to maintain downstream flows in order not to impair the water rights of farmers dependent on its waste stream. In all fairness, the same principal should be applied to the water interests of the Salton Sea, both for the wildlife and for the economic values the sea makes possible.

For every gallon of water that a farmer in Imperial Valley puts on his field, roughly three quarts evaporate or are taken up by plants. Only one quart of each gallon finds its way to the Salton Sea. Of that amount, about two-thirds (17 percent of the original gallon) flows off the field as tailwater. The rest percolates through the soil and is discharged as drainwater. Tailwaters, which dissolve fertilizers on their trip across the field, contribute much of the sea's nutrient load, while drainwaters become a vehicle for salt and selenium. Both sources help to maintain the sea.

Ignoring for the moment the idea of establishing water rights for the Salton Sea *in law,* such rights might come into being in a de facto sense as a condition of state or federal approval of a transfer agreement. Such a right would require that flows equivalent to drain and tailwater flows be maintained. That is to say, a farmer who sells water to San Diego would sell only three quarts from every gallon. He would sell, that is, only the portion he actually consumes of the water he diverts. He would allow the fourth quart to flow on to the Salton Sea, and it would make the trip directly and cleanly, without a contaminating detour across or under a field. This principle of respecting rights for the Salton Sea might be applied whether a farmer "harvested" water for sale through fallowing, tailwater pumpback, drip irrigation, or any other conservation method.[23]

The same principle ought also to be applied to the downstream *economic* effects of a water transfer. That is to say, a portion of the revenue generated by the sale of water should find its way to the communities whose economic activity declines thereby. Potential profits from water sales are high—according to one study, they could reach $85 an acre-foot, even after allowing for debt service on water conservation investments.[24] For both moral and pragmatic reasons, the Imperial Valley needs to distribute the benefits of water sales equitably. If farmers refuse to share the wealth, if they treat the water which the nation and its taxpayers provide them as if it were their exclusive property, if they refuse to acknowledge the collective interest of their communities in that water, then, in the words of IID director Don Cox, "you've got war."

In the Imperial Valley, the general population—not just farmers and landowners—elects the directors who run the IID. Cox explains the implications: "It really doesn't matter if a farmer owns that water or the public does because for every farm voter there are ten nonfarm voters. If you are making money profiteering from selling water, they are going to find some way to take that money or that water away from you."[25]

THE CONVERGENCE of multiple powerful interests in the waters of the Colorado guarantees that no solution to the predicaments of the former Colorado Desert will come easily. The interests exist at all levels of political organization from local to international, and in every possible arena: economic, environmental, social, and cultural. Still, a rescue of some kind for the Salton Sea, though difficult, is not as technically demanding as landing a man on the moon, nor is it as politically challenging as the pursuit of peace between Protestants and Catholics in northern Ireland or between Jews and Arabs in the Middle East. Even in these most difficult arenas, obstacles sometimes fall.

Nevertheless, the prospects are daunting. Efforts toward solution might falter for any number of reasons: the staggering costs, the impatience and short attention span of politicians, the difficulty of attaining—and maintaining—broad agreement on what should be done.

Whichever path is chosen, we may be sure of several things. Research will lag behind the need for knowledge. As a society, we will always be obliged to grope in the dark, for we can never know all we need to know, let alone all there is to know, about ecological systems and the effects our actions have on them. We should therefore hope hard, as well as take precautions, that we do no harm. This is easier said than done. The rescued Salton Sea, if such a creature comes to exist, will be a highly managed system. If past experience in other environmental venues holds true, the design and operation of the management system will be guided by notions of economic efficiency. The sea will be freshened and purified only so much—so as not to entail undue cost. The judgment of adequacy will be based on current knowledge, which is always imperfect. Since full agreement on the meaning of such knowledge rarely exists, management decisions will shift from the domain of science to the world of politics and economics, where expedience rules. Such shifts are not at all uncommon. A number of studies suggest that in virtually every instance where people have tried to manage complex systems for "sustainable" outputs, whether of cod or timber, grass or freshwater, the systems have crashed—or at least have departed from their natural range of variation.[26] The Salton Sea starts at a great disadvantage, being already far advanced in the process of collapse. No one with a serious appreciation of the odds should bet heavily on a happy outcome.

Yet American society will likely make a large bet on the Salton Sea, and it will do so, not least, because of fear of what will happen if it does not. Society's legitimate fear of the unknown will cause it to stumble into other unknown realms. The law of unintended consequences, which is as unappeasable as the law of evaporation, will extract its due. Elements of the desiderata of the Salton Sea will be saved—a species here, a shoreline there. Other elements will be lost, and the muddling through of trying to make things work, of placing Band-Aids upon Band-Aids, will continue ad infinitum. The hope must be to do no conscious harm, to provide as much buffer and backup as possible for all systems, economic and environmental, so that misfortune in one dimension of the problem does not spread to all the others.

A SOCIETY can rightly be judged by the dreams of its people. Every dream is a story about the future, a vision of how things *should be*. A dream necessarily mirrors the qualities of its dreamers; perhaps nothing mirrors them better. In the lands of the old Colorado Desert, the Quechan Indians dreamed of a world that had achieved its greatest perfection at the moment of its creation, a world complete in its unity and in the satisfaction of its gods, to which the people ever aspired to return. The forty-niners could not have been more different. They dreamed unabashedly of wealth, adventure, forward movement, and continental domination. They dreamed of a future unlike anything the past had seen. They were true to their time and true to the rash, bold spirit of a young America.

Reclamationists are more of a puzzle. Some, like William Ellsworth Smythe, drank deeply of the political and social idealism bequeathed them by Jefferson and earlier dreamers. If Jefferson grew tipsy at the thought of a yeoman democracy, the same vision, distilled through time and circumstance, gave Smythe the staggers. We may applaud his utopian hopefulness, but his gullibility was deplorable and dangerous. Smythe utterly misunderstood the relationships of power that large-scale reclamation would produce. To the degree that he failed to grasp that economics, more than ideals, would shape the irrigated West, men like Rockwood and Chaffey played him for a fool, even as their own dreams turned to folly.[27]

The dreams of our time may prove equally ill conceived, but we cannot live without them. The rescue of the Salton Sea, if it is attempted, will serve a vision of how the basin should be. It cannot be done otherwise. Unfortunately, the emergent rhetoric for that vision sounds too often like the discredited tinsel promises of M. Penn Phillips. It would be tragic to ransack the public treasury for the sake of floating marinas and shabby casinos.

The vision that guides restoration of the Salton Sea will be binocular. One lens will focus on the needs of the environment, the other on economic growth. In the first field of view, avoiding wildlife deaths will deservedly command priority, but mere avoidance of further loss is not half so inspiring or worthwhile as the idea of restoring habitats throughout the land, both U.S. and Mexican, that the lower Colorado River formerly sustained. The Colorado delta remains the greatest untapped environmental opportunity in the region, if not the continent.

Those who venture into the field of the second lens should be especially wary. Promises of economic growth are easily betrayed, and the record of dreamers like Charles Rockwood and Penn Phillips hardly inspires confidence. Now that the occupants of the Salton basin have appealed to the rest of the nation for rescue from the consequences of their use and misuse of water, the momentum will build for great projects and grand designs. The heirs of Rockwood and Phillips must relish the coming opportunities. They are a numerous breed; some lurk in the halls of Congress, some at companies like US Filter. They will beat the drum for restoring the sea in order to create a sump-side version of Las Vegas. Caveat emptor. Such a dream is fitter for the analyst's couch than for a legacy to history.

The tide of history, meanwhile, has shifted. The most powerful stories in the

North American past have concerned the efforts of European Americans to reshape the land to suit their needs and dreams. The spread of neo-European settlement, the subjugation of native people, the development of agriculture and industry, the growth of cities, the utilization of rivers, forests, rangelands, and other resources, these things and their kin have shaped the physical expression of history's passage on the continent. Both literally and figuratively, much of what has happened in this part of the world since 1492 has involved the breaking of new ground. But that story is now largely the story of the past. The story of the future will be more like the story of the Salton Sea. It will concern society's efforts to live with and at times ameliorate the consequences of what was broken. We have entered an age of obligatory adjustment and repair.

Twined into the story of the future will be an infinity of human lives and relations, which are the main cargo of history. Not a few of them will feature the same quiet heroism that Joe R. Hernandez and his mother, Amparo, showed in placing the good of others first, their private dreams second. Some will demonstrate the same dedication to duty and kindness that the staff and allies of the Salton Sea Wildlife Refuge have shown in battling through the ugly work of repeated epidemics. Somewhere in such episodes of selflessness may lie the elements of a vision for the region that will be true to basic good sense and worthy of effort and loyalty. Perhaps in years to come, the glitz and hype of dream hucksters may even be offset by small armies of down-to-earth people like Lester Murrah, whose heads are not easily turned and whose idea of the good life is modest and sane. One may so hope. One may even hope that the future of the basin will have room in it for a few stray cats and desert pupfish, for eccentrics like Leonard Knight and misfits like Mad Jack. The Colorado Desert has long attracted the restless and the original, drifters like Godfrey Sykes and John Van Dyke. May their kind not cease to wander through.

In the future of the old Colorado Desert, each new day will begin much like the one before. As the sky lightens, towns like Brawley and El Centro will stir to the sound of a guttural diesel roar. The throaty argument of engine against load, of gears under strain, will commence. Tractor trucks will pull their trailers onto the highway, bound for the far away with masses of produce and products. The trucks will gather speed as they head down the road to markets innumerable, each one like an individual life carrying the jumble of the present into an infinite and unknown future.

WHILE PREPARING to write this last chapter, I dreamed my own salt dream:

My wife and I were in Salton City, shopping for a house. The one we liked best and hoped to buy had formerly belonged to Sonny and Cher, early in their marriage before they were famous. (The Bonos, let it be said, never owned a house in Salton City, but this was, after all, a dream.) We found the house appealing but unremarkable, except for the last room. We were told that Sonny had designed and built the room himself.

It was a bright, airy meeting room that might easily have accommodated fifty people. The realtor told us that Sonny built the room in order to provide a venue for finding solutions to the problems of the Salton Sea.

Light poured in from windows on three sides. Thin, gauzy shades diffused the light and seemed to trap it in the room, so that anything that might take place within that space would be fully, even brilliantly, illuminated.

The fourth wall of the room, which adjoined the rest of the house, was the room's most extraordinary feature. It consisted entirely of doors, one after another with no space between them—doors that overlapped and rolled on tracks or swung open to reveal deep walk-in closets, each of which was crowded with shelves and cabinets from which books, files, and reports spilled. Each door and each trove of information to which it gave access was differently colored and configured. I wondered if there were as many of these doors as there were chapters in this book.

One of the doors led ultimately to the utility room of the house, which I was most curious to see. I followed a narrow corridor and came to a hot little room lit by a bare bulb and crammed with machinery. I identified the hot water heater but was puzzled that there should be so much additional equipment. I listened as pumps started up with an electric whir, or shut down. The room was never silent; the hum of pumps filled it. I felt their vibration in every tank or pipe I touched. And the many pipes that snaked around, connecting tank to pump to other pipes to tank again, created a baffling labyrinth of plumbing. A few pipes, thank goodness, had been labeled. Each label was like a sign at a desert crossroads: it named a faraway location, indicated its direction with an arrow, and gave a distance. One said, "Colorado River Aqueduct," another said, "Los Angeles," and a third, "Mexican Delta." In the moment of the dream, all the distances were illegible, and some of the labels were hard to see, but I was sure that if I could have squeezed to the far side of the largest tank, I would have found a pipe labeled, "San Diego."

I came back to the meeting room, where the hardwood floor gleamed as though it were a pool of light. Like the desert bowl outside, the room was shadowless with illumination, and it seemed to radiate an unusual spirit: it was a little like a ballroom, a little like a sports arena, a little like a conference hall. It had the spirit of a place where people would strive and struggle and perhaps celebrate. I stood there, feeling almost optimistic and hoping earnestly that in such a space, people might one day resolve the dilemmas of the Salton Sea—and of the many places that are its hydraulic co-dependents.

I cling to that hope even now, though it may indeed be a dream.

Notes

SOURCES FOR INVOCATIONS

E. W. Gifford, *The Kamia of the Imperial Valley,* 48; Herbert Eugene Bolton, *Rim of Christendom,* 594; Cormac McCarthy, *Blood Meridian,* 254–255; William E. Smythe, "An International Wedding," 286; D. T. MacDougal, "A Voyage below Sea Level on the Salton Sea"; J. Smeaton Chase, *California Desert Trails,* 194; Pelham D. Glassford, as quoted in Cletus E. Daniel, *Bitter Harvest,* 248–249; Guillermo Gómez-Peña, *Warrior for Gringostroika,* 44; Richard Shelton, *Going Back to Bisbee,* 181; Mary Belardo, personal communication, January 14, 1994 (at the time of the cited interview, Mary Belardo was chairman of the Torres-Martinez Desert Cahuilla).

CHAPTER 1. HEAD WATERS

1. University of California, Imperial County, Cooperative Extension, "Guide Lines to Production Costs and Practices, 1992–1993" (Holtville, Calif.); Agricultural Commissioner, Imperial County, "Imperial County Agricultural Crop and Livestock Report, 1992" (El Centro, Calif.).

2. María de la Paz Carpio-Obeso (Instituto de Ingeniería, Universidad Autónoma de Baja California, Mexicali), oral presentation at the workshop "Water and Environmental Issues of the Colorado River, Border Region," San Luís Río Colorado, Sonora, Mexico, April 30, 1998, sponsored by the Pacific Institute for Studies in Development, Environment, and Security and by Defenders of Wildlife.

3. Richard E. Lingenfelter, *Death Valley and the Amargosa: A Land of Illusion,* 366.

4. A facsimile can be found a hundred miles north of the Salton Sea in Bristol Dry Lake, where salt is still mined in the caustic bed of what was once a lake bottom, a place that replicates, in miniature, the sterility and shadelessness that once ruled the heart of the Colorado Desert. Bristol Dry Lake is a reminder, if one were needed, of how phenomenal was the transformation of settlement.

5. B. A. Cecil-Stephens, "The Colorado Desert and Its Recent Flooding," 372.

6. Marc Reisner, *Cadillac Desert,* 2 and *passim.*

7. Contemporary advocates of "sustainable development" would do well to compare their rhetoric with that of the reclamationists of a century ago. After allowing for differences in stylistic conventions, the promises of the two groups sometimes bear striking similarities.

8. William E. Smythe, *The Conquest of Arid America*, 43. Smythe was a Massachusetts Yankee whose uninspired career as a journalist flourished once he began to beat the drum for reclamation. His conversion to the cause came in 1889 in northern New Mexico when he encountered *acequia* irrigation along Vermejo Creek. The next year, drought struck the Great Plains, and Smythe, from his newspaper job in Omaha, watched as farms failed, homesteads were abandoned, and a stream of defeated prairie settlers retreated eastward. The spectacle caused him to reflect on the durable Hispanic hamlets of New Mexico and on their tradition of irrigation farming. Smythe soon quit his job and moved to the capital of irrigation, Salt Lake City, where he founded what was to become reclamation's most influential periodical, *Irrigation Age*. (See Martin E. Carlson, "William E. Smythe: Irrigation Crusader"; for Smythe's later life, after his disillusionment with large-scale irrigation, see Henry S. Anderson, "The Little Landers' Land Colonies.")

9. William E. Smythe, "An International Wedding," 286–300.

CHAPTER 2. DREAMS OF EARTH

1. Alfred L. Kroeber, *Handbook of the Indians of California*, 755–785. (This encyclopedia was originally published in 1925 as Bulletin 78 of the Bureau of American Ethnology of the Smithsonian Institution. The Dover edition that I cite is an unabridged re-publication of the original.) See also Jack D. Forbes, *Warriors of the Colorado*. The following excerpt, collected by Kroeber (773), gives some taste of the quality of Yuman tales and song cycles. It is, however, no more than a small fragment of a tale of epic length:

> In the morning Amainyavererkwa went up river once more. This time he went farther and traveled until he came to Asesmava. There he slept. In the morning he went on. He visited all his friends and received to eat whatever kinds of food they ate different from his own. He thought that perhaps his son had gone up for the purpose of eating these strange foods. Late in the afternoon he came to Amata-akwata, Kukake, Ahtanye-ha, and Avinyidho. He inquired there among his friends. It was now sunset and his friends gave him to eat, but told him they had not seen his son. He ate a little and slept there that night.
>
> In the morning he ate a little again. His friends gave him red paint and feathers which he packed and put on his shoulder and then he started (south) homeward.
>
> At sunset he came to Akwereha. He had [*sic*] no one living there, but lay and slept there. He thought: "I will call this place Akwereha, and all will call it by this name when they tell stories. And I will leave my paint and feathers here, and will call it also Amata-sivilya-kwidhaua (feather-having place)."
>
> In the morning he went on, and while it was still early came to Kwakitupeva and Kwasekelyekete (Union Pass, north of Fort Mohave). There he drank a little. He was now feeling very bad on account of his son. Then he began to run until he came to Amata-kamota'ara. There he drank again and then ran on until he came to Ammo-heva (Hardyville). Having drunk once more, he went on until he came to Ismavakoya and Mach-ho. There he looked to see how far he still had to go. Then he began running again. He ran until he came to Akweretonyeva. Then he thought: "I am nearly at my house now." So he walked fast until he came to Selya'aya-kumicha, and then to the top of Amai-kwitasa. Then he looked towards his house at Amata-tasilyka. As he stood, he saw it was dusty about his house as if there were wind there. The dust was from many people.

When he returned home he again thought of the north. He wanted to go north once more to look for his son. Then in the morning he took his sandals and started. He came to Oachavampeva, Asmalya-kuvachaka, Amata-kuma-ta'ara, and Avi-tunyora.

2. Robert L. Bee, "Quechan," 97; Bee, *The Yuma,* 13; Stephen Trimble, *The People,* 393, 410–415. Avikwame is Newberry Mountain, near the extreme southern tip of Nevada.

A word on nomenclature: the Quechan people are also widely known as the Yuma, a term that probably is a Spanish version of the name by which O'odham (Pima and Papago) people knew them, people who, incidentally, helped guide Spanish explorers from Kino's time to and through the area. *Yuma* and *Yuman* have since come to be applied in certain contexts to all members of the language group to which the Quechan belong. This large and heterogeneous group includes dryland, upcountry people east of the Colorado River (Yavapai, Walapai, and Havasupai), desert and mountain bands west of the river (Ipai, Paipai, and Tipai, who include Imperial Valley bands identified in ethnographic literature as Diegueño and Kamia), and, particularly, the tribal people inhabiting the riparian corridors of the lower Colorado and Gila rivers.

This last group will herein be referred to as river Yumans. They include the Cocopa of the Colorado delta, the Quechan of the area of Yuma Crossing, the Mohave of certain upstream canyons and valleys of the Colorado as far as present-day Nevada, and the Maricopa of the middle Gila. The river Yumans also included groups that are no longer tribally distinct, such as the Halchidhoma, who formerly dwelled along the Colorado between the Quechan and Mohave but were forced by them to abandon their lands and move up the Gila. Along with similarly uprooted tribes, the Halchidhoma were ultimately absorbed among the Maricopa. During historic times a non-Yuman, Shoshonean people, the Chemehuevi, moved into the area vacated by the Halchidhoma. The mosaic of Yuman people is more complex than can be described here. For more information, start with Kenneth M. Stewart, "Yumans: An Introduction," in *Handbook of North American Indians,* vol. 10. A thorough attempt to trace the shifting territories and allegiances of Colorado River tribes is in Forbes, *Warriors of the Colorado.*

Today, the Mohave possess a reservation upstream of Needles, California. Also, together with the Chemehuevi and relatively recently uprooted Navajos, they occupy the Colorado River Indian Reservation downstream of Parker. The Quechan Reservation lies across the Colorado from Yuma, while the Cocopa retain a small reservation near Somerton, Arizona, and maintain several small, impoverished communities in the Mexican state of Baja California del Norte.

3. Carroll L. Riley, *The Frontier People,* 141.

4. George P. Hammond and Agapito Rey, *Narratives of the Coronado Expedition, 1540–1542,* 146ff. See also Riley, *Frontier People,* 149.

5. Herbert Eugene Bolton, *Rim of Christendom,* 416–417.

6. Bolton, *Rim of Christendom,* 414.

7. Kroeber, *Handbook,* 784.

8. Bolton, *Rim of Christendom,* 418.

9. Later accounts would identify the Quíquima as the Halyikwamai, a tribe forced from the Colorado during historic times by warfare with the Quechan and their Mohave allies. The Halyikwamai moved up the Gila River, eventually to become incorporated among the Maricopa along with several other displaced Yuman groups. See Henry O. Harwell and Marsha C. S. Kelly, "Maricopa," 74.

10. Bolton, *Rim of Christendom,* 471.

11. J. N. Bowman and R. F. Heizer, *Anza and the Northwest Frontier of New Spain,* 32.

12. Stanley Noyes, *Los Comanches,* chapter 7; William deBuys, *Enchantment and Exploitation,* 79.

13. Bowman and Heizer, *Anza,* 111. Anza revealed slightly more of his emotional response in the description of the Quechan he confided to his diary:

> They are a heathen and barbarous people, but of this class they are the best that I have seen in their attitude toward the Spaniards, for in their affection and liberality they exceed all the tribes whom we have reduced. . . . They are not naturally ugly, but they make themselves so in a superlative degree with many and diverse paintings with which they cover the entire body. (Bolton, *Anza's California Expeditions,* vol. 2, 171)

14. The literature on Yuman subsistence is extensive. See Kroeber, *Handbook,* chapters 50, 52–53; Bee, *The Yuma,* 20ff.; Alfonso Ortiz, ed., *Handbook of North American Indians, vol. 10: Southwest, passim;* William H. Kelly, *Cocopa Ethnography;* Edward W. Gifford, *The Kamia of the Imperial Valley.*

15. The most detailed discussion of the use of Sonoran panic grass is in Gary Nabhan, *Gathering the Desert,* 151ff.

16. DeBuys, *Enchantment and Exploitation,* 17; Alfonso Ortiz, ed., *New Perspectives on the Pueblos,* 3–6, 143.

17. Bolton, *Anza's California Expeditions,* vol. 4, 101.

18. Bolton, *Anza's California Expeditions,* vol. 4, 103ff.

19. Kroeber, *Handbook,* 753; Bee, *The Yuma,* 53. See also Forbes, *Warriors of the Colorado,* 291–292 and *passim* for discussion of intertribal conflicts.

20. Bolton, *Anza's California Expeditions,* vol. 4, 119.

21. Bee, *The Yuma,* 42–47.

22. A monumental exception is Alvar Núñez Cabeza de Vaca's *Adventures in the Unknown Interior of America.* Written in 1542 under the title *La Relación* as an official report to the king of Spain, Cabeza de Vaca's story is truly a founding narrative of American literature. Rare among early chroniclers, Cabeza de Vaca wrote with sympathy for indigenous people and appreciation for their world and ways—no doubt because he sojourned with them so long and because throughout his years of wandering (1527–1536) he was obliged to meet the problems of existence on the same terms as they. One matter into which Cabeza de Vaca offers considerable insight is the ever-presence of hunger. Unlike Yuman river tribes, the Indians among whom Cabeza de Vaca traveled rarely knew the luxury of surplus. His narrative recounts many instances of extreme suffering from cold and hunger. Again and again, not only he but his Indian captors and later friends and followers starved or froze, only to be saved by the lucky discovery of a supply of edible cactus fruit or a few handfuls of pine nuts, which sustained them, if barely, while they journeyed to greener pastures or waited for some new crop to manifest in nature. This was indeed a balance with nature, but an unenviable one, and the penalty for stepping off the teeter-totter of survival was unambiguous and final.

Such was the probable condition of life for most North American tribes, including those who lived in the deserts and mountains bordering the fecund corridor of the Río Colorado. Said Anza of the Quemeya (the Kamia or eastern Tipai) of the Imperial Valley: "They are so hungry that, with the dirt and everything, they gathered up in a hurry some grains of maize which remained on the ground from that which was given to some mules. These [together with other Tipai or possibly Cahuillas or Cupeños] which I saw afterward at the Pass of San Carlos, appeared to me the most unhappy and unfortunate of all the people that I have seen" (Bolton, *Anza's California Expeditions*, vol. 2, 131–132).

23. Kroeber, *Handbook*, 798.

24. Hammond and Rey, *Narratives of the Coronado Expedition*, 146. Font, among others, mentions that the Quechan were well supplied with Hopi blankets; see Bolton, *Anza's California Expedition*, vol. 4, 103.

25. Boma Johnson, *Earth Figures of the Lower Colorado and Gila River Deserts*, 29. Certainly the diggers of the pit, as they proceeded, would not have thought to seek the advice of the Quechan. Considering that the U.S. Bureau of Reclamation never consulted the tribe before building the All-American Canal across Quechan land in 1938 (Bee, *The Yuma*, 94), a road builder or cement producer can hardly be held to a higher standard on land outside the reservation. For additional insight to the horse intaglio, see Barry Lopez, "The Stone Horse."

26. Johnson, *Earth Figures*, 14; see also *Imperial Valley Press*, April 19, 1994.

27. Forbes, *Warriors of the Colorado*, 129. Cf. Bee, *The Yuma*, 38.

28. Bolton, *Anza's California Expeditions*, vol. 2, 175; Bee, *The Yuma*, 38; Bolton, *Coronado: Knight of Pueblos and Plains*, chapter 15.

29. Kroeber, *Handbook*, 755 and chapter 51, generally.

30. The ethnographic literature is vague about the extent to which Yuman women shared, if at all, in the potential for "significant" dreaming.

31. Kenneth Stewart, "Mohave," 65. Also Kroeber, as in note 29.

SIDE TRIP. JACUMBA PASS

1. As quoted by Olivia Doughty, personal communication, July 1993. The source of the anecdote was Mrs. Doughty's father, Efrain "Blackie" Zambada, who came to the United States in 1919 from revolution-wracked Mexico. Aside from a stint working on the Golden Gate Bridge and a few other forays to distant lands, Mr. Zambada spent his entire adult life working in the fields of Imperial Valley.

3. DEAD MULES AND NIGHTMARES

1. Ralph P. Bieber, ed., *Southern Trails to California in 1849*, 380.

2. David Lavender, *California*, 165ff. J. E. Durivage offers specific information about the costs of one of the more economical routes to the gold fields. While still in northern Chihuahua and having "come thus far with as much ease and comfort as could be expected," Durivage estimated that "unless a man has four or five hundred dollars in his pocket he had better remain at home" (Bieber, *Southern Trails*, 198–200).

3. Anza's route crossed the Colorado Desert well to the south of the Southern Emigrant Trail. On his expedition of 1776 Anza's worst weather problem was a steady cold rain. When he distributed brandy to his shivering colonists to restore their spirits, Padre Font censured him for encouraging immorality. Anza, as usual, ignored the abstemious Jesuit (Bolton, *Anza's California Expeditions*, vol. 2).

4. Ralph P. Bieber, *Exploring Southwestern Trails, 1846–1854,* 217 n.220.

5. Bieber, *Exploring Southwestern Trails,* 216.

6. The route followed by the forty-niners was well known to river Yumans, whose best runners, according to oral tradition, could bring abalone from Pacific tides to Colorado floodplains before it spoiled (Boma Johnson, staff archaeologist, Yuma Office, Bureau of Land Management, personal communication, November 23, 1993). Such a feat required crossing nearly 200 miles of desert and mountains in no more than three days—an astounding performance but not out of keeping with others that have been reliably recorded. In 1886, Lieutenant John G. Bourke made note of a Mohave Indian, Panta-cha, who covered nearly 200 miles along the Colorado in less than 24 hours. Farther east in Arizona, a decade or so later, Charlie Talawepi, a Hopi, is said to have run the 156 miles from Tuba City to Flagstaff—and back—in the same amount of time (Peter Nabokov, *Indian Running,* 17–23). Such achievements are not entirely a thing of the past. In 1993, Victoriano Churro, a 55-year-old Tarahumara wearing huarache sandals, won the Leadville Trail 100, covering a grueling, high-altitude course of 100 miles through the Colorado Rockies in just 20 hours, 2 minutes, and 33 seconds (Don Jones, "Tarahumara: Running as an identity," *Santa Fe New Mexican,* August 21, 1994, C-3). Being Indian, however, conveyed no certain protection against the rigors of the desert. In 1773, Sebastian Taraval, a Cochimi Indian from Baja California who later served Anza as a guide, fled San Gabriel Mission (near Los Angeles) with his wife and attempted the bleak jornada of the Colorado desert from west to east in August heat. Ignorant of the location of water, he struck directly for the Colorado across the burning Algodones dunes, where his wife, after horrible suffering, perished of thirst (Forbes, *Warriors of the Colorado,* 149).

7. Bieber, *Southern Trails,* 159.

8. Ibid., 225.

9. Ibid., 226–227.

10. Ibid., 228.

11. Forbes, *Warriors of the Colorado,* 297ff.

12. Bieber, *Southern Trails,* 228.

13. Ibid.

14. David P. Robrock, ed., *Missouri Forty-Niner,* 188–189.

15. Terminology for the Kamia is imprecise. As used here, the term refers to the Indians of the Colorado Desert described in Edward W. Gifford's 1931 ethnography *The Kamia of the Imperial Valley.* These people appear never to have been numerous and to have intermarried and maintained fluid relations with neighboring groups including the Quechan and, more importantly, the mountain bands to the east and southeast, to whom the Kamia were most closely related. Essentially the Kamia of the Colorado Desert were a subset of the people variously called Diegueño (from the Spanish mission established to Christianize them), Tipai, or Kumeyaay (names by which the people have referred to themselves). Ronald May ("A Brief Survey of Kumeyaay Ethnography") reported in 1975 that members of the tribe preferred the name Kumeyaay, numerous variants of which appear in the anthropological literature. The Kumeyaay homeland includes most of the Imperial and Mexicali valleys and the coastal mountains between the Imperial Valley and the California coast southward, far into northern Baja California. See also Katherine Luomala, "Tipai and Ipai."

16. Robert F. Heizer and Adan E. Treganza, *Mines and Quarries of the Indians of California,* 359.

17. John W. Audubon, *Audubon's Western Journal,* 166; Robrock, *Missouri Forty-Niner,* 188.

18. Bieber, *Southern Trails,* 229.

19. Ibid., 230.

20. Ibid., 232–233. Durivage refers to a famous report written by William Emory as a lieutenant in the U.S. Army. Durivage was evidently aware that soon thereafter Emory received promotion to major. See William H. Emory, *Lieutenant Emory Reports: A Reprint of Lieutenant W. H. Emory's* Notes of a Military Reconnaissance.

21. Bieber, *Southern Trails,* 234.

22. Ibid.

23. "Carrizo creek is dry, except at occasional points where the water is forced to the surface by rock. There is a constant supply of water where it emerges from the hills to lose itself in the desert." R. S. Williamson in U.S. Senate, Exec. Doc. no. 78 (hereinafter abbreviated *Pacific Railroad Survey Reports),* 40.

24. About Palm Spring, William Hunter wrote on November 28, 1849, "Many of the trees had been cut down, no doubt by emigrants to feed their animals, a number of which lay in and about the springs" (Robrock, *Missouri Forty-Niner,* 193–194). See also William Phipps Blake, "Geological Report," part 2 of *Pacific Railroad Survey Reports.*

25. Hunter describes a "Laguna" formed by New River flows (probably that which was later known as Blue Lake, near the present site of Seeley): "On its bosom sported many water fowl, some of which we shot but did not get, as no one would venture in least [*sic*] he should mire. Here were the fragments of burned wagons and other wasted property, which some miserable looking Indians were collecting. Many of these are now lurking about the road, watching for exhausted oxen, or such as have died, from the carcass of which they speedily cut the flesh, and retire to glut themselves on their unsavory prize, or with it strung across their ponies, make their way to their wigwams to lay up a store for winter use. Some 5 or 6 miles off they have several broken down horses and mules, which I suppose they intend recruiting" (Robrock, *Missouri Forty-Niner,* 191).

26. Lavender, *California,* 149–150, 165.

27. Cf. Joan Didion, "The Golden Land."

28. Bieber, *Southern Trails,* 163, 168–169.

29. Audubon, *Audubon's Western Journal,* 167.

30. Ibid., 38.

31. Bieber, *Southern Trails,* 379.

32. The preeminent interpreter of the California dream is Kevin Starr, state librarian of California, whose serial history of the state includes *Americans and the California Dream, Inventing the Dream, Material Dreams, Endangered Dreams,* and *The Dream Endures.*

SIDE TRIP. YUMA CROSSING

1. As quoted in Arthur Woodward, *Feud on the Colorado,* 29. In addition to Woodward, valuable sources on this place and period include Forbes, *Warriors of the Colorado,* and Richard E. Lingenfelter, *Steamboats on the Colorado River 1852–1916.* A surviving member of

the Glanton gang, Samuel E. Chamberlain, gives his version of events in *My Confession*. No account of this troop of psychopaths, however, can be more vivid than Cormac McCarthy's fictionalized treatment, *Blood Meridian*, certainly one of most powerfully imagined novels of the American West.

CHAPTER 4. MEMORIES OF SEAS

1. Bieber, *Southern Trails*, 272.

2. Ibid.

3. Gifford, *The Kamia of Imperial Valley*.

4. Job Dye, a mountain man who first crossed the Colorado Desert with Ewing Young early in 1832, claimed credit in his memoirs for naming New River (*Recollections of a Pioneer*, 67–68). In February 1849 he sailed from California, where mules were in great demand, to Mazatlán. He gathered a herd of three hundred and drove them north across Sonora to Yuma Crossing. From there he followed the emigrant trail to Monterey, arriving on August 12, but he gives no other dates and so we may only suppose that he encountered the New River some time in early or mid-July. His comments about crossing the Colorado and the Gila, however, warrant reflection. He says, "I lost 80 odd head in crossing the Jila [*sic*] and the Colorado," and then later adds, "On our arrival at the Colorado river we found the stream very high and difficult to cross the mules and baggage. . . . The mules . . . had to swim half a mile or more, and in crossing fourteen head were carried down the stream and lost." Dye's account seems to suggest that he crossed both the Gila and the Colorado (losing mules in each instance) and that both rivers were in flood. Indeed, since Dye came from the south, it would have made no sense for him to pass up the usual fords and cross the Gila at all unless by so doing he avoided the even more dangerous, combined waters of both rivers, downstream of their confluence. If floods on the Gila in 1849 caused the appearance of the New River, it is not surprising. A flooding Gila River also triggered the Great Diversion of 1905. As to the New, the desert stream that Dye encountered was no trickle: "That evening we traveled up the New River about ten miles, but could not find a crossing to ford the stream, so we were compelled to camp for the night. The next day I made a raft and crossed the river" (69).

5. As quoted by Robrock, *Missouri Forty-Niner*, 267.

6. Couts also went into the ferry business at Yuma Crossing and collected customs duties from Sonorans returning from the gold fields (Forbes, *Warriors of the Colorado*, 308–309; Audubon, *Audubon's Western Journal*, 167). Audubon reported obtaining biscuits and rice at Camp Salvation from a "Col. Collins." This was probably Colonel James Collier, a newly appointed customs collector on his way to San Diego with thirty dragoons (Robrock, *Missouri Forty-Niner*, 187, 192–193, 269–270. Goetzmann, *Army Exploration*, 160–163).

7. Robrock, *Missouri Forty-Niner*, 190.

8. Abert to Francis Markoe, May 18, 1849, as quoted in William Goetzmann, *Army Exploration in the American West 1803–1963*, 209–210.

9. Goetzmann, *Army Exploration*, chapter 7, especially 262–266, 295.

10. Goetzmann theorizes that had Davis and high-ranking officers of the Topographical Engineers been less biased in favor of a route along the thirty-second parallel, its competitor

along the thirty-fifth might have marshaled support enough to win selection, thereby serving better the interests of the nation and the South. Ibid., 303–304.

11. *Pacific Railroad Survey Reports,* 2A.

12. David B. Dill, Jr., "William Phipps Blake."

13. William Phipps Blake, "The Cahuilla Basin and Desert of the Colorado," 1.

14. Blake, "Geological Report," 90.

15. Blake "Cahuilla Basin," 1.

16. Blake, "Geological Report," 93.

17. Philip J. Wilke and Harry W. Lawton, "Early Observations on the Cultural Geography of the Coachella Valley," 21–24. See also Lowell John Bean and William Marvin Mason, *Diaries and Accounts of the Romero Expedition in Arizona and California, 1823–1826.*

18. Blake, "Geological Report," 95.

19. Finis C. Farr, *The History of Imperial County, California,* p. 27. Farr's history treats the settlement of Imperial Valley as a heroic conquest by estimable men and women under the most trying natural circumstances. Its section titled "Indian Occupation" (25–28), however, is a litany of deprecation. Farr's thesis, belief in which we may assume was shared among much of the white community, was that the native dwellers of the desert so lacked skill and energy that they had failed in the Darwinian struggle for survival.

20. William Phipps Blake, "Ancient Lake in the Colorado Desert," 436.

21. Ibid. The Mohave also had a great fondness for tattooing, especially among women. According to Kenneth M. Stewart ("Mohave," 67), the Mohave believed that tattoos helped assure an appropriate afterlife: "Regardless of a person's behavior while alive, the soul went to [the land of the dead], with only a few exceptions, such as the victims of witchcraft, and those who died without having been tattooed, who were believed to pass down a rat hole at death." Perhaps, then, it was out of kindness to Olive Oatman, a white women held by Mohaves from 1852 until 1856, that her captors adorned her chin and lower cheeks with an array of dark ciphers. Miss Oatman's photograph after her return to white society (Stewart, "Mohave," 63), shows a dark-eyed, handsome woman in a pinched-waist gown, and the incongruous, indelible staining of her face renders it one of the most forlorn images in the history of the American West.

22. Blake, "Geological Report," 98.

23. Blake, "Ancient Lake," 436.

24. Blake, "Geological Report," 100.

25. Lowell John Bean, Sylvia Brakke Vane, and Jackson Young, *The Cahuilla Landscape,* 24, 26, 50; *Imperial Valley Press,* June 7, 1998, "Ancient fish traps, villages line shore of former Lake Cahuilla; protection sought." For more on the ancient lake and those who dwelled upon its shores, see David L. Weide, "Regional Environmental History of the Yuha Desert."

26. Blake, "Geological Report," 103.

27. Bieber, *Exploring Southwestern Trails,* 213. Early speculations about a prehistoric lake by Heintzelman and others are discussed in Wilke and Lawton, "Early Observations," 10n3.

28. W. G. Mendenhall, "The Colorado Desert," 687.

29. Blake's letter to the *Commercial Advertiser* was soon reprinted in the *American Journal of Science,* 2d series, vol. 17 (1854): 435–438.

30. Blake, "The Colorado Desert," in *Pacific Railroad Survey Reports,* 237.

31. Ibid., 236.

32. Blake, "Cahuilla Basin," 4.

33. Wilke and Lawton, "Early Observations," 11.

34. Fred B. Kniffen, *The Natural Landscape of the Colorado Delta,* 165; Norris Hundley, "The Politics of Reclamation: California, the Federal Government, and the Origins of the Boulder Canyon Act—A Second Look."

35. Donald L. Baars, *The Colorado Plateau,* 32–33.

36. Blake, "Colorado Desert," *Pacific Railroad Survey Reports,* 249–250.

CHAPTER 5. LOOMINGS

1. Daniel Hillel, "Lash of the Dragon." The Yellow River again served as a weapon of war in 1938, when Chiang Kai-shek breached the dikes to flood the path of an advancing Japanese army. The ensuing inundation, thought to be the worst in China's history, slowed the invaders for only three months. Its greater legacy was the death of an estimated 890,000 of Chiang's countrymen. See also Patrick E. Tyler, "China's Endless Task to Stem Centuries of Floods," *New York Times,* September 15, 1996, A-1, 6.

2. On the readiness of the Colorado to begin moving westward, see Godfrey Sykes, *The Colorado Delta,* 169ff. On the magnitude of the flood of 1905, see David Meko and Donald A. Graybill, "Tree-Ring Reconstruction of Upper Gila River Discharge."

3. Mahlon Dickerson Fairchild, "A Trip to the Colorado Mines in 1862"; Francis J. Johnston, "Stagecoach Travel through the San Gorgonio Pass"; Hubert Howe Bancroft, "Guide to the Colorado Mines."

4. Lester Reed, *Old Time Cattlemen and Other Pioneers of the Anza-Borrego Area.*

5. Passenger Department, Southern Pacific Company, "Imperial Valley, California" (1908 pamphlet).

6. Sykes, *Colorado Delta,* 110. Additional accounts of the salt works may be found in Charles F. Holder, "A Remarkable Salt Deposit," 224; Arthur J. Burdick, *The Mystic Mid-Region,* 162–171; and George Wharton James, *The Wonders of the Colorado Desert,* 324–329.

7. Burdick, *Mystic Mid-Region,* 162.

8. Wilke and Lawton, "Early Observations," 12.

9. Godfrey Sykes, *A Westerly Trend,* 224.

10. Not until the herd reached the Arkansas River did Sykes learn that the previous owner of his blankets had come down with smallpox. As Sykes later explained, neither the owner nor the foreman "thought it necessary to enquire whether I had ever had small-pox—which I had not!—but relied on the prophylactic air of the plains, and the bean-pot and corn-bread of our cook, to ward off infection. These measures had proved to be so effective, and the treatment so pleasant in this early experience of mine, that I forthwith adopted them as remedial agents and have placed great reliance upon them ever since." *A Westerly Trend,* 73.

11. Ibid., 208–209.

12. Sykes, *Colorado Delta,* 49 and map, flyleaf.

13. Sykes, *Westerly Trend,* 220.

14. Cf. Cecil-Stephens, "The Colorado Desert and Its Recent Flooding," 376.

15. Sykes, *Colorado Delta,* 40. Volcano was near the location shown as Wister on some contemporary maps. Sykes gives the trip no further mention in either *The Colorado Delta,* a

superb physiographic monograph, or *A Westerly Trend,* his arch and fascinating autobiography, an overlooked western classic.

16. Farr, *History of Imperial County,* 157.

17. Cecil-Stephens, "The Colorado Desert and Its Recent Flooding," 374.

18. Blake, "Geological Report," 250.

CHAPTER 6. NATURE REDREAMT AND REDRAWN

1. Harold Bell Wright, *The Winning of Barbara Worth,* 79.

2. Franklin Walker, *A Literary History of Southern California,* 214–217.

3. Vilma Banky played Barbara and Ronald Colman took the role of Willard Holmes, the young engineer patterned after Henry Cory, who ultimately wins Barbara's hand. Henry King directed (Donald Worster, *Rivers of Empire,* 197–198.)

4. Wright, *Barbara Worth,* 85. Near the bottom of Wright's social hierarchy are dark-skinned primitives including Mexicans, whom Wright usually presents as compliant laborers who speak childlike English and smile obsequiously—such as "dark Pablo, softly touching his guitar, representing a people still far down on the ladder of the world's upward climb" (290).

5. Farr, *History of Imperial County,* 311–313.

6. Rockwood, "The Early History of Imperial County," 103.

7. Ibid., 98.

8. Ibid., 98.

9. Ibid., 99. Sykes, *Colorado Delta,* 110.

10. Rockwood, "Early History of Imperial County," 100.

11. Ibid., 102.

12. Like Rockwood, Wozencraft drew encouragement from William Blake's researches. All he needed to launch the project, he boldly explained to the California legislature in 1859, was a grant of some sixteen hundred square miles of desert. The sale of land to settlers would generate capital for building the irrigation infrastructure, as well as compensate him for his labors.

Previously Wozencraft had served as one of the state's three Indian commissioners and negotiated a range of treaties and advocated a system of reservations encompassing almost seven and a half million acres. California's white citizens, however, begrudged assigning so much territory to the land's first inhabitants, and neither the treaties nor the reservations were ever approved. Wozencraft turned to other dispositions of land. He lobbied the legislature on behalf of the Atlantic-Pacific railroad, a task that further versed him in the rhetoric and legalities of giant land grants. Then he turned to his own proposal.

In the early days of statehood, Californians were not averse to making vast grants of land, so long as the right people benefited. In 1859 the legislature approved Wozencraft's plan for colonizing the desert and ceded him all the state's rights in the domain he sought. Wozencraft's victory, however, was not as grand as it sounds, for the state's rights were few. Nearly all the land lay in the public domain of the United States. Only Congress might give it away.

And so Oliver Wozencraft went to Washington, where timing undid him. The Civil War soon outweighed all other public business. Nevertheless, Wozencraft doggedly pressed his case with the Public Lands Committee of the House of Representatives. In May 1862, Congress at last considered his proposal—and rejected it. Only a week earlier, it had approved the famous Homestead Act, which promised ownership of 160 acres, free, to any family with the tenacity to live on the land for five years. With the ink hardly dry on enactment of that central, democratic dream, Congress was indisposed to enrich a single individual with a million-acre gift. Wozencraft pressed on for another fifteen years. He died in 1887 while a Congressional committee, behind closed doors, once more considered his proposal. See James J. Rawls, *Indians of California,* 141–148: Helen Hosmer, "Triumph and Failure in the Imperial Valley," 205–221; Hundley, "Politics of Reclamation," 292–325.

13. Rockwood "Early History," 112.

14. J. A. Alexander, *The Life of George Chaffey,* 40. Also George Kennan, *The Salton Sea,* 22ff.

15. Alexander, *Life of George Chaffey,* chapters 14–18.

16. Smythe, "International Wedding." See also discussion of Smythe in chapter 1.

17. These assessments are a distillation of Rockwood's "Early History," Alexander on Chaffey, and Kennan on Harriman, plus syntheses by Worster, Hosmer, and Cory. *Material Dreams,* by Kevin Starr, though inaccurate in a few details, provides an excellent overview.

18. J. A. Alexander places Chaffey's solo trip in December 1899, whereas Rockwood places the start of the earlier group expedition in that month. Since Rockwood was an actual participant in the events and wrote closer to them in time, I have accepted his date, which places Chaffey's solitary inspection probably in February 1900.

19. See especially Alexander, *Life of George Chaffey,* 283–287; Starr, *Material Dreams,* 24.

20. Kennan, *Salton Sea,* 25.

21. Alexander, *Life of George Chaffey,* 293.

22. Ibid., 294.

23. The mechanics of the water stock transaction deserve examination. The settler paid his money to the Imperial Land Company and received stock in a "mutual" company—that is, a company nominally owned by those whom it served. But in fact, most settlers did not receive their stock outright, because they lacked the funds to pay $25 per acre in addition to all the other expenses they faced. So they bought their water stock on credit at 6 percent, and the Imperial Land Company held the stock as security for the loan. Naturally, if the settler defaulted on his loan, the land company became the owner of the deed to his land, if any had been issued, as well as his water stock, without which the land was worthless. Because failures were frequent, this process was repeated again and again, with the result that the company fairly rapidly became what its name described: a land company of imperial scope and dimension.

24. Hosmer, "Triumph and Failure," 211.

25. The foregoing account of the trades and manipulations of stock by the CDC, Delta Investments, and the Imperial Land Company is distilled mainly from Starr, Hosmer, Alexander, and the 1908 promotional pamphlet of the Southern Pacific Railroad, "Imperial Valley, California."

SIDE TRIP. THE SHIMMERING DESERT

1. Farr, *History of Imperial County,* 25.

1. As quoted by Richard Shelton in his introduction to John C. Van Dyke, *The Desert*, xiv. For more on Van Dyke see Patricia Nelson Limerick, *Desert Passages*. For another view of the Salton Sea at this time, see articles by Frances Anthony for *Land of Sunshine*, 1900–1901: "To Palm Canyon," "At Indian Well," and "Below Sea Level."

2. Van Dyke, *The Desert*, 56–57.

3. Ibid., 60.

4. Farr, *History of Imperial County*, 433; Starr, *Material Dreams*, 28.

5. Burdick, *Mystic Mid-Region*, 230.

6. J. Smeaton Chase, *California Desert Trails*, 291.

7. Among those "who came to help hew a piece of destiny out of the raw material, one sometimes caught a glimpse of a tear on a face set with fortitude" (Farr, *History of Imperial County*, 266).

8. Chase, *California Desert Trails*, 286.

9. Margaret Romer, *A History of Calexico*, and Farr, *History of Imperial County*, generally. Also see Allen Day, "Irrigation in Southern California," 439; and Karen J. Frisby, *Imperial Valley: The Greening of a Desert*.

10. Burdick, *Mystic Mid-Region*, 217.

11. Romer, *History of Calexico*, 19

12. Sykes, *Westerly Trend*, 257. Also Romer, *History of Calexico*, 18.

13. Farr, *History of Imperial County*, 212.

14. Romer, *History of Calexico*, 21–22.

15. As quoted in Starr, *Material Dreams*, 33.

16. How much the soils report influenced Chaffey's rapid departure from the CDC remains a matter of debate—his defenders accord it little importance.

17. Cf. Starr, *Material Dreams*, 27.

18. Hundley, "Politics of Reclamation," 302. Helen Hosmer, "Triumph and Failure," 213, gives a figure of 1.2 million acres, which can hardly be reconciled with Hundley's.

19. Twenty thousand second-feet indeed represents a controlling portion of the Colorado's flow, which averaged 23,683 cubic feet per second at Yuma in the years 1905–1924. (See note 11, chapter 8, for the source of river flow information.) If validated, the CDC's claim would have greatly limited other withdrawals from the Colorado at times of low or even average flow. Water rights today are more normally expressed in acre-feet per year. The Imperial Irrigation District, for instance, holds claim to upwards of 3 million acre-feet (maf) of Colorado River water per year—more than one-fifth of the lower river's average annual flow—and these claims derive from the CDC's original filings.

20. Hosmer, "Triumph and Failure," 213.

21. A member of the committee asked Heber if he thought "there should not be any limit to the profits a private corporation should be permitted to earn while taking the public waters of the river and irrigating and controlling largely the public lands?" Heber answered with a vehemence not unknown among similar pleaders today: "I am opposed to the government interfering in every instance with the private property and . . . profits of any private corporations." He left no doubt that he considered the claimed twenty thousand second-feet

of continuous flow of the Colorado River to be the exclusive property of the California Development Company. See Hosmer, "Triumph and Failure," 213; Starr, *Material Dreams,* 35.

22. Kennan, *Salton Sea,* 33; Hundley, "Politics of Reclamation," 302. Rockwood ("Early History," 129) asserted that the claims were settled for "less than $35,000, paid entirely in water and water stock," which hardly sounds like an attractive settlement for the plaintiffs.

23. Rockwood, "Early History," 128.

24. James, *Wonders of the Colorado Desert,* 536.

25. The agreement specified that half the water diverted should be used in Mexico; not coincidentally, it was also at this time that Harry Chandler, who, with his father-in-law, Harrison Gray Otis, controlled the *Los Angeles Times,* acquired 862,000 (or 840,000—accounts vary) acres in northern Baja California—Andrade's remaining estate.

26. Rockwood, "Early History," 137. Also Kennan, *Salton Sea,* 36.

27. Hosmer, "Triumph and Failure," 217.

28. Rockwood, "Early History," 137ff. Also Kennan, *Salton Sea,* 38.

29. Rockwood ("Early History," 133–134) argued that the company was fatally handicapped by the government's delay in correcting its survey of the valley, which slowed the completion of land filings and hence the sale of water stock. No doubt the General Land Office dragged its heels, but as Hosmer suggests ("Triumph and Failure," 212), it was probably for good reason: the government's agents were in no hurry to help the CDC subvert the Homestead and Desert Land Acts.

DIMENSIONS. THE RIVER

1. Some "unofficial" estimates based on tree-ring reconstruction of long-term river flow run as low as 13.5 maf per year. Jason I. Morrison, Sandra L. Postel, and Peter H. Gleick, *The Sustainable Use of Water in the Lower Colorado River Basin,* 9.

2. The Supreme Court has confirmed the allocation to Arizona of tributary flows downstream of Lees Ferry. The 1.0 maf thus allocated are separate from the 2.8 maf otherwise allocated to Arizona under the Colorado River Compact. Very little of the tributary water actually reaches the main stem. Sources for this section include U.S. Geological Survey river gauge records at http:/waterdata.usgs.gov/nwis; Hundley, "Politics of Reclamation," 299; Thomas H. Watkins, ed., *The Grand Colorado,* 198, 206; Morrison, Postel, and Gleick, *Sustainable Use of Water,* 9–10 and *passim;* and Jason I. Morrison and Michael Cohen, Pacific Institute for Studies in Development, Environment, and Security, personal communication.

CHAPTER 8. A SEA OF UNINTENTION

1. Meko and Graybill, "Tree-Ring Reconstruction of Upper Gila River Discharge."

2. Shifting channels is basic behavior for any delta-building river. Were it not for colossal resistance provided by the U.S. Army Corps of Engineers, the Mississippi would long ago have redirected its main flow through the Atchafalaya Basin to the Gulf of Mexico, well to the west of its present channel, leaving New Orleans all but riverless.

3. Sykes, *Colorado Delta,* 7–18, 169; Odie B. Faulk, *Derby's Report on Opening the Colorado, 1850–51,* 43; James Ohio Pattie, *The Personal Narrative of James O. Pattie of Kentucky,* 217ff.

4. The following account of the great desert flood of 1905–1907 is drawn from a variety of sources. Many of the principal actors in the drama of the flood produced memoirs of the events. These include Charles R. Rockwood's *Born of the Desert,* which appeared as a special edition of the *Calexico Chronicle* in 1909. The *Chronicle* reissued it as a booklet in 1930 under the supervision of its editor, Randall Henderson. Included in the latter publication was "Personal Recollections of the Early History of Imperial Valley" by Dr. W. T. Heffernan, a former surgeon at Fort Yuma, who befriended Rockwood when he first arrived in the region and became one of the Colorado Development Company's earliest investors. Finis Farr incorporated Rockwood's essay in his 1918 *History of Imperial Country,* where it constitutes a lengthy chapter (pp. 97–153) titled "The Early History of Imperial County." As one would expect, Rockwood attempted to justify his actions and minimize his responsibility for the flood, although not very convincingly, in this writer's opinion. H. T. Cory rendered his account for posterity in *The Imperial Valley and the Salton Sink.* George Kennan presented Edward Harriman's role in a favorable light in *The Salton Sea: An Account of Harriman's Fight with the Colorado River.* Other useful portraits of the flood include Margaret Romer's *A History of Calexico* and various period accounts from *Scientific American* (April 14, April 21, November 24, December 15, and December 22, 1906, and August 10, 1907). The April 21 report by George Wharton James is especially useful. James describes his descent by boat from Yuma into the Salton Sink during the height of the flood in *The Wonders of the Colorado Desert,* and Godfrey Sykes relates his experiences crossing the flooded valley by wagon in *A Westerly Trend.* Sykes also provides a very balanced appraisal of the flood and efforts to stem it in *The Colorado Delta.* Additional primary sources concerning the flood include records of the litigation associated with it, especially those in the *Federal Reporter* (Annotated), October–November, 1909 (St. Paul, Minn.: West Publishing, 1910), 792–825. Secondary sources include Kevin Starr, *Material Dreams;* Helen Hosmer, "Triumph and Failure"; and Norris Hundley, "Politics of Reclamation." None of these sources agrees entirely with any of the others, yet each casts light on the overall picture. I have drawn on all of them while trying to distill a defensible and convincing version of events that remains consistent with a majority of sources.

5. Rockwood, "Early History," 139.

6. Kennan, *Salton Sea,* 42.

7. Sykes, *Colorado Delta,* 115–116.

8. *Federal Reporter,* October–November 1909, 794ff. Sykes, *Colorado Delta,* 115.

9. Hosmer, "Triumph and Failure," 218.

10. As quoted in Kennan, *Salton Sea,* 44.

11. U.S. Geological Survey data at http:/waterdata.usgs.gov/nwis. The government's official records differ from Kennan (*Salton Sea,* 47, 50), who had access to Southern Pacific's records taken from the Yuma gauge and who gives two figures, 102,000 and 115,000 cfs, for the same highwater event.

12. *The Wonders of the Colorado Desert* was initially published in two volumes in 1906 by Little, Brown and Company. Page references given here refer to a one-volume edition from the same publishers that appeared in 1911. For more on James, see his profile in Limerick, *Desert Passages.*

13. James, *Wonders,* 493–494.

14. Ibid., 496–497.

15. Ibid., 497. The volcanic butte was probably Mullet Island, now submerged.

16. Sykes, *Colorado Delta,* 55. Although Sykes states that the flow reached 110,000 cfs, official U.S. Geological Survey records show that it approached only 100,000. The difference is academic. In either case the flood was an awesome force of water, particularly for the channel of a river where the mean yearly flow was only one-fifth as great.

17. Allen Day, "The Inundation of the Salton Basin by the Colorado River and How It Was Caused," *Scientific American* (April 14, 1906): 310.

18. Rockwood, *Born of the Desert,* 40. As Rockwood put it, with habitual self-inflation, "I had found, at this time, that it was impossible for me to manage the affairs of the company in accordance with my ideas, and unless I could do so, I believed that it was best for the stockholders of the company that I should resign as assistant general manager."

19. Elizabeth Harris, compiler, "Townsite of Silsbee and Indian Well," pamphlet produced by the Imperial Valley Pioneers, 1980. Also Randall Stocker, director of environmental affairs for the Imperial Irrigation District, Imperial, Calif., personal communication, November 13, 1991. Some accounts of heroism during the height of the flood may be fanciful, but not the dropping of explosives from an aerial cable, which showed clearly in a period photograph Mr. Stocker showed me. Stocker also mentioned reports of the hurried blasting of a channel between the Alamo River and the New, upstream of Calexico, in order to divert water from the former to the latter for the sake of the town. Although the story sounds credible, I was unable to confirm it.

20. Dr. F. W. Peterson, "Medical History," in Farr, *History of Imperial County,* 212. The peripatetic Godfrey Sykes, traveling by wagon with his son from San Diego to Tucson, also confirmed that in the aftermath of the summer flood no small number of settlers hoped fervently to "unsettle" from the valley as quickly as their means allowed: "There were weak-kneed brethren here and there who were more than anxious to trade their embryonic farms and prospects of future prosperity and affluence for a good sturdy team with appurtenant waggon [*sic*] and camp-duffle, such as my small boy and I were piloting deviously around and amongst the muddy streaks in the valley" (*Westerly Trend,* 256). Sykes, being little interested in acquiring farmland, whether submerged or dry, declined the offers and kept driving.

21. As quoted in Kennan, *Salton Sea,* 60.

22. Kennan, *Salton Sea,* 61–62.

23. Ibid., 62.

24. Ibid., 66. Cory's authority to mobilize the tribes was apparently not questioned.

25. Ibid., 66.

26. *Scientific American,* December 15, 1906.

27. Kennan, *Salton Sea,* 76.

28. Ibid., 78.

29. U.S. Geological Survey data for the Yuma gauge at http:/waterdata.usgs.gov/nwis.

30. See M. J. Dowd, "History of Imperial Irrigation District and the Development of Imperial Valley," for details of postflood damages, legal proceedings, and establishment of the IID.

1. *Scientific American* (August 10, 1907): 94.

2. Passenger Department, Southern Pacific Company, "Imperial Valley California" (1908), 15.

3. Ibid., 8–9.

4. Farr, *History of Imperial County*, 16; Hundley, "Politics of Reclamation," 306.

5. "U.S. v. Imperial Irrigation District," 799 Federal Supplement 1052 (Southern District of California 1992), 1057. Hereinafter identified as "Keep Decision." Lands awarded to the Cahuila were generally "checkerboarded," i.e. interspersed with previously awarded railroad sections. The resulting scatter makes it impossible to depict tribal lands at the scale of the maps included in this volume.

6. Ibid. See also Frank H. Bigelow, "Studies on the Rate of Evaporation at Reno, Nevada, and in the Salton Sink."

7. Hundley, "Politics of Reclamation," 307.

8. The rate of evaporation at the Salton Sea is very high and fairly constant, about 5.8 feet per year. At this rate, if you filled a six-foot-deep swimming pool in Salton City on January 1 and added no more water for the rest of the year, the pool would be just about empty by the end of the following December. Throughout most of the 1990s, inflows held relatively steady at 1.3 million acre-feet per year and were offset by the evaporation of 230,000 acres of surface area (230,000 acres x 5.8 feet per year = approx. 1.3 million acre-feet per year). See Salton Sea Authority, *Proceedings of the Salton Sea Symposium, January 13, 1994*.

9. J. Penn Carter, chief counsel, to board of directors, Imperial Irrigation District, memorandum: "Formation of IID," October 3, 1989. Also Dowd, "History of Imperial Irrigation District."

10. Keep Decision, 1061.

11. Formed in 1918, CVWD pumped only groundwater to client farmers until 1950, when it began to distribute Colorado River water via the newly completed Coachella Canal. Connection to this new source of water triggered the expansion of agriculture within the valley, which in turn increased drainage flows from Coachella fields to the Salton Sea.

12. The figure of $30 million in damages was compiled by Alec Rosenberg, a reporter for the *Imperial Valley Press* who tracked the IID's affairs for a number of years (personal communication, March 23, 1998). For information on the suits filed against IID, see *John Elmore v. Imperial Irrigation District*, 205 California Reporter 433 (Cal. App. 4 Dist. 1984); *Imperial Irrigation District v. State Water Resources Control Board*, 231 California Reporter 283 (Cal. App. 4 Dist. 1986); and *Salton Bay Marina Inc., et al. v. Imperial Irrigation District*, 218 California Reporter 839 (Cal. App. 4 Dist. 1985). This last decision notes (p. 846) that for the period from 1967 to 1976, the "amount of spillwaste was 556,509 acre feet, or 20 percent of the total amount the District delivered to its farmers. The general managers of the District testified five percent (127,150 acre feet) would be a reasonable amount for operation spill."

Considerable irony attends the damage suits against the IID and CVWD. To a large degree, the suits that have been successful were triggered by the weather. For a few years in the 1970s and again in the mid-1980s, heavy winters, influenced by El Niño, drenched the

Salton Basin with unexpected and largely unwanted rain. More to the point, the weather of those wet years deposited unusually deep snowpacks in the intermountain West, and the Colorado flowed with abundance: reservoirs filled, and spillways and floodgates were obliged to open. Flows that exceeded the treaty-bound minimum surged across the international border, and Mexican farmers happily put the additional water to use in the Mexicali valley, much of which drains northward via the Alamo and New rivers. The IID and its member farmers in Imperial Valley seem also to have treated themselves to greater than usual profligacy in their use of water. As irrigation on both sides of the border increased, the level of the sea dutifully rose, leading to the damage actions and settlements.

The successful damage suits arose in response to unusual weather years. The Cahuilla's grievances, which arise entirely from normal operations, remain uncompensated. Viewed in the simplest terms, the situation suggests that the two districts may safely ignore injuries caused by their day-to-day operations, but they become liable for acts of God and El Niño.

13. Keep Decision, 1066. The total of damages given in the published decision is $69,563,213—a few dollars less than the sum of the alleged constituent damages.

14. Ibid., 1069.

15. Mary Belardo, personal communication, January 14, 1994.

16. Thomas Luebben, personal communication, March 15, 1997.

SIDE TRIP. PORT ISABEL

1. Godfrey Sykes, *A Westerly Trend,* 216–217. See also Glenton G. Sykes, "Five Walked Out! The Search for Port Isabel."

CHAPTER 10. THE DELTA, HUNG OUT TO DRY

1. Joseph Grinnell, "Birds of a Voyage on Salton Sea"; James, *Wonders,* 518.

2. Sykes, *Colorado Delta,* 87; Pattie, *Personal Narrative,* 241–242.

3. Aldo Leopold, "The Green Lagoons," in *A Sand County Almanac;* Luna B. Leopold, *Round River,* 10ff.; Curt Meine, *Aldo Leopold,* 207–209.

4. Leopold, "Green Lagoons," 142.

5. See Meine, *Aldo Leopold,* 453, on the writing of "The Green Lagoons."

6. A few outsiders, such as Godfrey Sykes, who were better acquainted than Leopold with the place and its people, knew that the mazelike twists and turns of delta waterways were hardly a mystery to the Cocopa. On one occasion, Sykes traveled an inundated delta in the company of a Cocopa guide who knew its waters even to the point of recognizing old channels by the merest show of treetops above the floodwaters (*Colorado Delta,* 55–56).

7. Kelly, *Cocopa Ethnography;* Nabhan, *Gathering the Desert,* 158.

8. U.S. Geological Survey data at http:/waterdata.usgs.gov/nwis. See also Sykes, *Colorado Delta,* 80.

9. David Harris, "Recent Plant Invasions in the Arid and Semi-Arid Southwest of the United States," gives a figure of 723,000 acres lost to tamarisk by 1961. Tamarisk has hardly halted its spread since then, and continued dam construction has increased the availability of suitable habitat. See also Nabhan, *Gathering the Desert.* Among possible points of entry to

the literature on tamarisk spread and control are Michael R. Kunzmann et al., "Tamarisk Control in Southwestern United States"; and U.S. Bureau of Reclamation, "Vegetation Management Study: Lower Colorado River."

10. Sykes, *Colorado Delta,* 76ff.

11. Lewis R. Freeman, *Down the Grand Canyon,* 68.

12. Angela Moyano Pahissa, *California y sus relaciones con Baja California,* 97–98, 114.

13. Jorge A. Calderón Salazar, *Reforma agraria y colectivización ejidal en México,* 84–86.

14. See both Moyano Pahissa and Calderón Salazar as cited above.

15. Peter Singelmann, "Rural Collectivization and Dependent Capitalism: The Mexican Collective Ejido"; Marilyn Gates, "Codifying Marginality: The Evolution of Mexican Agricultural Policy and its Impact on the Peasantry."

16. The salinity of waters delivered to Mexico has sparked considerable debate and negotiation. See chapter 15, note 12.

17. Morrison, Postel, and Gleick, *Sustainable Use of Water,* 22; Edward P. Glenn, et al., "Water Management Impacts on the Wetlands of the Colorado River Delta, Mexico."

18. Glenn et al., "Water Management Impacts." See also Mark K. Briggs and Steve Cornelius, "Opportunities for Ecological Improvement along the Lower Colorado River and Delta."

19. *Arizona Republic,* February 23, 1993. *Arizona Daily Star,* February 27, 1993.

20. Both the channel and the railroad represent important milestones in the history of the delta. After the lower Pescadero clogged with silt in 1922, the plumbers of the delta built several generations more of levees and drains, including, in 1929, the Vacanora Canal, which was intended primarily for the distribution of irrigation water. The Vacanora ran from the Rio Abejas southward to a point just downstream from present-day Colonias Nuevas and the railroad bridge. The river first shunned the canal, moving its channel so that the Vacanora's intakes were high and dry. Engineers then made deeper and wider cuts to bring the river to the project. Once this was done, the river appropriated the Vacanora as its principal channel, which it has since remained (Sykes, *Colorado Delta,* 82; Nabhan, *Gathering the Desert,* 158–159). The Ferrocarril Inter-California del Sur, the first direct rail link between Baja and the rest of Mexico, was completed in 1948. Previously, branch lines of the Southern Pacific had provided the Mexicali Valley's only rail connection to the rest of the world.

21. See chapter 15, note 12.

22. United States Bureau of Reclamation, "Yuma Desalting Plant: A Status Report."

23. Glenn et al., "Water Management Impacts"; Edward P. Glenn et al., "Cienega de Santa Clara: Endangered Wetland in the Colorado River Delta, Sonora, Mexico."

24. Martin Van Der Werf, "Draining the Budget to Desalt the Colorado." Various observers have pointed out that it would have been cheaper for the U.S. to buy up the Wellton-Mohawk agricultural area and retire it from production than to build the Yuma desalting plant.

25. U.S. State Department, Bureau of Oceans and International Environmental Affairs, "International Sonoran Desert Alliance"; Peggy J. Turk-Boyer, "Reserves in Action." The commitment of the Mexican government also increases the odds that when (not if) problems of excessive salinity develop within the cienega, the United States and Mexico will cooperate to solve them.

26. Glenn et al. "Water Management Impacts." Glenn identifies a third area, the El Doctor wetlands at the foot of Sonora Mesa on the east side of the delta. These spring-fed wetlands are small, and most of them are burdened by heavy local grazing and other uses. Nevertheless, they possess considerable biological significance, as well as high potential for restoration should they receive protection. Another natural area, encompassing many thousands of acres, lies downstream of the low-water crossing between Colonia Carranza and Héroes de la Patria, nourished in recent years by flood control releases from Hoover Dam (Edward Glenn, personal communication, April 29, 1998).

27. Kelly, *Cocopa Ethnography*, 13.

28. Glenn et al., "Water Management Impacts."

29. The fabulous early history of the Volcano Lake area is recounted in Carl L. Strand, "Mud Volcanoes, Faults, and Earthquakes of the Colorado Delta Before the Twentieth Century."

30. Frank Clifford, "Efforts Focus on Dried-Up Colorado Delta"; Edward Glenn, workshop presentation, "Water and Environmental Issues of the Colorado River Border Region," San Luís Río Colorado, April 30, 1998. Freshwater flows appear to enhance estuarine fisheries by reducing salinity in nursery areas to a degree that discourages entry of marine predators.

31. One (admittedly preliminary) estimate of needed flows is surprisingly modest. Carlos Valdes of the Instituto Technológico y de Estudios Superiores de Monterrey (Guaymas), Edward Glenn of the Environmental Research Laboratory (Tucson), and a host of colleagues have suggested that base flows of about 32,000 cfs, supplemented by surges of 300,000 cfs or more every four years, would nourish and sustain much of the potential natural habitat of the lower delta. This regime roughly averages to the annual allocation of 130,000 cfs mentioned below. It would represent the dedication of less than 1 percent of the river's natural flow to the support of its delta.

32. Ernesto Reynoso Nuño, Centro Regional de Estudios Ambientales y Socioeconómicos, personal communication, April 30, 1998. The people of El Mayor refer to themselves not as Cocopa but as Wua-pah.

CHAPTER 11. UPHILL TOWARD MONEY

1. Dale Pontius, *Colorado River Basin Study: Final Report*.

2. *Imperial Valley Press*, October 14, 1996.

3. Under a more severe accounting, the sale price of the water was calculated to be $78 per acre-foot (*Imperial Valley Press*, June 11, 1997). Either way, payment for the water took the form of installation, at MWD's expense, of water conservation structures (canal lining, intercept ditches, etc.), as well as the adoption of procedures to avoid waste, thanks to a new, high-tech command station, which MWD also paid for. The difference in the calculation of the cost of water saved depends on whether a $23-million loan fund for mitigating the effects of water conservation is included.

4. Farr, *History of Imperial County*.

5. Norris Hundley, *The Great Thirst*, 206.

6. U.S. Geological Survey data at http://www.waterdata.usgs.gov/nwis.

7. The story of Boulder Dam has been told many times. See Norris Hundley, *The Great Thirst,* 202–232; Hundley, "Politics of Reclamation," 292–325; Philip L. Fradkin, *A River No More,* 269–287; Reisner, *Cadillac Desert,* chapter 4 and *passim;* Watkins, *The Grand Colorado,* part 2 and *passim;* and Worster, *Rivers of Empire,* 194–212.

8. Hundley, *The Great Thirst,* 220.

9. See Hundley, "Politics of Reclamation," for a complete discussion of the origins of the Boulder Canyon Act.

10. The decade 1914–1923, during which the river averaged an annual discharge of 18.8 million acre-feet, is the wettest ten-year period on record. The degree to which the river is overallocated depends on whose math and climate data one accepts. In dry years, the vast storage capacity of the river's chain of reservoirs somewhat compensates for this colossal oversight: the cumulative volume of the man-made lakes is roughly equal to four times the river's flow (Pontius, *Colorado River Basin Study,* 10). Nevertheless, even the average flow of the past 90 years (about 15 million acre-feet) may be 10–15 percent higher than the average flow of the past 1,000 years (*U.S. Water News,* July 1998). See also "Dimensions: The River," note 1.

11. As quoted in Bogan et al., "Southwest." Also Dale Pontius, personal communication, April 1998 (undated letter).

12. Smythe, "International Wedding," 286–300.

13. California Health and Welfare Agency, "Annual Planning Information, Imperial County, 1993."

14. Fradkin, *A River No More,* 272–287. Dr. Ben Yellen of El Centro, long a gadfly to the IID, filed the suit on behalf of "small farmers." Yellen died in 1994, one day before his eighty-seventh birthday, with yet another suit pending—this one against Attorney General Janet Reno for allowing Mexicans to enter the country illegally and take jobs from U.S. citizens (*Imperial Valley Press,* July 3, 1994).

15. *Imperial Valley Press,* May 18, 1995.

16. A case study of the IID-MWD transfer agreement appears in Marc Reisner and Sarah Bates, *Overtapped Oasis,* Appendix A.

17. *Imperial Valley Press,* June 19, 1997.

18. *Imperial Valley Press,* June 9, 1995, December 29, 1996, September 10, 1997, September 19, 1997. The jury's report caused no action to be brought against Western Farms or IID, but the temperature of the war of words over southern California water increased by several degrees. The top management of the MWD soon called IID a water "cartel" controlled by the Basses. One IID director, in his highest headline dudgeon, retorted that he would not trust the MWD general manager "in an outhouse with a muzzle on"—a species of invective that left partisans on both sides scratching their heads (*Imperial Valley Press,* September 10, 1997).

19. In 1995 San Diego consumed about 200,000 acre-feet per year. SDCWA anticipated water demand's rising to 500,000 acre-feet by 2020 and to 1 million acre-feet by 2045 (*Imperial Valley Press,* July 13, 1998).

20. *Imperial Valley Press,* June 30, 1998. The lone holdout on the IID board was Don Cox, who was due to become IID board president in January 1996. Three of the remaining four members of the board, however, voted to deny him his normal rotation as president. Said Cox, "It's sad to me that since I refused to join the Bass brothers team, I have been locked in a

closet by this board" (*Imperial Valley Press,* September 6, 1998). Information about the Western Farms–SDCWA negotiations came to light during the summer of 1998 as a result of a leak of confidential documents to the Imperial Valley Press (see *Imperial Valley Press,* June 30, 1998, July 6, 1998, July 8, 1998, July 13, 1998, September 6, 1998; also *U.S. Water News,* August 1998).

21. *Imperial Valley Press,* July 16, 1998.

22. *Imperial Valley Press,* August 4, 1997, August 5, 1997, August 13, 1997, August 15, 1997, September 17, 1997. A year and a half later, US Filter itself became the target of an acquisition. Vivendi, a French environmental services firm, bought US Filter for $6.2 billion in what was the largest French acquisition ever made in the United States (*Imperial Valley Press,* March 23, 1999).

23. *Imperial Valley Press,* December 11, 1997, April 26, 1998. Exhibit *E* of the transfer agreement sets forth the procedures for price adjustments, which include factors for water quality, water supply reliability, and time elapsed since the previous adjustment. Exhibit *E* generated much consternation during public review of the agreement prior to its approval, as the following headline attests: "Math professor flown in to explain transfer's exhibit E at workshop" (*Imperial Valley Press,* April 21, 1998).

24. Pontius, *Colorado River Basin Study,* 42–45; *Albuquerque Journal,* December 18, 1997.

25. *Imperial Valley Press,* March 31, 1998.

26. Don Cox, personal communication, March 25, 1998. See *Imperial Valley Press,* September 23, October 7, and expecially September 20, 1998, for reviews of the controversy over fallowing and its many definitions. Even the most straightforward of those definitions—that fallowing be considered "the non-use of farmland for the purpose of creating water for transfer"—would prompt difficult questions regarding a farmer's intent. While on the surface a prohibition against fallowing is a black-and-white issue, in actual practice and enforcement it becomes irreducibly gray.

27. According to Paul Cunningham, a principal spokesperson for IID, "We have an open-ended water right. What we did was entirely legal. Our farmers were simply responding to market forces" (personal communication, April 30, 1998).

28. Reisner and Bates, "Overtapped Oasis," 66, 120, and Appendix A *passim;* Pontius, *Colorado River Basin Study,* 37; *Torres-Martinez Desert Cahuilla Indians v. IID,* Civil Case no. 91–1670, U.S. District Court, Southern District of California, Complaint, 12. See also chapter 9, note 12.

29. US Filter launched a pilot program along these lines in 1998 (*Imperial Valley Press,* June 5, 98).

30. In late 1997, MWD reached a new high in terms of irritating the IID by filing with the state for rights to 500,000 acre-feet of water in the Alamo and Whitewater rivers, whose flows consist mainly of drainwater from the Imperial and Coachella valleys, respectively. *Imperial Valley Press,* August 18, 1997, September 11, 1997, September 12, 1997.

31. Don Cox, personal communication, March 25, 1998.

32. Douglas L. Hayes, "The All-American Canal Lining Project." See also U.S. Department of the Interior, "Draft Environmental Impact Statement/Environmental Impact Report."

33. *Imperial Valley Press,* June 23, 1997.

34. The secretary of interior has threatened to intervene if CVWD and IID cannot come to agreement (*Imperial Valley Press*, September 13, 1998, September 15, 1998, September 17, 1998).

35. *Imperial Valley Press*, August 11, 1998, August 12, 1998, August 30, 1998, September 1, 1998.

36. *Imperial Valley Press*, December 11, 1997, April 26, 1998, April 29, 1998, April 30, 1998.

37. State of California, Colorado River Board, "Colorado River Board 4.4 Plan," draft of December 17, 1997.

CHAPTER 12. THE THEORY AND PRACTICE OF BORDERS

1. Paul Ganster, "Environmental Issues of the California–Baja California Border Region," 2.

2. Cletus E. Daniel, *Bitter Harvest*, 228–249. See also Carey McWilliams, *Factories in the Field: The Story of Migratory Farm Labor in California* and Mark Day, *Forty Acres*.

3. Victor Salandini, *The Confessions of the Tortilla Priest*. See also obituaries of Salandini in the *Santa Fe New Mexican*, August 31, 1994, and the *New York Times*, August 6, 1994.

4. Ernesto Reynoso Nuño, personal communication, April 30, 1998.

5. Ganster, "Environmental Issues."

6. *Imperial Valley Press*, May 8, 1998 ("23 Injured, deaths likely in van wreck)," June 9, 1998 ("Border crossers avoid beefed-up patrol)," August 16, 1998 ("Bounty offered for smuggler who left 9 to die").

7. *Imperial Valley Press*, August 13, 1998 ("Border Patrol apprehensions top 200,000").

8. *Imperial Valley Press*, April 22, 1993, February 14, 1995, March 30, 1997, April 3, 1997.

9. *New York Times*, June 23, 1994; *Imperial Valley Press*, April 26, 1994, July 4, 1994, November 9, 1994.

SIDE TRIP. CALIPATRIA

1. Mike Davis, "Hell Factories in the Field."

CHAPTER 13. HOME BY THE RANGE

1. See, for example, Marian Seddon, *What Ever Happened to the Chocolate Mountain Gang*.

2. *Imperial Valley Press*, February 26, 1993.

3. J. Hector St. John de Crèvecoeur, "Letters from an American Farmer," 8.

4. Ibid., 42–43.

5. Steve Sorenson, "Heaven Sent: Shrapnel, Shells, Bomb Casings, Machine Gun Clips, Missile Parts." See also Sorenson's other excellent pieces in *San Diego's Weekly Reader*: "Hot Place in the Sun: A Different Breed of Snowbird Finds a Town with No City Limits" (on Slab City); "Mission over the Mountains: Mock War in a True Wilderness" (on the gunnery range); and "Bury My Bait at Bombay Beach."

6. *Imperial Valley Press*, June 12, 1992, December 30, 1996, January 9, 1997; Sorenson, "Heaven Sent."

7. *Imperial Valley Press,* October 4, 1993, July 21, 1994, December 28, 1994.

8. *Imperial Valley Press,* June 23, 1994, June 28, 1994, June 30, 1994, July 1, 1994, July 18, 1994, July 24, 1994, July 27, 1994, August 5, 1994, August 31, 1994, September 30, 1994; Roger Manley and Mark Sloan, *Self-Made Worlds,* 78–79.

9. The Final Rule on endangerment of *Cyprinodon macularius* appeared in the *Federal Register* 51, 61 (March 31, 1986): 10842–10851. See also Peter B. Moyle, *Inland Fishes of California,* 252–254.

CHAPTER 14. HAVE WE GOT A DEAL FOR YOU

1. In 1889, for instance, Congress briefly closed the public domain to homestead entries so that carefully designed settlement plans, based on detailed surveys, might be put into effect. Protest was widespread and vehement, and the closure was rescinded the following year. See Wallace Stegner, *Beyond the Hundredth Meridian,* 316–324, 328–338.

2. This account of development efforts in and around Salton City draws heavily on a curious book by David J. Levenkron, *Sand and Rubble: The Salton City Story,* which Levenkron published, evidently at his own expense, in Los Angeles in 1982. Levenkron, a successful entrepreneur in the "wall covering" business, was an early investor in Salton City, and when it foundered, he invested again—in an ill-advised effort to attract jobs and residents to the area by establishing a scientific research center there. *Sand and Rubble* is partly a litany of complaint, partly a source-book for a class action suit against various developers for fraud, partly a scrapbook of Salton City brochures and newsclips, and partly an exculpatory explanation of how a wise and experienced businessman like David J. Levenkron was taken to the cleaners. At the outset of researching *Salt Dreams,* I contacted Levenkron with a request for access to his considerable collection of promotional and other material relating to Salton City. Levenkron replied that the cost for such access would be no less than $10,000—a measure, one might say, of the primacy of the Salton City story in his life and imagination. The invitation to pay for access to Levenkron's files was declined, and, failing to find any other detailed source, I have relied on *Sand and Rubble,* notwithstanding its heavy bias. A leading virtue of the book is its unedited reproduction of many photographs, brochures, and news clippings dealing with Salton City. These documentary materials provide a strong foundation for understanding the "rise and fall" of the Salton Riviera.

3. *Los Angeles Times,* December 11, 1970, as quoted in Levenkron, *Sand and Rubble,* 193.

4. As quoted in Levenkron, *Sand and Rubble,* 51.

5. Levenkron, *Sand and Rubble,* 74. Phillips would live another nineteen years, succumbing to a stroke at the age of 92 (*Los Angeles Times,* June 3, 1979).

6. Levenkron, *Sand and Rubble,* 92.

7. Levenkron's data are inconsistent on these points, and so I have used the most conservative numbers he offers. Compare pp. 130 and 192–193 in *Sand and Rubble.*

8. See, for instance, *Imperial Valley Press,* June 12, 1994, "Services district draws battle lines." The copy begins, "Joe Martinez, a Salton City real estate agent, refuses to go to Salton Community Services District meetings anymore. 'They're too violent,' Martinez says. A recent special meeting was no exception." See also *Imperial Valley Press,* December 1, 1993.

1. Sykes, *Colorado Delta,* 118. Other sources for this section include Salton Sea Authority, *Proceedings;* and Salton Sea Authority, "The Salton Sea: A Brief Description of Its Current Condition and Potential Remediation Projects." See chapter 16, notes 16 and 17, for source information on the Sonny Bono Memorial Salton Sea Restoration Act.

CHAPTER 15. A SEA OF TROUBLES

1. Clark Bloom, personal communication, March 24, 1998.
2. William Radke and refuge staff, internal refuge report on the 1991–1992 eared grebe die-off. Unless otherwise noted, information on the grebe die-off is taken from this report.
3. Water Education Foundation, "The Salton Sea," 7. Salinity is also frequently expressed in milligrams per liter (mg/L), which is equivalent to parts per million.
4. See chapter 8, notes 14, 16.
5. See chapter 10, note 1.
6. Manuel Lujan, "Report to Congress on Salton Sea National Wildlife Refuge."
7. Ibid.; Gary W. Page et al., "Shorebird Numbers in Wetlands of the Pacific Flyway." Also Clark Bloom, personal communication, June 1998.
8. *Imperial Valley Press,* April 23, 1995; William Radke, Salton Sea National Wildlife Refuge, personal communication, July 23, 1993.
9. Boyd W. Walker, ed., *The Ecology of the Salton Sea, California, in Relation to the Sportfishery,* 77, 83.
10. Ibid., 77–91, 185ff.; Richard G. Thiery, "The Aquatic Ecosystem of the Salton Sea," in Salton Sea Authority, *Proceedings.*
11. Ibid.; Gary Kramer, "North America's Most Unique Fishery," 47–48.
12. Reisner, *Cadillac Desert,* 478; Salton Sea Authority, *Proceedings.* Downstream of Imperial Dam, the Colorado grows rapidly saltier due to return flows from the Yuma area and, especially, drainwaters from the Wellton-Mohawk Irrigation District forty miles upstream along the Gila River. In 1961, Wellton-Mohawk flows boosted the salinity of Colorado River water at the border to 2,700 ppm, crippling agriculture in the Mexicali Valley and threatening the permanent poisoning of its fields. Mexico protested, but the United States merely asserted that the 1944 treaty apportioning water between the two nations was mute on the question of quality. The dispute continued for more than a decade. In the early 1970s, President Echeverría of Mexico threatened action in the World Court at The Hague. Perhaps more impressed by Mexico's newly found oil reserves than by its legal position, the United States finally agreed to negotiate the salinity issue. In 1973, an agreement was at last embodied in Minute 242 of the International Boundary Waters Commission, under which the United States is obliged to deliver Colorado River water to Mexico with an average annual salinity of not more than 115 ppm over the salinity of water diverted in 1976 at Imperial Dam—which turned out to be 879 ppm (Reisner, *Cadillac Desert,* 6–9, 474–482; David C. Sweigert, "Relining Canals in the Border Region: Can the U.S. Ignore impacts on Mexico?"; Bureau of Reclamation, "Yuma Desalting Plant"; Van Der Werf, "Draining the budget").
13. Salton Sea Authority, *Proceedings.* See especially Colorado Regional Water Quality

Control Board, "Salton Sea Briefing Paper—December 31, 1993" and Gary Polakovic, "Farm runoff: a challenge," *Riverside Press-Enterprise,* January 3–10, 1993, reproduced therein. Also see Morrison, Postel, and Gleick, *Sustainable Use of Water,* 42.

14. Walker, *Ecology of the Salton Sea,* 187.

15. There are periods of exception in the long-term decline of the Salton Sea fishery, including, for instance, 1997, which provided exceptionally good corvina fishing. The wet, El Niño winter of 1993 may offer a partial explanation: high freshwater inflows to the sea may have created areas of lower salinity where corvina successfully spawned. By 1997, the young of that year had grown large enough to attract the attention of fishermen and to be attracted to the fishermen's baits and lures.

16. *Imperial Valley Press,* October 3, 1997, October 6, 1997, October 28, 1997, May 13, 1997.

17. Tom Harris, *Death in the Marsh,* 21, 119ff. See also Tom Harris et al., "David Love," and Jane E. Brody, "Hope Rising for Selenium."

18. The selenium of the Salton Sea is not merely a consequence of geology and water transport. Some of it flows down to the sea because of what people and their animals have done to increase the profusion of selenium-fixing plants. The species that most effectively tolerate and utilize selenium include various members of the woody aster (*Xylorrhiza*) and poisonvetch or locoweed (*Astragalus*) families. Cattle, sheep, and horses generally find these plants unpalatable but eat them when other feed is unavailable. Where soils are heavily seleniferous, the results can be disastrous. In the 1890s, for instance, a couple of Mormon herders driving four thousand head of sheep happened upon a little-used range in the heart of Wyoming. In those days of range wars and Mormon persecution, such an expanse of unoccupied land, well covered with vegetation, looked like a gift from God. The sheepherders probably did not notice that the dominant plant in that lush-looking landscape was woody aster, or if they did, it meant nothing to them. When cowboys rode by and told them that the land was poisoned, the sheepherders dismissed the warning as a lie intended to scare them off. They made their camp and bedded down. In the morning, they awoke to find nearly three-quarters of their sheep as dead as stones. By the time they escaped the toxic plain, only a tenth of their original herd survived (Harris, *Death in the Marsh,* 129–130).

The names stockmen have given to the symptoms of selenium poisoning conjure a vision of sluggish wandering: the blind staggers, cracker-heel, knocking disease (from heels clacking together in a stumbling gait), roaring disease (from wheezing induced by respiratory paralysis), mountain disease, and timber trouble, timber paralysis (USDA Forest Service, *Range Plant Handbook,* W31). The problems increase where the range is overused. Animals first eat the palatable, nutritious forage, especially grasses. Then they turn to asters and astragaluses, specimens of which have been found to consist of from 0.1 to 1.0 percent selenium (1,000 to 10,000 ppm)—enough, as the saying goes, to kill a horse (Harris, *Death in the Marsh,* 78). As the pattern of selective herbivory persists, desired grasses die back, and undesirable species, which suffer less grazing pressure, increase. In this way, overgrazing fosters the spread of asters and locoweed, guaranteeing multiplication of the number of flowering selenium pumps operating in the landscape and intensification of the synthesis of water-soluble selenate. Thus the waters of the West become charged with ever more of the toxic mineral.

Grazing is not the only process contributing to increasing concentrations of selenium in the environment. Coal-fired power plants discharge selenium in their fly ash, which eventually settles back to earth. Selenium comes to the surface of the earth in the tailings of mines, from which the weather washes it downstream along with other unwanted metals and compounds. Selenium also travels from lithosphere to ecosphere in the manufacture and use of phosphate fertilizers (Harris, *Death in the Marsh*, 134–135).

19. Harris, *Death in the Marsh*, 10–13.

20. *Brawley News*, May 8, 1986; *Los Angeles Times*, May 8, 1986.

21. *Imperial Valley Press*, May 1, 1995; Water Education Foundation, "Salton Sea," 8–9; Salton Sea National Wildlife Refuge, Annual Narrative Report, 1992, 15.

22. Rich Thiery, Coachella Valley Water District, personal communication, March 20, 1998; Salton Sea Authority, *Proceedings* (see contribution by Jim Setmire, USGS). See also Jim Setmire et al., "Detailed Study of Water Quality, Bottom Sediment, and Biota Associated with Irrigation Drainage in the Salton Sea Area, California, 1988–90."

23. *Imperial Valley Press*, March 28, 1995; Wayne W. Carmichael, "The Toxins of Cyanobacteria."

24. *Imperial Valley Press*, September 18, 1995, June 30, 1994, September 29, 1993, September 16, 1993; *Los Angeles Times*, May 18, 1994; *New York Times*, January 27, 1997.

25. Farr, *History of Imperial County*, 218; *Imperial Valley Press*, April 10, 1994.

26. *Imperial Valley Press*, October 1, 1996, October 12, 1995, April 10, 1994, September 16, 1993, September 20, 1993; *Los Angeles Times*, May 18, 1994; Salton Sea Authority, *Proceedings;* Colorado Regional Water Quality Control Board, "Salton Sea Briefing Paper." The health effects of the river on people who live close to the New River but farther upstream are another matter, which has long provoked debate but little real study. Doctors in both the Mexicali and Imperial valleys report a higher than normal incidence of eye, skin, and respiratory infections, as well as persistent hair loss, among people living along the banks of the river (*Imperial Valley Press*, October 1, 1996). But these reports, for want of detailed epidemiological study, remain more anecdotal than systematic. The paranoid among us might say that such studies have not been undertaken because the authorities who might commission them do not want to know what the studies would reveal. A kinder explanation holds that those authorities despair that bureaucratic inertia and international complexities prevent doing more than is already being done. Mexicali discharges approximately 23 million gallons of raw sewage and industrial waste into the New River every day. That's more than enough slop to fill 1,100 diving-depth backyard swimming pools, provided, of course, that the Mexicali sewage treatment plant operates roughly according to plan. When it suffers a major breakdown, which it does with regularity, the New River's daily freight of sewage increases by about half. Remedy, however, is promised. Under side accords of the 1994 North American Free Trade Agreement, the U.S. and Mexico will jointly fund construction of a new $70 million sewage treatment plant for Mexicali (*Imperial Valley Press*, March 2, 1997, March 28, 1997; *New York Times*, January 26, 1997). Skeptics might argue that this and other remedies have been in the works, off and on, for decades, and that nothing in that time has actually changed. Such a point of view may be ungenerous, but where the New River is concerned, skepticism has never proved unjustified.

27. Fish kills from algal blooms occur in warm water. Winter kills of fish in the Salton Sea

can also occur when water temperature drops to about 50 degrees F or lower, at which point tilapia succumb to hypothermia. Tilapia are a tropical fish, and prolonged cold weather, though rare in the Salton basin, can drive large schools of them into shallows where they try to burrow into the marginally warmer sands and sediments, there to die and rot. Widespread suffocation arising from algal blooms is a more recent and increasingly frequent phenomenon arising from continued eutrophication of the lake.

28. Unless otherwise noted, information on the bird die-offs of 1996 and subsequent years is drawn from Clark Bloom, "The Salton Sea Experience," and from my interviews with Bloom on March 24, 1998, and May 21, 1998.

29. William Booth, "Seeking Salvation for a Sick Salton Sea"; Clark Bloom, personal communication, June 1998.

30. *Imperial Valley Press,* May 13, 1997, approx. June 10, 1997 ("Salton Sea bird deaths linked to Newcastle virus"), October 5, 1997, October 6, 1997, October 28, 1997; Biological Resources, United States Geological Survey, "Lethal Parasite Prime Suspect in Fish Kills at Salton Sea," press release, September 10, 1997. Also Bloom, "Salton Sea Experience" and personal communication, June 1998.

31. *Imperial Valley Press,* July 23, 1998.

32. *Imperial Valley Press,* October 14, 1997; *Washington Post,* August 1, 1997; U.S. Department of the Interior, Fish and Wildlife Service, "Saving the Salton Sea: A Research Needs Assessment"; Clark Bloom, personal communication, March 24, 1998.

CHAPTER 16. PIPE DREAMS

1. Salton Sea Authority, "The Salton Sea: A Brief Description."

2. William Radke, U.S. Fish and Wildlife Service, Salton Sea National Wildlife Refuge, personal communication, July 23, 1993; Stuart Hurlbert, Department of Biology, San Diego State University, personal communication, April 30, 1998.

3. The effects of wind-borne lake sediments on human health can be deadly, as people downwind of the shrunken Aral Sea have discovered. See Murray Feshback and Alfred Friendly, Jr., *Ecocide in the USSR;* "Pollution of the Aral sea has increased illness in region," *US Water News,* March 1997; and William S. Ellis, "The Aral: A Soviet Sea Lies Dying."

4. Walker, *Ecology of the Salton Sea; Desert News* (Indio, Calif.), "Salton Sea dike plan loses priority label," December 22, 1974.

5. *Imperial Valley Press,* September 27, 1996. For a record of the sea's elevation, see Salton Sea Authority, *Proceedings;* and Sykes, *Colorado Delta,* 118.

6. *Imperial Valley Press,* September 29, 1996.

7. Don Cox, personal communication, March 25, 1998.

8. Brian Nestande, oral presentation at Salton Sea Symposium II, January 12, 1998.

9. *Imperial Valley Press,* October 5, 1998. A gross cost estimate of a pipeline plan may be found in R. Wayne Hardie, Los Alamos National Laboratory, testimony before the Subcommittee on Water and Power, U.S. House of Representatives Committee on Resources, March 12, 1998.

10. H. Martin Jessen, senior vice president, US Filter, oral presentation at Salton Sea

Symposium II, January 12, 1998; Michael Bazdarich, "An Economic Analysis of the Benefits of Rehabilitating the Salton Sea." The Bazdarich study claimed that cleanup of the sea would produce economic benefits with a capitalized value of $4.6 billion, equivalent to increased regional cash flow of $160 million a year. Not everyone was impressed. Critics questioned the assumptions on which the study was built and waved off the analysis of "data" as "smoke and mirrors." Don Cox, the lone political opponent of the Basses and US Filter on the IID board, succinctly called the study "bullshit" (personal communication, March 25, 1998).

11. Roughly 150,000 acre-feet a year flow across the border in the New River channel. If and when Mexicali's sewage treatment plants are rebuilt, expanded, and enter into reliable service, it is expected that their relatively clean outflows will be retained in Mexico, reducing the flow of the New River at the border to about 50,000 acre-feet. Coupled with water transfers, the net reduction in inflow to the Salton Sea might cause a precipitous drop in the sea's surface elevation, depending on the degree to which the effects of the water transfers are mitigated. The drop in elevation might strand present "shoreline" property far from water and, by uncovering many square miles of former lake bottom, expose sediments with potentially dangerous contents to wind transport. See note 3 above.

12. Richard Thiery, personal communication, March 20, 1998.

13. Jessen, oral presentation at Salton Sea Symposium II.

14. *Imperial Valley Press,* January 11, 1998, January 13, 1998.

15. Frank Cullen, Jr., oral presentation at Salton Sea Symposium II, January 12, 1998.

16. H.R. 3267, 105th Congress, 2d Session. Senators Barbara Boxer and Diane Feinstein introduced a similar bill in the U.S. Senate permitting a slightly longer study period.

17. The legislative history of H.R. 3267, 2d Session, 105th Congress, may be found at http:/ thomas.loc.gov.

18. See the discussion in chapter 12. Some of California's water experts take exception to the phrase "4.4 plan" because California's allocation also includes one-half of any year's surplus, including the unused allocations of other states. Exception noted.

19. As approved, the plan holds IID responsible for environmental compliance but sets a limit of $15 million on the costs it is obliged to pay. If the costs of compliance rise above $15 million, San Diego and IID may voluntarily make up the difference, or the entire agreement becomes null and void, and the transfer dies.

20. See chapter 15, note 32.

21. A "minute" is essentially an amendment to the treaty that is adopted by executive agreement of the signatories. It acquires the force of treaty, but because it expands the existing treaty relationship rather than creating a new one, it does not require ratification under the constitutions of either the U.S. or Mexico. See William J. Snape, "Adding an Environmental Minute to the 1944 Water Treaty: Impossible or Inevitable?" See also, in the same volume, the summary of oral comments by David Getches, Robert Ybarra, and Luís Antonio Rascón concerning the Snape paper.

22. The needs of the delta had little influence on the inclusion of this provision. The nominal author of the Senate version of the bill was Senator Jon Kyl of Arizona, and the traditional jealousy of Arizona over California's use of the Colorado entirely explains the prohibition against considering use of surplus river flows for Salton Sea rescue.

23. So far as I know, the idea of requiring water roughly equivalent to drain and tailwater to flow through to the Salton Sea is original with Don Cox (personal communication, March 25, 1998).

24. Ibid. The mentioned economic study, by James Merchant of the San Francisco–based consulting firm Dornbush and Company, is summarized in *Imperial Valley Press,* February 13, 1998. Some of the issues in "keeping communities whole" are explored in Santos Gomez and Anna Steding, "California Water Transfers: An Evaluation of the Economic Framework and a Spatial Analysis of the Potential Impacts."

25. Don Cox, personal communication, March 25, 1998.

26. Donald Ludwig, Ray Hilborn, and Carl Waters, "Uncertainty, Resource Exploitation and Conservation: Lessons from History"; C. S. Holling, "What Barriers, What Bridges?"

27. The most extensive study of power relationships in western water is Donald Worster's magisterial *Rivers of Empire.* Readers may also wish to examine Norris Hundley's superb history of California water development in *The Great Thirst,* which reaches conclusions different from Worster's.

References Cited

Agricultural Commissioner, Imperial County. "Imperial County Agricultural Crop and Livestock Report." El Centro, Calif., 1992.

Alexander, J. A. *The Life of George Chaffey.* New York: MacMillan, 1928.

Anderson, Henry S. "The Little Landers' Land Colonies: A Unique Agricultural Experiment in California." *Agricultural History* 5, 4 (October 1931): 139–150.

Anonymous. "Controlling the Colorado River and the Salton Sea." *Scientific American,* December 22, 1906: 467–469.

Anthony, Frances. "To Palm Canyon." *Land of Sunshine* 13 (October 1900): 235–240.

——. "At Indian Well." *Land of Sunshine* 14 (February 1901): 121–125.

——. "Below Sea Level." *Land of Sunshine* 15 (July 1901): 22–26.

Audubon, John Woodhouse. *Audubon's Western Journal: 1849–1850.* Glorieta, N.M.: Rio Grande Press, 1969. (Originally published 1906.)

Baars, Donald L. *The Colorado Plateau: A Geologic History.* Albuquerque: University of New Mexico Press, 1972.

Bancroft, Hubert Howe. "Guide to the Colorado Mines." *California Historical Society Quarterly* 12, 1 (1933): 3–10. (Originally published 1863.)

Bazdarich, Michael. "An Economic Analysis of the Benefits of Rehabilitating the Salton Sea." Inland Empire Economic Databank and Forecasting Center, University of California, Riverside. January 16, 1998.

Bean, Lowell John, and William Marvin Mason. *Diaries and Accounts of the Romero Expedition in Arizona and California, 1823–1826.* Palm Springs, Calif.: Palm Springs Desert Museum, 1962.

Bean, Lowell John, Sylvia Brakke Vane, and Jackson Young. *The Cahuilla Landscape: The Santa Rosa and San Jacinto Mountains.* Menlo Park, Calif.: Ballena Press, 1991.

Bee, Robert L. "Quechan." In *Handbook of North American Indians, vol. 10: Southwest,* Alfonso Ortiz, ed. Washington, D.C.: Smithsonian Institution, 1983.

——. *The Yuma.* New York: Chelsea House, 1989.

Bieber, Ralph P., ed. *Southern Trails to California in 1849.* Glendale, Calif.: Arthur H. Clark, 1937.

——. *Exploring Southwestern Trails, 1846–1854.* Glendale, Calif.: Arthur H. Clark, 1938.

Bigelow, F. H. "Studies on the Rate of Evaporation at Reno, Nevada, and in the Salton Sea." *National Geographic* 19 (1908): 20–28.

Blake, William Phipps. "Ancient Lake in the Colorado Desert." *American Journal of Science* (2d series) 17 (1854): 435–438.

——. "Geological Report." Part 2 in *Pacific Railroad Survey Reports,* 1857. (See below: United States Senate, Exec. Doc. no. 78.)

——. "The Cahuilla Basin and Desert of the Colorado." In *The Salton Sea: A Study of the Geography, the Geology, the Floristics, and the Ecology of a Desert Basin,* Daniel Trembly MacDougal, ed. Washington, D.C.: Carnegie Institution, 1914.

Bloom, Clark. "The Salton Sea Experience." Unpublished paper delivered at a joint meeting of the California Department of Fish and Game and the U.S. Fish and Wildlife Service, Sacramento, Calif., August 19, 1997.

Bogan, M. A., C. D. Allen, E. H. Muldavein, S. P. Platania, J. N. Stuart, G. H. Farley, P. Melhop, and J. Belnap. "Southwest." In *National Status and Trends Report,* M. J. Mac, P. A. Opler, and P. D. Doran, eds. Washington: BRO-USGS, 1998.

Bolton, Herbert Eugene. *Anza's California Expeditions.* 5 vols. Berkeley: University of California Press, 1930.

——. *Rim of Christendom: A Biography of Eusebio Francisco Kino, Pacific Coast Pioneer.* Tucson: University of Arizona Press, 1936, 1984.

——. *Coronado: Knight of Pueblos and Plains.* Albuquerque: University of New Mexico Press, 1949, 1990.

Booth, William. "Seeking Salvation for a Sick Salton Sea." *Washington Post,* August 1, 1997.

Bowman, J. N., and R. F. Heizer. *Anza and the Northwest Frontier of New Spain.* Los Angeles: Southwest Museum, 1967.

Briggs, Mark K., and Steve Cornelius. "Opportunities for Ecological Improvement along the Lower Colorado River and Delta." Final Report to Defenders of Wildlife, Washington, D.C., and National Park Service, US/Mexico Affairs Office, Las Cruces, N.M., July 24, 1997.

Brody, Jane E. "Hope Rising for Selenium." *New York Times,* February 19, 1997.

Burdick, Arthur J. *The Mystic Mid-Region: The Deserts of the Southwest.* New York: G. P. Putnam's Sons, 1904.

Cabeza de Vaca, Alvar Núñez. *Adventures in the Unkown Interior of America.* Cyclone Covey, ed. and translator. New York: Collier Books, 1961. Reprint, 1983, Albuquerque: University of New Mexico Press.

Calderón Salazar, Jorge A. *Reforma agraria y colectivización ejidal en México: La experiencia cardenista.* Culiacán: Universidad Autónoma de Sinaloa, 1990.

California Health and Welfare Agency. "Annual Planning Information, Imperial County, 1993." Labor Market Information Division, Southern Area Information Group, Los Angeles.

Carlson, Martin E. "William E. Smythe: Irrigation Crusader." *Journal of the West* 8 (January 1968): 41–47.

Carmichael, Wayne W. "The Toxins of Cyanobacteria." *Scientific American* 270, 1 (January 1994): 78–86.

Cecil-Stephens, B. A. "The Colorado Desert and Its Recent Flooding." *Bulletin of the American Geographical Society* 23 (September 1891): 367–377.

Chamberlain, Samuel E. *My Confession: The Recollections of a Rogue.* New York: Harper and Brothers, 1956.

Chase, J. Smeaton. *California Desert Trails*. Boston: Houghton Mifflin, 1919.
Clifford, Frank. "Efforts Focus on Dried-Up Colorado Delta." *Albuquerque Journal*, March 30, 1997.
Cory, Harry Thomas. *The Imperial Valley and the Salton Sink*. San Francisco: J. J. Newbegin, 1915.
Crèvecoeur, J. Hector St. John de. "Letters from an American Farmer." New York: E. P. Dutton, 1957. (Originally published 1782.)
Daniel, Cletus E. *Bitter Harvest: A History of California Farmworkers, 1870–1941*. Ithaca, N.Y.: Cornell University Press, 1981.
Davis, Mike. "Hell Factories in the Field." *The Nation*, February 25, 1995: 229–234.
Day, Allen. "Irrigation in Southern California." *Scientific American*, December 17, 1904: 439.
——. "The Inundation of the Salton Basin by the Colorado River and How It Was Caused." *Scientific American*, April 14, 1906: 310–312.
Day, Mark. *Forty Acres*. New York: Praeger, 1971.
deBuys, William. *Enchantment and Exploitation: The Life and Hard Times of a New Mexico Mountain Range*. Albuquerque: University of New Mexico Press, 1985.
Didion, Joan. "The Golden Land." *New York Review of Books*, October 21, 1993: 85–95.
Dill, David B., Jr. "William Phipps Blake: Yankee Gentleman and Pioneer Geologist of the Far West." *Journal of Arizona History* 32 (1991): 385–412.
Dowd, M. J. "History of Imperial Irrigation District and the Development of Imperial Valley." Manuscript in the possession of the Imperial Irrigation District, Imperial, Calif., 1956.
Dye, Job F. *Recollections of a Pioneer.* Los Angeles: Glen Dawson, 1951.
Ellis, William S. "The Aral: A Soviet Sea Lies Dying." *National Geographic*, February 1990: 73–92.
Emory, William H. *Lieutenant Emory Reports: A Reprint of Lieutenant W. H. Emory's* Notes of a Military Reconnaissance. Albuquerque: University of New Mexico Press, 1951, 1968.
Fairchild, Mahlon Dickerson. "A Trip to the Colorado Mines in 1862." *California Historical Society Quarterly* 12 (1933): 11–17.
Farr, Finis C., editor and principal author. *The History of Imperial County, California*. Berkeley: Elms and Franks, 1918.
Faulk, Odie B., ed. *Derby's Report on Opening the Colorado, 1850–51*. Albuquerque: University of New Mexico Press, 1969.
Feshback, Murray, and Alfred Friendly, Jr. *Ecocide in the USSR: Health and Nature under Siege*. New York: BasicBooks, 1992.
Forbes, Jack D. *Warriors of the Colorado: The Yumas of the Quechan Nation and Their Neighbors*. Norman: University of Oklahoma Press, 1965.
Fradkin, Philip L. *A River No More: The Colorado River and the West*. New York: Alfred A. Knopf, 1981.
Freeman, Lewis R. *Down the Grand Canyon*. New York: Dodd, Mead, 1930.
Frisby, Karen J. *Imperial Valley: The Greening of a Desert*. Occasional Paper no. 9, Imperial Valley College, Desert Museum Society, 1992.
Ganster, Paul. "Environmental Issues of the California–Baja California Border Region." Border Environment Research Reports, no. 1, Southwest Center for Environmental Research and Policy, San Diego State University, June 1996.

Gates, Marilyn. "Codifying Marginality: The Evolution of Mexican Agricultural Policy and Its Impact on the Peasantry." *Journal of Latin American Studies* 20 (November 1988): 277–311.

Gifford, Edward W. *The Kamia of the Imperial Valley.* Bureau of American Ethnology Bulletin 97. Washington, D.C., 1931.

Glenn, Edward P., Richard S. Felger, Alberto Búrquez, and Dale S. Turner. "Cienega de Santa Clara: Endangered Wetland in the Colorado River Delta, Sonora, Mexico." *Natural Resources Journal* 32 (Fall 1992): 817–824.

Glenn, Edward P., Christopher Lee, Richard Felger, and Scott Zengel. "Water Management Impacts on the Wetlands of the Colorado River Delta, Mexico." *Conservation Biology* 10, 4 (August 1996): 1175–1186.

Goetzmann, William H. *Army Exploration in the American West 1803–1863.* Lincoln: University of Nebraska Press, 1979.

Gomez, Santos and Anna Steding. "California Water Transfers: An Evaluation of the Economic Framework and a Spatial Analysis of the Potential Impacts." Pacific Institute for Studies in Development, Environment, and Security, Oakland, California, April 1998.

Gómez-Peña, Guillermo. *Warrior for Gringostroika.* St. Paul, Minn.: Graywolf Press, 1993.

Grinnell, Joseph. "Birds of a Voyage on Salton Sea." *The Condor,* 10 (1908): 185–191.

Hammond, George P., and Agapito Rey. *Narratives of the Coronado Expedition, 1540–1542.* Albuquerque: University of New Mexico, 1940.

Harris, David. "Recent Plant Invasions in the Arid and Semi-Arid Southwest of the United States." In *Man's Impact on Environment,* Thomas R. Detwyler, ed. New York: McGraw-Hill, 1971.

Harris, Tom. *Death in the Marsh.* Washington, D.C.: Island Press, 1991.

Harris, Tom, et al. "David Love" and assorted other articles relating to selenium. *High Country News* 24, 2 (February 10, 1992).

Harwell, Henry O., and Marsha C. S. Kelly. "Maricopa." In *Handbook of North American Indians, vol. 10: Southwest,* Alfonso Ortiz, ed. Washington, D.C.: Smithsonian Institution, 1983.

Hayes, Douglas L. "The All-American Canal Lining Project: A Catalyst for Rational and Comprehensive Groundwater Management on the United States-Mexico Border?" *Transboundary Reources Report* 5, 1: 1–3. Albuquerque: University of New Mexico School of Law, International Transboundary Resources Center.

Heffernan, W. T. *Personal Recollections of the Early History of Imperial Valley.* Calexico, Calif.: *Calexico Chronicle,* 1928, reprint 1930.

Heizer, Robert F., and Adan E. Treganza. *Mines and Quarries of the Indians of California.* Ramona, Calif.: Ballena Press, 1972.

Hillel, Daniel. "Lash of the Dragon." *Natural History,* August 1991, 29–37.

Holder, Charles F. "A Remarkable Salt Deposit." *Scientific American* 74, 14 (April 6, 1901): 224.

Holling, C. S. "What Barriers, What Bridges?" In *Barriers and Bridges to the Renewal of Ecosystems and Institutions,* S. Light, L. Gunderson, and C. S. Holling, eds. New York: Columbia University Press, 1995.

Hosmer, Helen. "Triumph and Failure in the Imperial Valley." In *The Grand Colorado: The*

Story of a River and Its Canyons, Thomas H. Watkins, ed. Palo Alto, Calif.: American West, 1969, 205–221.

Hundley, Norris. "The Politics of Reclamation: California, the Federal Government, and the Origins of the Boulder Canyon Act—A Second Look." *California Historical Quarterly* 52, 4 (Winter 1973): 292–325.

——. *The Great Thirst: Californians and Water, 1770s–1990s.* Berkeley: University of California Press, 1992.

James, George Wharton. "The Overflow of the Colorado and the Salton Sea—II. *Scientific American,* April 21, 1906: 328–329.

——. *The Wonders of the Colorado Desert.* Boston: Little, Brown, 1911.

Johnson, Boma. *Earth Figures of the Lower Colorado and Gila River Deserts: A Functional Analysis.* Arizona Archeological Society Publication no. 20 (*Arizona Archaeologist*), November 1985.

Johnston, Francis J. "Stagecoach Travel through the San Gorgonio Pass." *Journal of the West* 11, 4 (1972): 616–635.

Kelly, William H. *Cocopa Ethnography.* Anthropological Papers of the University of Arizona, no. 29, Tucson, 1977.

Kennan, George. *The Salton Sea: An Account of Harriman's Fight with the Colorado River.* New York: Macmillan, 1917.

Kniffen, Fred Bowerman. *The Natural Landscape of the Colorado Delta.* Lower California Studies, 4. Berkeley: University of California Press, 1932.

Kramer, Gary. "North America's Most Unique Fishery." *Petersen's Fishing,* September 1987: 47–48.

Kroeber, Alfred L. *Handbook of the Indians of California.* New York: Dover, 1925, 1976.

Kunzmann, Michael R., et al. "Tamarisk Control in Southwestern United States." Special Report no. 9, Cooperative National Park Resources Studies Unit, University of Arizona School of Renewable Natural Resources, Tucson, 1990.

Lavender, David. *California: Land of New Beginnings.* Lincoln: University of Nebraska Press, 1972, 1987.

Leopold, Aldo. *A Sand County Almanac, and Sketches Here and There.* New York: Oxford Univerity Press, 1949, 1968.

Leopold, Luna B., ed. *Round River: From the Journals of Aldo Leopold.* New York: Oxford University Press, 1983.

Limerick, Patricia Nelson. *Desert Passages: Encounters with the American Deserts.* Niwot, Colo.: University Press of Colorado, 1989.

Lingenfelter, Richard E. *Death Valley and the Amargosa: A Land of Illusion.* Berkeley: University of California Press, 1986.

——. *Steamboats on the Colorado River 1852–1916.* Tucson: University of Arizona Press, 1978.

Lopez, Barry. "The Stone Horse." In *Crossing Open Ground,* 1–17. New York: Vintage Books, 1989.

Ludwig, Donald, Ray Hilborn, and Carl Waters. "Uncertainty, Resource Exploitation, and Conservation: Lessons from History." *Science* 260 (April 2, 1993): 17, 36.

Lujan, Manuel, Secretary of the Interior. "Report to Congress on Salton Sea National Wildlife Refuge." Washington, D.C., December 14, 1989.

Luomala, Katherine. "Tipai and Ipai." In *Handbook of North American Indians, vol. 8: California,* Robert F. Heizer, ed. Washington, D.C.: Smithsonian Institution, 1978.

MacDougal, Daniel Trembly. "A Voyage below Sea Level on the Salton Sea." *Outing,* 1908: 592–601.

Manley, Roger, and Mark Sloan. *Self-Made Worlds: Visionary Folk Art Environments.* New York: Aperture, 1997.

May, Ronald V. "A Brief Survey of Kumeyaay Ethnography: Correlations Between Environmental Land-Use Patterns, Material Culture, and Social Organization." *Pacific Coast Archaeological Society Quarterly* 11, 4 (October 1975): 1–25.

McCarthy, Cormac. *Blood Meridian, or the Evening Redness in the West.* New York: Random House, 1985.

McWilliams, Carey. *Factories in the Field: The Story of Migratory Farm Labor in California.* Santa Barbara, Calif.: Peregrine Publishers, 1935, 1971.

Meine, Curt. *Aldo Leopold: His Life and Work.* Madison: University of Wisconsin Press, 1988.

Meko, David, and Donald A. Graybill. "Tree-Ring Reconstruction of Upper Gila River Discharge." *Water Resources Bulletin* (American Water Resource Association) 31, 4 (August 1995): 605–616.

Mendenhall, W. G. "The Colorado Desert." *National Geographic* 20 (1909): 681–701.

Morrison, Jason I., Sandra L. Postel, and Peter H. Gleick. *The Sustainable Use of Water in the Lower Colorado River Basin.* Oakland, Calif.: Pacific Institute for Studies in Development, Environment, and Security, 1996.

Moyano Pahissa, Angela. *California y sus relaciones con Baja California.* Mexico, D.F.: Fondo de Cultura Económica, 1983.

Moyle, Peter B. *Inland Fishes of California.* Berkeley: University of California Press, 1976.

Nabhan, Gary Paul. *Gathering the Desert.* Tucson: University of Arizona Press, 1985.

Nabokov, Peter. *Indian Running: Native American History and Tradition.* Santa Fe, N.M.: Ancient City Press, 1987.

Noyes, Stanley. *Los Comanches: The Horse People, 1751–1845.* Albuquerque: University of New Mexico, 1993.

Ortiz, Alfonso, ed. *New Perspectives on the Pueblos.* Albuquerque: University of New Mexico Press, 1972.

——, ed. *Handbook of North American Indians, vol. 10: Southwest.* William C. Sturtevant, general editor. Washington, D.C.: Smithsonian Institution, 1983.

Page, Gary W., Lynne E. Stenzel, Janet E. Kjelmyr, and W. David Shuford. "Shorebird Numbers in Wetlands of the Pacific Flyway. A Summary of Spring and Fall Counts in 1988 and 1989." Point Reyes Bird Observatory, February 1990.

Pattie, James Ohio. *The Personal Narrative of James O. Pattie of Kentucky.* Milton Milo Quaife, ed. Chicago: Lakeside Press, 1930 (a reissue of the edition edited by Timothy Flint and published in Cincinnati by John H. Wood in 1831).

Pontius, Dale. *Colorado River Basin Study: Final Report.* Western Water Policy Review Advisory Commission, June 1997.

Rawls, James J. *Indians of California: The Changing Image.* Norman: University of Oklahoma Press, 1984.

Reed, Lester. *Old Time Cattlemen and Other Pioneers of the Anza-Borrego Area.* Benson, Ariz.: Border-Mountain Press, 1963, 1977.
Reisner, Marc. *Cadillac Desert: The American West and Its Disappearing Water.* New York: Penguin, 1986.
Reisner, Marc, and Sarah Bates. *Overtapped Oasis: Reform or Revolution for Western Water.* Washington, D.C.: Island Press, 1990.
Riley, Carroll L. *The Frontier People: The Greater Southwest in the Protohistoric Period.* Albuquerque: University of New Mexico Press, 1987.
Robrock, David P., ed. *Missouri Forty-Niner: The Journal of WIlliam W. Hunter on the Southern Gold Trail.* Albuquerque: University of New Mexico Press, 1992.
Rockwood, Charles R. *Born of the Desert.* Calexico, Calif.: *Calexico Chronicle,* 1909, 1930.
——. "The Early History of Imperial County." In *The History of Imperial County, California,* Finis C. Farr, ed. Berkeley: Elms and Franks, 1918.
Romer, Margaret. *A History of Calexico.* Annual Publication of the Historical Society of Southern California, 1922.
Salandini, Victor. *The Confessions of the Tortilla Priest. San Diego Review,* 1992.
Salton Sea Authority. *Proceedings of the Salton Sea Symposium, January 13, 1994.* Indian Wells, Calif.
——. "The Salton Sea: A Brief Description of Its Current Condition and Potential Remediation Projects." Indio, Calif., January 22, 1998.
Seddon, Marian. *What Ever Happened to the Chocolate Mountain Gang.* Yuma, Calif.: Sun Graphics, 1982.
Setmire, J. G., R. A. Schroeder, J. N. Densmore, S. L. Goodbred, D. J. Auder, and W. R. Radke. "Detailed Study of Water Quality, Bottom Sediment, and Biota Associated with Irrigation Drainage in the Salton Sea Area, California, 1988 to 1989." U.S. Geological Survey, Water Resources Investigations, report no. 93–4014, 1993.
Shelton, Richard. *Going Back to Bisbee.* Tucson: University of Arizona Press, 1992.
Singelmann, Peter. "Rural Collectivization and Dependent Captialism: The Mexican Collective Ejido." *Latin American Perspectives* 5 (Summer 1978): 38–61.
Smythe, William E. "An International Wedding." *Sunset* 5 (October 1900): 286–300.
——. *The Conquest of Arid America.* Seattle: University of Washington Press, 1899, 1905, 1969.
Snape, William. "Adding an Environmental Minute to the 1944 Water Treaty: Impossible or Inevitable?" Appendix D in *Workshop Proceedings, Water and Environmental Issues of the Colorado River Border Region,* San Luís Río Colorado, Sonora, Mexico, April 30, 1998. Defenders of Wildlife and the Pacific Institute for Studies in Development, Environment, and Security.
Sorenson, Steve. "Heaven Sent: Shrapnel, Shells, Bomb Casings, Machine Gun Clips, Missile Parts." *San Diego's Weekly Reader* 20, 8 (February 28, 1991).
——. "Hot Place in the Sun: A Different Breed of Snowbird Finds a Town with no City Limits." *San Diego's Weekly Reader* 17, 9 (March 10, 1988).
——. "Mission over the Mountains: Mock War in a True Wilderness." *San Diego's Weekly Reader* 16, 43 (October 29, 1987).
——. "Bury My Bait at Bombay Beach." *San Diego's Weekly Reader* 15, 17 (May 1, 1986).

Southern Pacific Railroad Company, Passenger Department. "Imperial Valley, California" (pamphlet). San Francisco, 1908. (Collection of Joan Myers.)

Starr, Kevin. *Americans and the California Dream, 1850–1915.* New York: Oxford University Press, 1973.

——. *Inventing the Dream: California through the Progresssive Era.* New York: Oxford University Press, 1985.

——. *Material Dreams: Southern California through the 1920s.* New York: Oxford University Press, 1990.

——. *Endangered Dreams: The Great Depression in California.* New York: Oxford University Press, 1996.

——. *The Dream Endures: California Enters the 1940s.* New York: Oxford University Press, 1997.

Stegner, Wallace. *Beyond the Hundredth Meridian: John Wesley Powell and the Second Opening of the West.* New York: Penguin Books, 1954, 1992.

Stewart, Kenneth M. "Yumans: An Introduction." In *Handbook of North American Indians, vol. 10: Southwest,* Alfonso Ortiz, ed. Washington, D.C.: Smithsonian Institution, 1983.

Strand, Carl L. "Mud Volcanoes, Faults, and Earthquakes of the Colorado Delta Before the Twentieth Century." *San Diego Journal of History* 27 (1981): 43–63.

Sweigert, David C. "Relining Canals in the Border Region: Can the U.S. Ignore impacts on Mexico?" Manuscript. School of Law, University of California, Davis, 1990.

Sykes, Glenton G. "Five Walked Out! The Search for Port Isabel." *Journal of Arizona History,* Summer 1976.

Sykes, Godfrey. *A Westerly Trend.* Tucson: Arizona Historical Society and University of Arizona Press, 1945, 1984.

——. *The Colorado Delta.* Washington and New York: Carnegie Institution and American Geographical Society, 1937.

Trimble, Stephen. *The People.* Santa Fe, N.M.: School of American Research Press, 1993.

Turk-Boyer, Peggy J. "Reserves in Action." *CEDO News/Noticias del CEDO* (Lukeville, Ariz.), Fall–Winter 1996.

United States Bureau of Reclamation. "Yuma Desalting Plant: A Status Report." Yuma, Ariz., August 1990.

——. "Vegetation Management Study: Lower Colorado River." Boulder City, Nev., 1995.

United States Department of the Interior. "Draft Enviromental Impact Statement/Enviromental Impact Report, All-American Canal Lining Project Imperial County, Calif." Bureau of Reclamation, Lower Colorado Region, Boulder City, May 1991.

United States Department of State, Bureau of Oceans and International Environmental Affairs. "International Sonoran Desert Alliance: A Participatory Process to Support Conservation and Sustainable Use in a Tri-National Region." In *Biosphere Reserves in Action: Case Studies of the American Experience,* Department of State Publication 10241, June 1995: 37–44.

United States Fish and Wildlife Service. "Saving the Salton Sea: A Research Needs Assessment." October 1997.

United States Senate. Exec. Doc. no. 78. 32d Congress, 2d Session. "Reports of Explorations and Surveys to Ascertain the Most Practicable and Economical Route for a Railroad from the Mississippi River to the Pacific Ocean," vol. 5: "Reports on Routes in California to connect with the routes near the thirty-fifth and thirty-second parallels explored by Lt. R. S. Williamson, Corps of Topographical Engineers, in 1853." Washington, D.C., 1855. (Abbreviated in notes as *Pacific Railroad Survey Reports.*)

University of California, Imperial County, Cooperative Extension. "Guide Lines to Production Costs and Practices, 1992–1993." Holtville, Calif.

USDA Forest Service. *Range Plant Handbook.* Washington, D.C., 1937.

Van Der Werf, Martin. "Draining the Budget to Desalt the Colorado." *High Country News,* February 21, 1994.

Van Dyke, John C. *The Desert.* Park City, Utah: Peregrine Smith, 1901, 1980.

Walker, Boyd W., ed. *The Ecology of the Salton Sea, California, in Relation to the Sportfishery.* Fish Bulletin no. 113, Department of Fish and Game, State of California, 1961.

Walker, Franklin. *A Literary History of Southern California.* Berkeley: University of California Press, 1950.

Water Education Foundation. "The Salton Sea." *Western Water,* March–April 1994.

Watkins, Thomas H., editor. *The Grand Colorado: The Story of a River and Its Canyons.* Palo Alto, Calif.: American West, 1969.

Weide, David L. "Regional Environmental History of the Yuha Desert." In *Background to Prehistory of the Yuha Desert Region,* Philip J. Wilke, Harry W. Lawton, Thomas F. King, and Stephen Hammond, eds. Ramona, Calif.: Ballena Press, 1976.

Wilke, Philip J., and Harry W. Lawton. "Early Observations on the Cultural Geography of the Coachella Valley." In *The Cahuilla Indians of the Colorado Desert: Ethnohistory and Prehistory,* by Philip J. Wilke, Harry W. Lawton, Thomas F. King, and Stephen Hammond. Ramona, Calif.: Ballena Press, 1975.

Woodward, Arthur. *Feud on the Colorado.* Los Angeles: Westernlore Press, 1955.

Worster, Donald. *Rivers of Empire: Water, Aridity, and the Growth of the American West.* New York: Pantheon, 1985.

Wright, Harold Bell. *The Winning of Barbara Worth.* Chicago: Book Supply Company, 1911.

Index

S A L T   D R E A M S

Text set in Berkeley Old Style Medium with Goudy Sans display
by Keystone Typesetting, Orwigsburg, Pennsylvania
Text printed by Thomson Shore, Inc., Dexter, Michigan
on fifty-pound Glatfelter Supple Opaque
Duotones printed by Southeastern Printing, Stuart, Florida
on eighty-pound Celesta Dull
Bound by Thomson Shore, Inc., Dexter, Michigan
Maps by Deborah Reade
Designed by Kristina Kachele